P9-AEZ-573

Interfacial Phenomena in Apolar Media

SURFACTANT SCIENCE SERIES

Volume 1: NONIONIC SURFACTANTS, edited by Martin J. Schick

Volume 2: SOLVENT PROPERTIES OF SURFACTANT SOLUTIONS, edited by Kozo Shinoda (*out of print*)

Volume 3: SURFACTANT BIODEGRADATION, by R. D. Swisher (*out of print*)

Volume 4: CATIONIC SURFACTANTS, edited by Eric Jungermann

Volume 5: DETERGENCY: THEORY AND TEST METHODS (*in three parts*), edited by W. G. Cutler and R. C. Davis

Volume 6: EMULSIONS AND EMULSION TECHNOLOGY (*in three parts*), edited by Kenneth J. Lissant

Volume 7: ANIONIC SURFACTANTS (*in two parts*), edited by Warner M. Linfield

Volume 8: ANIONIC SURFACTANTS–CHEMICAL ANALYSIS, edited by John Cross

Volume 9: STABILIZATION OF COLLOIDAL DISPERSIONS BY POLYMER ADSORPTION, by Tatsuo Sato and Richard Ruch

Volume 10: ANIONIC SURFACTANTS–BIOCHEMISTRY, TOXICOLOGY, DERMATOLOGY, edited by Christian Gloxhuber

Volume 11: ANIONIC SURFACTANTS–PHYSICAL CHEMISTRY OF SURFACTANT ACTION, edited by E. H. Lucassen-Reynders

Volume 12: AMPHOTERIC SURFACTANTS, edited by B. R. Bluestein and Clifford L. Hilton

Volume 13: DEMULSIFICATION: INDUSTRIAL APPLICATIONS, by Kenneth J. Lissant

Volume 14: SURFACTANTS IN TEXTILE PROCESSING, by Arved Datyner

Volume 15: ELECTRICAL PHENOMENA AT INTERFACES: FUNDAMENTALS, MEASUREMENTS, AND APPLICATIONS, edited by Ayao Kitahara and Akira Watanabe

Volume 16: SURFACTANTS IN COSMETICS, edited by Martin M. Rieger

Volume 17: INTERFACIAL PHENOMENA: EQUILIBRIUM AND DYNAMIC EFFECTS, by Clarence A. Miller and P. Neogi

Volume 18: SURFACTANT BIODEGRADATION, Second Edition, Revised and Expanded, by R. D. Swisher

Volume 19: NONIONIC SURFACTANTS: CHEMICAL ANALYSIS, edited by John Cross

Volume 20: DETERGENCY: THEORY AND TECHNOLOGY, edited by W. Gale Cutler and Erik Kissa

Volume 21: INTERFACIAL PHENOMENA IN APOLAR MEDIA, edited by Hans-Friedrich Eicke and Geoffrey D. Parfitt

Volume 22: SURFACTANT SOLUTIONS: NEW METHODS OF INVESTIGATION, edited by Raoul Zana

OTHER VOLUMES IN PREPARATION

NONIONIC SURFACTANTS: PHYSICAL CHEMISTRY, edited by Martin J. Schick

IN MEMORIAM

GEOFFREY D. PARFITT

The sudden death of Geoffrey Parfitt, in July 1985, interrupted a life of intense research and editorial work and ended his endeavors to promote the International Association of Colloid and Interface Scientists, of which he was one of the founders.

Personally, I lost my coeditor, an invaluable and competent advisor in editing this volume on Interfacial Phenomena in Apolar Media, which turned out to be his last contribution to interfacial science. I remember with great pleasure the days Geoff spent in Basel where our cooperation started and the plan of this book was settled. This volume is thus devoted to his memory.

Hans-Friedrich Eicke

Interfacial Phenomena in Apolar Media

edited by

HANS-FRIEDRICH EICKE
Institut für Physikalische Chemie der
Universität Basel
Basel, Switzerland

GEOFFREY D. PARFITT
Department of Chemical Engineering
Carnegie-Mellon University
Pittsburgh, Pennsylvania

MARCEL DEKKER, INC. New York and Basel

Library of Congress Cataloging-in-Publication Data

Interfacial phenomena in apolar media.

(Surfactant science series ; v. 21)
Includes bibliographies and index.
1. Surface active agents. 2. Surface chemistry.
I. Eicke, Hans-Friedrich II. Parfitt, G. D.
III. Series.
TP994.I56 1987 541.3'453 86024018
ISBN 0-8247-7506-6

MARCEL DEKKER, INC.
270 Madison Avenue, New York, New York 10016

Current printing (last digit):
10 9 8 7 6 5 4 3 2 1

PRINTED IN THE UNITED STATES OF AMERICA

Preface

This is the first comprehensive account of interfacial phenomena in apolar media. The volume covers areas of surprising variety where interfacial phenomena in liquid hydrocarbons are highly significant. Ranging broadly, the exposition covers such topics as foaminess and capillarity in apolar solutions, stabilization of inverse micelles as well as aqueous nanophases and their applications in enhanced oil recovery, interfacial catalysis, and other more practical aspects of general water/oil emulsions. The treatment includes the dispersion of pigments in such media and deals with adsorption mechanisms and the important area of lubrication. The nine chapters each present the current state of the art. They have been written by specialists in separate fields and provide reliable information in each of the areas covered.

Thus, research workers in industrial, governmental, and university laboratories as well as technicians who are presented with problems of any kind concerning interfacial phenomena in hydrocarbon media may profitably resort to this up-to-date source book; graduate students of colloid and surface science and of related fields such as engineering, environmental, pharmaceutical, and life sciences will also benefit from the broad coverage it presents.

In addition, an extensive reference list at the end of every chapter provides readers with information about the most relevant publications in their respective fields to help in the formation of independent opinions on particular aspects of the general subject.

Hans-Friedrich Eicke
Geoffrey D. Parfitt

Contents

Contributors

PAUL BECHER Paul Becher Associates Ltd., Wilmington, Delaware

B. BRISCOE Department of Chemical Engineering and Chemical Technology, Imperial College, London, England

HANS-FRIEDRICH EICKE Institut für Physikalische Chemie der Universität Basel, Basel, Switzerland

STIG E. FRIBERG Chemistry Department, University of Missouri at Rolla, Rolla, Missouri

GORDON J. HOWARD Visiting Scientist, 1983-85, Marshall Research and Development Laboratory, E. I. Dupont de Nemours & Company, Philadelphia, Pennsylvania

R. B. McKAY Ciba-Geigy Pigments, Paisley, Scotland

C. A. MILLER Chemical Engineering Department, Rice University, Houston, Texas

CHARMIAN J. O'CONNOR Department of Chemistry, University of Auckland, Auckland, New Zealand

S. QUTUBUDDIN Chemical Engineering Department, Case Western Reserve University, Cleveland, Ohio

SIDNEY ROSS Department of Chemistry, Rensselaer Polytechnic Institute, Troy, New York

D. TABOR Department of Physics, Cavendish Laboratory, University of Cambridge, Cambridge, England

Interfacial Phenomena in Apolar Media

1

Foaminess and Capillarity in Apolar Solutions

SYDNEY ROSS Department of Chemistry, Rensselear Polytechnic Institute, Troy, New York

I. THEORIES OF FOAM STABILITY

Pure liquids do not form stable foams, but they allow entrained air to escape with no delay other than that inseparable from the Stokesian rate of rise, which is controlled by the diameter of the bubble of dispersed air and the viscosity of the bulk liquid. Certain solutes are able to stabilize thin sheets (or lamellae) of liquid; if these solutes are present the escape of entrained bubbles is more or less retarded, and a head of foam is produced. Theories of foam postulate plausible mechanisms to account for this behavior, with the ultimate objective of understanding the phenomenon so thoroughly that predictions about the behavior of a given solute can be made prior to actual observation. One may say at the outset that this final goal has not yet been completely attained.

A. Rayleigh-Gibbs Theory of Foam

The earliest of these theories, the one usually designated the Rayleigh-Gibbs theory [1,2], has best withstood criticism through the years. This theory refers the stability of foam to an elasticity or restoration of liquid lamellae, which depends on the existence of an adsorbed layer of solute at the liquid surface and the effect of this adsorbed layer in lowering the surface tension of the solution below that of the solvent. The two effects, surface segregation, or adsorption, and the lowering of the surface tension, are concomitant: a reduction of surface tension due to the addition of a solute is evidence, admittedly indirect but no less certain than were it given by direct observation, that the solute is segregated at the surface. The degree of the segregation is measured as excess moles of solute per square centimeter of surface, designated Γ_2^G, and is proportional to the variation of the surface tension lowering with concentration of solute, i.e.,

$$\Gamma_2^G = \frac{1}{RT}\,\frac{d\pi}{d\ln a_2} \tag{1}$$

where π is the lowering of the surface tension caused by a thermodynamic activity a_2 of solute in the solution. Equation (1) is based on thermodynamics, derived for a two-component system. In this chapter the term *surface-active solute* denotes a solute that reduces the surface tension of a liquid to any appreciable extent, even by as little as a few tenths of a millinewton per meter.

When local areas of a foam lamella are expanded, as would happen, for example, when a bubble of air pushes through a liquid surface, new areas of surface are created where the instantaneous surface tension is large, because the adsorbed layer has not had sufficient time to form. The greater surface tension in these new areas of surface exerts a pull on the adjoining areas of lower tension, causing the surface to flow toward the region of greater tension. The viscous drag of the moving surface carries an appreciable volume of underlying liquid along with it, thus offsetting the effects of both hydrodynamic and capillary drainage and restoring the thickness of the lamella. The same mechanism explains how liquid lamellae withstand mechanical shocks, such as the passage through the foam of lead granules, mercury drops, iron fillings, or steel spheres, all of which have been used by one or another investigator to test the resilience of lamellae [3]. The lamella survives because the local increase of surface tension, where the impinging solid deforms the surface, causes flow toward the weakened region. Very thin lamellae are less fluid than thick ones and so are more readily broken, as are lamellae made from solutions in which the surface tension gradient is small, e.g., oil solutions or very dilute aqueous solutions of surface-active agents.

In general, the elasticity arising from the variation of the surface tension during deformation of a liquid lamella may be manifested both in equilibrium (when a surface layer under forces leading to deformation is in equilibrium with its bulk phase) and in nonequilibrium conditions. The first case refers to the Gibbs elasticity and the second to the Marangoni elasticity [4]. The Marangoni elasticity is a dynamic, nonequilibrium property, normally larger in value than the Gibbs elasticity that could be obtained in the same system. Gibbs elasticity is defined as the ratio of the increase in the tension resulting from an infinitesimal increase in the area and the relative increase of the area. For a lamella with adsorbed solute on both sides, the elasticity E is given by [5]:

$$E = 2 \frac{d\sigma}{d \ln A} \tag{2}$$

where σ is the surface tension and A is the area of the liquid surface. The factor 2 is required because the stretching of the lamella increases the area on both of its sides.

Attempts to test Eq. (2) have been made by measuring the dynamic (i.e., nonequilibrium) surface tension as the surface of a solution is abruptly extended, or pulsated. Some investigators have found dynamic surface tensions occurring at rather low frequencies of dilatation-compression cycles, from one per minute to one every 30 min; others [6] have used frequencies as high as 15 to 135 Hz (cycles per second), although such disturbances are far from corresponding to the extension-contraction cycles occurring in an actual foam. The measurement must be made coincidentally with the changes of the surface area. The modulus of surface elasticity of a solution can be determined by measuring the damping of transverse ripples as a function of their frequency. The surface of a wave is contracted at the crest and extended at the trough. In the absence of any surface-active material, this contraction and extension does not change the surface tension; but if an adsorbed layer is present, it is compressed on the contraction and dilated on the extension, causing the surface tension to decline at the crest and to increase at the trough. These local differences of tension alter the pattern of subsurface flow giving rise to a greater rate of energy dissipation by viscous friction, and consequently to a greater damping of the waves than would otherwise occur. The distance-damping coefficient β may be measured by the logarithmic decrease of the wave amplitude, a, with distance x from the source:

$$\beta = -\frac{d \ln a}{dx} \tag{3}$$

The coefficient β is constant for Newtonian fluids: the surface of a solution contains adsorbed solute if the solute is surface active, and such a surface may not be Newtonian. The plot of ln a versus x would then not be linear. The testing of Eq. (3) therefore is informative about the presence of a non-Newtonian shear viscosity at the surface of a solution. If, however, the surface should be Newtonian, the data have a further use: the relative change of the damping coefficient compared to pure solvent ($\beta = \beta_0$), as a function of wave frequency, is directly related to the surface elasticity [7], assuming that diffusional interchange between bulk and surface is negligible during dilatation and compression of the adsorbed layer [8].

The effects described by the Rayleigh-Gibbs theory depend therefore on a combination of two physical properties of the solution: the solute should be capable of lowering the surface tension of the medium, but this alone is not enough: a rate process is also required by which a freshly created liquid surface retains its initial, high, nonequilibrium surface tension long enough for surface flow to occur. Many instances are known in which the mere reduction of surface tension by the solute does not lead to the stabilization of foam, presumably because it is not accompanied by the relatively slow attainment of equilibrium after a fresh surface is made that is the second requirement for the ability to stabilize bubbles.

In general, the surface tension of a freshly formed solution of a surface-active solute changes with time until it reaches a final equilibrium value. Equilibrium may be reached in a fraction of a second or it may require several days. Adsorption may be considered as a two-step process involving (1) diffusion of the solute molecules from the bulk phase to the subsurface (i.e., the layer immediately below the surface); and (2) adsorption of the solute molecules from the subsurface to the surface.

Recently, Borwankar and Wasan [9] developed a mathematical model of adsorption of surface-active solutes at a gas-water interface that takes into account both the diffusion in the bulk phase and the energy barrier to adsorption. Not all adsorptions require both steps: some systems are diffusion-controlled, and the activation energy barrier to adsorption is negligible in such cases. The rate of diffusion is much higher than the rate at which the activation energy barrier is overcome. A diffusion-controlled rate of adsorption makes itself evident, therefore, by a rapid approach to surface tension equilibrium (on the order of seconds or less). Surfaces that age more slowly imply a kinetically controlled rate of adsorption. The different types of instrument developed to measure dynamic surface tension vary with respect to the age of the surface that they are designed to handle: the vibrating jet deals with surface ages of a few milliseconds, the damping of capillary ripples deals with surface ages of seconds, and conventional techniques for measuring surface tension can detect surface aging of several minutes, hours, or days.

The above considerations apply equally to aqueous and apolar solutions. A major difference between the two lies, however, in the magnitude of the effects produced by surface-active solutes. The surface tension of water is reduced from 73 to 25 mN/m quite readily by amphipathic organic solutes; but the surface tension of most organic solvents is already in the low range of 25 to 30 mN/m, so that only a small reduction can be achieved by an organic solute. Thus, although Marangoni effects may arise in apolar solutions, they are usually much less pronounced than those in aqueous solutions of soaps or detergents. Oil lamellae, therefore, have a relatively low resistance to mechanical shock [3]; consequently oil foams are transient or evanescent, resembling the foam produced from very dilute aqueous solutions of detergents or more concentrated solutions of weakly surface-active solutes. Of course, special solutes have been developed to stabilize apolar foams, for application in the field of cellular plastics. These solutes incorporate poly(alkylsiloxane) or perfluoroalkyl moieties in solute molecules, which are able to reduce the surface tension of organic liquid monomers by 6 to 10 mN/m to boost the Marangoni effects; and the result is obvious in an increased stability of the foam.

B. Enhanced Viscosity or Plasticity at the Liquid-Gas Interface

A single surface-active species in solution does not usually confer any increase of the viscosity, much less rigidity, in the surface layer of the solution. Although foam is capable of being produced by such a solute, the foam is of brief duration. That kind of foam is called *evanescent foam*, but it can nevertheless be a cause of concern, because if produced rapidly it can reach a large expansion ratio and so flood any container. Much more stable foam is created if, in addition to the Rayleigh-Gibbs effect described above, the surface layer of the solution has an enhanced viscosity or, especially, plasticity [10]. This phenomenon is known to occur in water with certain mixtures of solutes or with certain polymers, both natural and synthetic. The best known examples in aqueous systems are solutions of water-soluble proteins, such as casein or albumen. Common examples are the stable foams produced with whipping cream, egg white, beer, or rubber latex. In many other examples the highly viscous surface layer is made by having present one or more additional components in the solution. An example is the increase in surface viscosity of a mixture of tannin and heptanoic acid in aqueous solution, compared to the effect of the two constituents separately.

In apolar liquids, particularly in bunker oils and crude oils, plastic surface layers have been observed; porphyrins of high molecular weight have been indicated as a possible source of this effect. In a hydrocarbon lubricant, additives, such as calcium sulfonate, have been identified as creating a plastic skin (or two-dimensional Bingham body) at the air interface; they may also act as foam stabilizers [11]. These viscous or rigid layers in apolar liquids enhance the stability of foam, just as they do in aqueous solutions [12–14]. Cellular plastics are made from a monomer foam, which on polymerization displays high viscosity, finally becoming solid; but the growth of the viscosity is not confined to the surface, nor does it even show preferential development at the surface.

Different kinds of surface viscosity are also distinguished:

1. Innate surface viscosity. This is the resistance to flow that is innately associated with the presence of a liquid surface, whether or not there are additional sources of resistance such as those described below [15].
2. Surface shear viscosity. This is associated with the presence of a pellicle or skin, such as an insoluble monolayer, but not restricted to that example, at the undisturbed, or static, liquid surface. A layer of denatured protein that stabilizes the foam of meringue, or of whipped cream, or of beer is a common example.
3. Dilatational (or compressional) viscosity. The surface elasticity that arises from local differences of surface tension is simultaneously associated with a resistance to surface flow. The local difference of surface tension is caused by dilatation (or compression) of the surface of the solution, so the resistance to flow that results from Marangoni counterflow is known as dilatational (or compressional) surface viscosity.

C. Mutual Repulsion of Overlapping Double Layers

Adsorption of ionic surfactants into the surface layer is evident in aqueous solutions and readily leads to the formation of charged surfaces of the lamellae in foams [16]. The counterions in the liquid interlayer of the lamella are the compensating charges. When the thickness of the lamella is of the order of magnitude of 20 times the Debye thickness of the electrical double layer, the counterions adjacent to the two opposite surfaces repel each other more or less according to an exponential increase of electrical potential with decreasing distance. This repulsion prevents further thinning of the lamella and so preserves it from imminent rupture.

The mechanism of charge separation that operates in water does not function in nonionizing solvents. Until relatively recently it was believed, therefore, that electrostatic repulsion of overlapping electrical double layers could not be a factor in stabilizing liquid lamellae in oil or other apolar foams. But we now recognize that mechanisms of charge separation other than electrolytic dissociation are possible, and indeed must operate; for zeta potentials of 25 to 125 mV have been observed for various kinds of particle dispersed in apolar media of low conductivity. Nevertheless, no evidence has yet been reported to suggest that foam may be stabilized by electrostatic repulsion in apolar solutions.

D. Theory of Foam Inhibition

The mechanism of foam inhibition by an insoluble liquid added for that purpose depends on a Marangoni effect, if that term is defined to refer to flow of liquid due to local differences in surface tension. The action of such an agent arises from its ability to spread spontaneously over the surface of the foam lamella. The lateral flow of the spreading liquid communicates a shearing stress to the liquid underneath so that the substrate, to a depth of several millimeters, is carried away by viscous drag as the liquid on top advances. A single drop of the agent, once it has arrived at the surface of the liquid lamella, performs in effect like a Venturi pump, ejecting on every side all the liquid lying beneath it as it advances over the surface, and causing the lamella to break by the agitation produced by its action. If the substrate is a foamable liquid and has formed a stable liquid lamella, the spreading action of the agent destroys it. The process just described can occur only when a certain relation obtains between the various surface and interfacial tensions. The agent will spread spontaneously if the value of the spreading coefficient S is positive, where S is given by

$$S = \sigma_1 - \sigma_2 - \sigma_{12} \tag{4}$$

where σ_1 is the surface tension of the foamy liquid, σ_2 is the surface tension of the foam-inhibiting agent, and σ_{12} is the interfacial tension between them. This principle is made use of in the application of dispersed droplets of poly(dimethylsiloxane) as a foam inhibitor in lubricating oils. The surface tensions of liquid hydrocarbons are between 25 and 30 mN/m; a liquid able to spread on such a low-energy substrate must have a surface tension lower by several millinewtons per meter. Such liquids are usually volatile, which makes them unsuitable in many applications. Only special polymers such as poly(dimethylsiloxane) or perfluorinated hydrocarbons combine the usually disparate properties of low surface tension and low volatility.

If S is negative, the insoluble drop of liquid may enter the surface without spreading on it, or may not even be able to enter the surface. This happens if the interfacial tension between the two immiscible liquids is too great. Foam inhibition is then greatly curtailed or does not occur at all.

Another action takes place when particles of a solid are introduced, on the surface of which the liquid medium makes a contact angle larger than 90° [17]. The substrate liquid withdraws from the solid; the process is described at "dewetting" and is analogous to the withdrawing or dewetting of an oily patch when paint is brushed over it. A liquid of higher surface tension cannot be made to coat a substrate of lower surface energy; it spontaneously withdraws if the attempt is made. The dewetting action of the liquid lamella on the surface of a hydrophobic solid provides sufficient mechanical shock to cause rupture of the lamella, the magnitude of the shock being proportional to the size of the particle. The difficulties to be overcome in putting this mechanism into effect are (1) to prevent hydrophobic particles from flocculating in the aqueous medium and (2) to avoid gravitational settling of the dispersed particles. The former difficulty is taken care of by introducing the solid encapsulated in oil, which oil is itself in the form of an emulsified droplet in the aqueous medium. The oil droplet enters the surface of the lamella, spreads away from its encapsulated solid, and so leaves the hydrophobic particle precisely where it can do the most harm to its supporting substrate, i.e., the liquid foam lamella, which then destroys itself by its dewetting action. The other difficulty is met by control of the particle size, so that the particle is as large as it may be without settling out of the system.

II. RELATION OF CAPILLARITY TO PHASE DIAGRAMS

A. Introduction

An amphipathic solute can be so called only with respect to a particular solvent. Such a solute contains in its molecular structure some moiety that interacts strongly with the solvent, whether it be by solvation, hydrogen bonding, or acid-base interaction (these may be merely different names for the same effect), and another moiety that has no specific interaction with the solvent. The former moiety is termed *lyophilic* and the latter *lyophobic*. The lyophilic moiety confers solubility, and the lyophobic moiety ensures that the solubility is limited. An example of an amphipathic solute for a hydrocarbon solvent is a molecule containing a hydrocarbon moiety, which interacts with the solvent by dispersion force attraction, and a polyhydroxy moiety, which, in spite of indiscriminate dispersion force attraction with solvent, has a stronger specific interaction by hydrogen bonding, probably involving traces of water in the system, with solute, thus reducing the solubility of the whole molecule. Such a solute would be recognized as a typical oil-soluble detergent, e.g., the Spans of I.C.I. Americas, Inc.

If such a solute and solvent are liquids, their combination in a two-component system is likely to lead to partial miscibility. The moieties need not even be chemically linked in one molecule; a combination of two solutes, incompatible when together, may be brought into solution by a third amphipathic component, as, for example, a mixture of water and benzene may be dissolved by a cosolvent such as ethanol. Again, such a combination is likely to include partial miscibility at some part of its (ternary) phase diagram. The position of a solution of a given concentration on the phase diagram may indicate degrees of heteromolecular interaction: compositions near a phase boundary have a weaker solute-solvent interaction than those farther from a phase boundary. Adsorption is a precursor of imminent phase separation: the surface offers a region for partial segregation of molecules prior to their more complete separation as a bulk phase. As compositions approach those of a phase boundary, therefore, adsorption increases and other interfacial phenomena associated with surface activity begin to occur. A propensity toward phase separation is therefore a general guide to surface-active behavior. A prescient but passing and incidental remark to this effect was made by Langmuir 50 years ago. Langmuir wrote: "In mutually saturated liquids, especially near the critical temperature, the conditions are favorable for orientation and segregation of the molecules in the liquid" [18]. Adsorption of solute is usually accompanied by micellization, and micelles are well known to occur near the critical temperature.

Industrial foaming problems were the stimulus that provided specific examples to confirm Langmuir's insight. Foaming in distillation and fractionation towers, for example, by which the liquid is carried into spaces intended for the vapor, is called *foam flooding* and is often encountered. Degasification after gas absorption or during "stripping" of a monomer in emulsion polymerization is a further example of the same effect. The cause of the foam stability may occasionally be traced to the presence of unintended contamination by minute concentrations of substances of strong surface activity; but usually the profoaming solute is a legitimate component of the system, present in substantial concentration. Such a component is not a general surfactant, but becomes surface active under certain conditions of temperature and concentration in a medium in which these conditions are conducive to decreasing its solubility. These conditions may occur in the process of fractionation, stripping, etc.; even so, the surface activity thus elicited is minimal, capable of stabilizing foam for very short times. Although such evanescent foams last no more than several seconds, they are the source of severe foam-flooding problems in distillation or fractionation towers, where rapid evolution of vapor may build up a large volume of dynamic foam. Gas-liuqid contacting is the basis of many production processes. For example, in the Girdler sulfide (GS) process for producing heavy water, liquid water and gaseous hydrogen sulfide are brought into contact at high temperature and at pressures of several atmospheres over sieve trays [19–21]. In all such processes, an adequate interfacial area is maintained for good mass transport but excessive foaming may lead to instability, especially in equipment such as sieve trays. Foaming is indeed so prevalent in this type of industrial equipment that design engineers routinely provide extra capacity in gas-liquid contact towers.

Hitherto the accepted theory of foaming in fractionation or distillation towers has traced the cause to Marangoni flow, induced by a difference in composition, and hence in surface tension, between thin films of liquid at the bubble caps and in the bulk liquid phase from which they originate [22]. Liquid films on walls of the container, or foam lamellae, because of their extended surfaces, evaporate faster than does a bulk liquid phase; if the loss of the more volatile constituent causes the surface tension of the residual solution to increase, then the liquid of greater surface tension draws liquid away from that of smaller surface tension, and so the stability of the liquid lamella is maintained.

This rule is far too inclusive: in normal liquids, volatility and small surface tension stem from a common cause, namely relatively small forces of intermolecular attraction, and therefore occur together, so that, with certain exceptions such as the poly(dimethylsiloxane)s, the usual behavior of solutions leads to the surface tension of the residual solution becoming greater after the loss of its more volatile components. According to this rule, therefore, foaming within a fractionation tower would be an almost universal condition. In practice, the problem of foaming is less prevalent than this rule predicts.

A more selective prediction can be obtained by studying the phase diagram of the system. Surface activity, and hence a propensity toward foaminess, is specifically inherent in solutions only at certain temperatures and compositions that are related to solubility curves and other features of the phase diagram, as well as to the relative surface tensions of the components. Although Marangoni effects may arise as a result of the volatility of a component of small surface tension, these are less significant in stabilizing foam than are Marangoni effects derived from adsorption of solute at the liquid-gas surface, i.e., surface activity.

The surface activity of a solute is not primarily due to an amphipathic molecular structure, but to a weak interaction with the solvent. This interaction must still be sufficient to dissolve the solute, but need be no greater than the least degree required to do so. A tendency toward phase separation is a general indicator of surface activity. This statement is akin to Lundelius' rule [23] that the least soluble materials are the most readily adsorbed. Thus, for example, the foam stability of gelatin solutions is greatest at the isoelectric point, where the gelatin is least soluble. Similarly, the foaminess of polymer solutions is maximum in poorer solvents, declining again, however, when insolubility supervenes [24]. Not surprisingly, therefore, the surface activity displayed by solutions or mixtures can be related to the phase diagram.

B. Experimental Observations

Experimental observations of the foaminess of binary and ternary solutions in systems in which a miscibility gap exists show that the foaming of a solution reaches it maximum under conditions of temperature and concentration where a transition into two separate liquid phases is imminent. Figure 1 shows one such diagram, for the system 2,6-dimethyl-4-heptanol and ethylene glycol [25]. Superimposed on the diagram by means of dashed lines are interpolated contours of equal foam stability (isaphroic lines, from *aphros*, the Greek word for foam). The isaphroic lines center about a point (the epicenter) close to the critical point as a maximum and decrease in value the farther they are from it.

Two-component systems show maximum foam stability at a temperature and composition near that of the critical point, and three-component systems show maximum foam stability at compositions near that of the plait point; but only as long as the systems are maintained as homogeneous one-phase solutions. The slightest degree of separation of liquid phases produces a conjugate solution that can defoam its foamy conjugate. Both these effects, the foaming enhancement in the one-phase solution and the foam inhibition in the two-phase solutions, can be ascribed to the surface activity of the component of lower surface tension in the system, which reveals itself in the one-phase solution by adsorption at the surface, and in the two-phase solutions by the positive spreading of one conjugate on the other. The occurrence of surface activity in the vicinity of critical or consolute points is well established experimentally by these and many other similar types of observation.

Foam stability is an indirect indication of surface activity: it is affected by other factors such as viscosity, and often fails to make itself manifest even when surface tension of the solution is less than that of the solvent. The fundamental index of the surface activity of a solute in any solvent is positive adsorption, which is measured quantitatively by the excess surface concentration of solute. For a binary solution, this quantity is given by $\Gamma_2{}^G$ in the well-known adsorption-isotherm equation of Gibbs [see Eq. (1)]. If Eq. (1) is applied to treat experimental data for a binary system at concentrations greater than about 0.01 M, activities rather than concentrations must be used. Nishioka et al. [26] used available data on surface

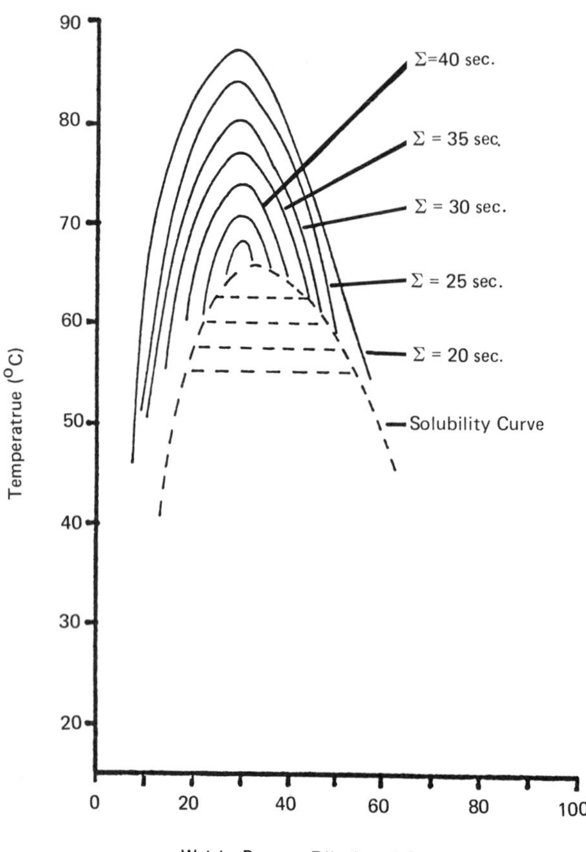

FIG. 1 Phase diagram and interpolated isaphroic contours of the two-component system 2,6-dimethyl-4-heptanol (diisobutyl carbinol) and ethylene glycol, showing an epicenter of maximum foaminess near the critical solution temperature. Foaminess is measured in terms of Σ, the lifetime of a bubble in the foam. (From Ref. [25] by permission.)

tensions and activity coefficients as functions of composition and temperature to calculate Gibbs excess concentrations of solute for the binary system diethylene glycol and ethyl salicylate. The Gibbs excess concentrations were then plotted as cosorption contours superimposed on the phase diagram for the system (see Fig. 2). The thesis that surface activity is the precursor of phase separation is amply confirmed by these observations of increasing surface activity of the unsaturated solutions as they approach saturation, with maximum surface activity near the critical solution point. In all solutions, ethyl salicylate, the component of lower surface tension, is the surface-active component. The maximum surface excess has an epicenter that does not correspond exactly to the consolute point, but is shifted toward the component of higher surface tension (diethylene glycol).

The resemblance of the foaminess (isaphroic) contours shown on Fig. 1 and the cosorption contours shown on Fig. 2 is striking, even though the diagrams pertain to different systems. Later work [27] compared the cosorption contours of Fig. 2 with isaphroic contours for the same system, and the resemblance confirmed the relation of surface activity and its manifestation in foaminess to the character of the phase diagram.

C. Regular Solution Theory

Nishioka et al. [26] compared these experimental results with a theoretical model of the same kind of system, using the "two-surface-layer" regular-solution model of Defay and Prigogine [28]. In this model only the top two monomolecular layers are considered to differ in composition from the bulk phase, arising from molecular coordination numbers less than those of molecules in the bulk phase. Consider a molecule in the bulk phase: it has z nearest neighbors, of which jz are in the same lattice plane, where j is the fraction of nearest neighbors in that plane (e.g., 6/12 for a close-packed lattice, 4/6 for a cubic lattice), and mz are in either contiguous plane, where m is the fraction of nearest neighbors in that plane (e.g., 3/12 for a close-packed lattice, 1/6 for a cubic lattice). Then

$$j + 2m = 1 \tag{5}$$

A molecule in the surface layer has a smaller number, z', of nearest neighbors given by:

$$z' = (j + m)z = (1 - m)z \tag{6}$$

Employing the Bragg-Williams approximation and assuming that the molecular surface areas of the two components are the same, the

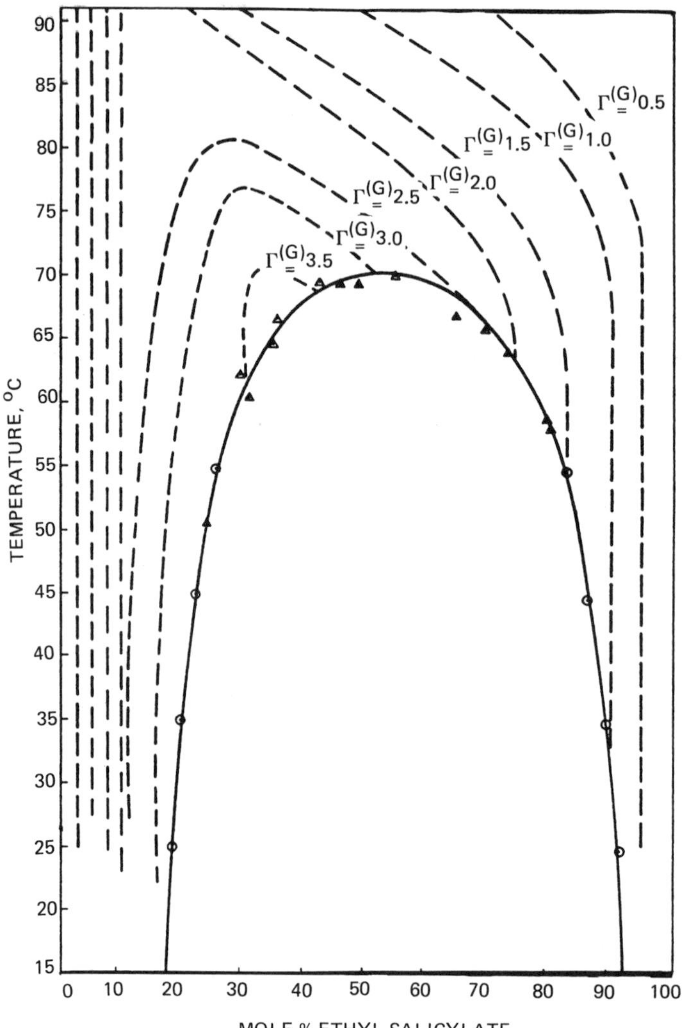

FIG. 2 Phase diagram and cosorption contours for the diethylene glycol-ethyl salicylate system. All values of surface excess concentration are × 10^{10} mole ethyl salicylate/cm^2. (From Ref. [26] by permission.)

compositions in the upper two layers are given by the following two equations:

$$\log \frac{x_1 x_2''}{x_1'' x_2} - \frac{2\alpha}{RT}(x_2'' - x_2) - \frac{4\alpha m}{RT} \frac{(x_2 + x_2' - 2x_2'')}{2} = 0 \tag{7}$$

$$\frac{RT}{\alpha} \log \frac{x_1 x_2'}{x_2 x_1'} + \frac{a(\sigma_2 - \sigma_1)}{\alpha} + 2j(x_2 - x_2') + m(x_2 - x_1)$$

$$+ 2m(x_2 - x_2'') = 0 \tag{8}$$

where x_1 and x_2 are the mole fractions of components 1 and 2 in the bulk phase; x_1' and x_2' the mole fractions of 1 and 2 in the top monolayer; x_1'' and x_2'' the mole fractions of 1 and 2 in the intermediate monolayer; α the interaction constant (2RT for a regular solution); σ_1 and σ_2 the surface tensions of pure components 1 and 2; and a the area per molecule.

Equations (7) and (8) were solved numerically for x_2' and x_2'' as a function of the composition of the solution (x_1). A close-packed lattice was assumed and the term $a(\sigma_2 - \sigma_1)/\alpha$ in Eq. (7) was assumed to be constant and equal to -0.5. This assumption was based on reasonable values of $a = 0.30\ nm^2$ per molecule, $T = 300$ K, and $(\sigma_2 - \sigma_1) = -6.9$ mN/m. The surface excess concentration of component 2 is then:

$$\Gamma_2^G = \frac{x_2' - x_2 + x_2'' - x_2}{\omega x_1} \tag{9}$$

where ω is the surface occupancy in square centimeters per mole.

The dimensionless surface excess Γ_2^G calculated for a regular solution is shown in Fig. 3 as a series of contours of equal surface concentration, called cosorption lines. In this hypothetical system the component of lower surface tension is the surface-active component at all points in the phase diagram. The maximum in surface activity, the epicenter, occurs at a point on the solubility curve, thus agreeing with Lundelius' rule; but it does not coincide with the critical point, being biased toward a higher concentration of the component of higher surface tension. Calculations based on the theory disclose that the greater the difference in surface tension between the two components, the greater is the bias toward that side of the composition scale, and also that the greater the bias, the greater is the surface activity.

Figure 3 shows that at higher temperatures a regular solution tends toward having maximum surface activity at a mole fraction of

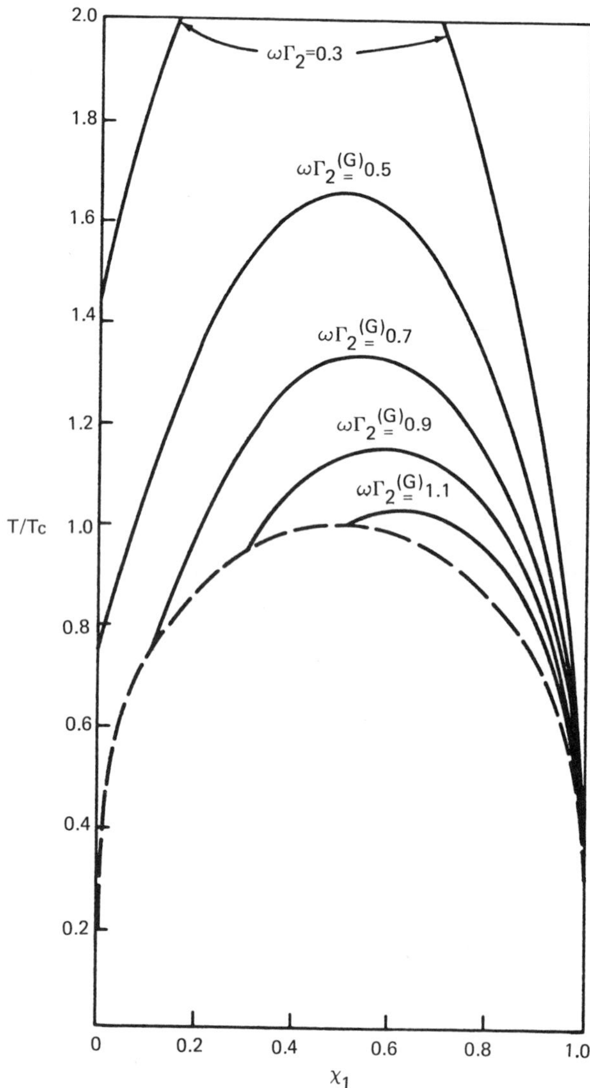

FIG. 3 Dimensionless surface excess concentration of component 2 in a two-layer regular-solution model. Component 1 is assumed to have an appreciably larger surface tension than component 2. (From Ref. [26] by permission.)

one-half, but experimental data do not confirm this prediction: they show the maximum shifting toward still higher concentrations of the component of higher surface tension. This behavior is also typically observed for solutions at temperatures far above a critical point, such as water-ethanol solutions at room temperature, which have a maximum surface activity at 5 to 10% alcohol. Nevertheless, Fig. 3 still bears a marked resemblance to published diagrams showing observed foam stabilities of two-component systems as functions of composition and temperature.

D. General Similarity of Cosorption and Isaphroic Contours

If the surface of a solution containing a surface-active solute is expanded suddenly, then, before adsorption from the bulk phase to the newly created surface has had time to take place, the same number of adsorbed molecules remain on the surface, so that the differential increase in area is the same as the differential increase in area per molecule, i.e.,

$$d\alpha = d\,\frac{A}{n_s} = \frac{dA}{n_s} \tag{10}$$

where α is the area per adsorbed molecule, A is the total area, and n_s is the number of molecules of adsorbed solute on the total area.

The suddenly expanded surface is not a stable state, but subsequent changes are so slow on a molecular scale that one can assume it to be at thermal equilibrium. Thermal equilibrium implies that thermodynamics is applicable even though the states of the system are not in mechanical equilibrium. We may therefore use equilibrium equations of state for the surface film at each stage of its path to final equilibrium. Let us suppose initially that the adsorbed surface film is described by the two-dimensional ideal equation of state:

$$\pi\alpha = (\sigma_0 - \sigma)\alpha = kT \tag{11}$$

or

$$(\sigma_0 - \sigma) = RT\Gamma_2^G \tag{12}$$

where $\Gamma_2^G = n_s/N_0A$ mole/cm^2

Gibbs elasticity is defined as the ratio of the increase in the surface tension resulting from an infinitesimal increase in area and the relative increment of the area. For a lamella with adsorbed films on both sides, the elasticity E is given by:

$$E = 2 \frac{d\sigma}{d \ln A} \tag{2}$$

Differentiating Eq. (11) and substituting in Eq. (2) gives

$$E = 2RT\Gamma_2^G \tag{13}$$

Equation (13) says that at a given temperature the Gibbs elasticity is proportional to the Gibbs excess surface concentration in the case of a two-dimensional ideal equation of state. It is not likely, however, that solutions that are sufficiently surface active to show foaminess would have adsorbed films of solute so dilute as to be described by the ideal equation of state. For more compressed films the elasticity is given by a few terms of a power series in Γ_2^G:

$$E = 2RT\Gamma_2^G - b\left(\Gamma_2^G\right)^2 + c\left(\Gamma_2^G\right)^3 \tag{14}$$

The direct proportionality between E and Γ_2^G is not maintained for high levels of surface activity; yet it remains a sufficiently good approximation to account for the similarity between the isaphroic contours and the cosorption contours, based on the entirely reasonable supposition that the foams we are discussing are stabilized by the Marangoni effect alone, which is measured by the Gibbs elasticity.

Gas-liquid contacting is the basis of many production processes, in which an adequate interfacial area is required for good mass transport, but excessive foaming leads to instability, especially in equipment such as sieve trays. Problems of this kind may be anticipated and avoided by studying the phase diagram of the system.

III. THREE REGIMES OF CONCENTRATION OF SURFACE-ACTIVE SOLUTES

A. Introduction

The presence of a surface-active solute in an apolar solvent may cause effects such as emulsified water, air entrapment, or foam. These effects are pernicious in many applications, for example, in lubrication. Such striking effects, when they occur, clearly imply a condition of surface activity as their necessary precursor; but surface activity in an apolar solution does not always make itself manifest by conspicuous phenomena: it may be subtly present. Detection of the condition, in the absence of the clear evidence afforded by foaming or emulsification, is not as straightforward in an apolar solvent as it is in aqueous solution. Most organic liquids have low surface tensions, usually less than 30 mN/m, and surface activity in such solvents may be a matter of lowering the surface tension by only a few tenths of a millinewton per meter. Observing a reduction of the surface tension by the solvent, which is the direct method for detecting the presence of a surface-active solute, is not a feasible procedure with solvents that initially are not chemically pure. Methods for detecting dynamic surface tension are better suited to reveal surface activity in the solvent.

Many interfacial phenomena, for example, emulsification, dispersion, foaming, and wetting, depend chiefly on a complex interaction of dynamic or nonequilibrium properties of the interface, such as dynamic surface tension and dilatational elasticity. The velocity of rise of a single bubble through a solution containing a surface-active solute is the simplest and best understood example of the influence of these properties of the interface. One limitation, however, common to all methods, must be remarked: they can all definitively demonstrate the *presence* of surface activity in a solution; but the absence of a positive response, by whatever method used, is a necessary but not sufficient reason to conclude that a surface-active solute is *not* present. The disequilibrium, on which dynamic surface tension depends, may be over before it is detected.

Several studies of this subject have been reported pertaining to aqueous solutions of surface-active solutes, but few studies of apolar systems are on record. Water is an atypical solvent, however, and offers special difficulties owing to adventitious contamination: one part of a surface-active solute in 10 million parts of water is enough to stabilize a single bubble and to retard its rate of rise through the solution. Nonaqueous solvents are much less susceptible to these effects of contaminants and besides are worth study on their own account and for comparison with what is known of the behavior of aqueous systems.

The rate of rise of single gas bubbles in a solution is determined by the behavior of newly formed interfaces that are subject to forces of dilatation and compression. In a pure liquid, under conditions of laminar flow, a rising bubble moves faster than predicted by Stokes' law, since the mobility of its interface allows lower velocity gradients in the liquid than those that develop at an immobile, or rigid, interface. When surface-active solute is adsorbed at the interface, however, the movement of the interface and of air inside the bubble is restricted; the velocity gradient in the outer fluid is increased, until at the limit the bubble acquires the properties of a rigid sphere, and its rate of rise is reduced to that given by Stokes' law.

B. Theories of Rate of Bubble Rise at Moderate Reynolds Numbers

1. Spherical Bubble

A perfectly spherical bubble (the Platonic ideal) is the form assumed at the mathematical limit of a Weber number of zero, where the Weber number W is defined as:

$$W = \frac{2a\rho V_c^2}{\sigma} \tag{15}$$

where σ is the surface tension of the liquid; ρ is its density; a is the radius of curvature at the top of the rising bubble; and V_c is the corrected rate of rise of the bubble.

The quantitative relation between the eccentricity of the bubble's shape and the Weber number was investigated experimentally by Wellek et al. [29], who showed, in particular, that at $W \leqslant 0.1$ the deviation of the eccentricity from unity does not exceed 0.01. For the application of this method, that condition is a requirement, to ensure that the deviation from the sphere is negligible.

2. Rise in Pure Liquids—Hadamard Regime

Stokes' theory for the terminal velocity of a solid sphere in a viscous medium was extended by Rybczynski [30] and Hadamard [31] to fluid spheres. For a liquid drop or a gas bubble of radius a, density ρ_1, and viscosity η_1, moving at constant velocity, V_c, through an infinite volume of a medium of density ρ_2 and viscosity η_2, the drag coefficient C_D is related to the Reynolds number Re as follows:

$$C_D Re = \kappa 24 \tag{16}$$

where

$$C_D = \frac{8ag}{3V_c^2} \quad \text{and} \quad R_E = \frac{2a(\rho_2 - \rho_1)V_c}{\eta_2} \tag{17}$$

g is the gravitational constant and κ, the Rybczynski-Hadamard correction factor, has the value:

$$\kappa = \frac{3\eta_1 + 2\eta_2}{3\eta_1 + 3\eta_2} \tag{18}$$

The derivation of Eq. (18) postulates that the medium exerts a viscous drag on the surface of the bubble or liquid drop, and so sets up a circulation of the fluid contained inside, whether gas or liquid. According to the theory, a bubble containing a circulating gas, with $\eta_1 \ll \eta_2$, would move 50% faster ($\kappa = 2/3$) than one in which the gas, for any reason, does not circulate; for in the latter case the Rybczynski-Hadamard factor does not apply, and the velocity of the bubble is given by the unmodified form of Stokes' law:

$$C_D Re = 24 \tag{19}$$

Garner and co-workers [32,33] demonstrated experimentally the existence of the circulation inside air bubbles and examined the effects of bubble size and shape. They showed that the validity of the above equations is limited to the range of Reynolds number less than about one ($Re < 1$) and to the same conditions as for Stokes' law to hold, including the requirement that the gas bubbles be spheres. Within these conditions the Rybczynski-Hadamard expression for an ascending gas bubble whose gas-liquid interface is characterized by a constant surface tension is:

$$C_D Re = 16 \tag{20}$$

Ryskin and Leal [34] recently presented a numerical solution of the problem of a rising bubble in a liquid by integrating the forces at the surface, for Reynolds numbers in the range $0.5 < Re < 200$ and for Weber numbers up to 20. Ryskin and Leal take the dynamic air-liquid interface to be completely characterized by a constant surface tension, that is, to be spatially uniform, which is to assume, in effect, that it is free of adsorbed surface-active solute and is isothermal.

Theoretically computed rates of rise of bubbles with fluid interfaces at high Reynolds numbers are tabulated by Ryskin and Leal. Of these computations we require for this application only those that refer to spherical bubbles, that is, to Weber numbers of zero. The numbers from Table 1 of [34] for $W = 0$ and $0.75 \leqslant Re \leqslant 20$ are fitted to the following algebraic interpolation:

$$C_D = -0.023 + \frac{16.780}{Re} - \frac{0.244}{Re^2} + 1.021 \tag{21}$$

and for $Re < 0.75$ the Rybczynski-Hadamard equation, Eq. (20), applies.

3. Effect of Surface-Active Solutes—Stokes Regime

Frumkin and Levich [35] first provided a satisfactory explanation of the retardation of the velocity of a rising bubble caused by surface-active solutes in the medium. They postulated that adsorbed solute is not uniformly distributed on the surface of a moving bubble. The surface concentration on the upstream part of the bubble is less than the equilibrium concentration, while that on the downstream part is greater than equilibrium. This disequilibration of the concentrations is brought about by the viscous drag of the medium acting on the interface, which in turn creates a disequilibrium of surface tensions, with the lower tension where the concentration of adsorbate is greater. The liquid interface then flows (Marangoni flow) from the region of lower tension to that of higher tension, and the direction of this flow offsets the flow induced by the shear stress in the outer fluid acting on the interface. This action restrains the net flow of the interface; its consequent loss of fluidity, or, if you will, its increasing rigidity, reduces the circulation of internal gas to a greater or lesser degree. When the interface is completely rigidified, by a sufficiently large gradient of surface tension, the circulation of internal gas in the bubble is inhibited; both the velocity gradient in the medium and the terminal velocity of rise then become indistinguishable from those of a rigid sphere of the same density.

The criterion for a rigid interface on a rising bubble is that at Re less than approximately 1 its velocity of rise be described by Stokes' law. Kuerten et al. [36] give empirical formulas for rates of motion of solid spherical particles at higher Reynolds numbers. The following equation is based on numerous experimental measurements of the drag coefficients of solid spheres in fluid media at $0.75 < \mathrm{Re} < 20$:

$$C_D = \frac{24}{\mathrm{Re}} + 2 \tag{22}$$

Boussinesq [37] offered a different explanation for the same effect, in terms of an enhanced viscosity of the interface when a surface-active solute is present in the medium. In such a case a reduced rate of internal circulation would occur on both sides of the interface. But a gaseous adsorbed monolayer is adequate to prevent surface flow by the Frumkin-Levich mechanism, while the surface shear viscosity of such a monolayer is low. The Boussinesq and Frumkin-Levich explanations are identical if, for the static surface shear viscosity that Boussinesq had in mind, the concept of a dynamic dilatational viscosity is substituted.

C. Experimental Observations

1. Apparatus and Procedure for Measuring Rate of Bubble Rise

By means of an extended syringe, single bubbles of gas are released at a capillary orifice (see Fig. 4) into oil contained in a graduated

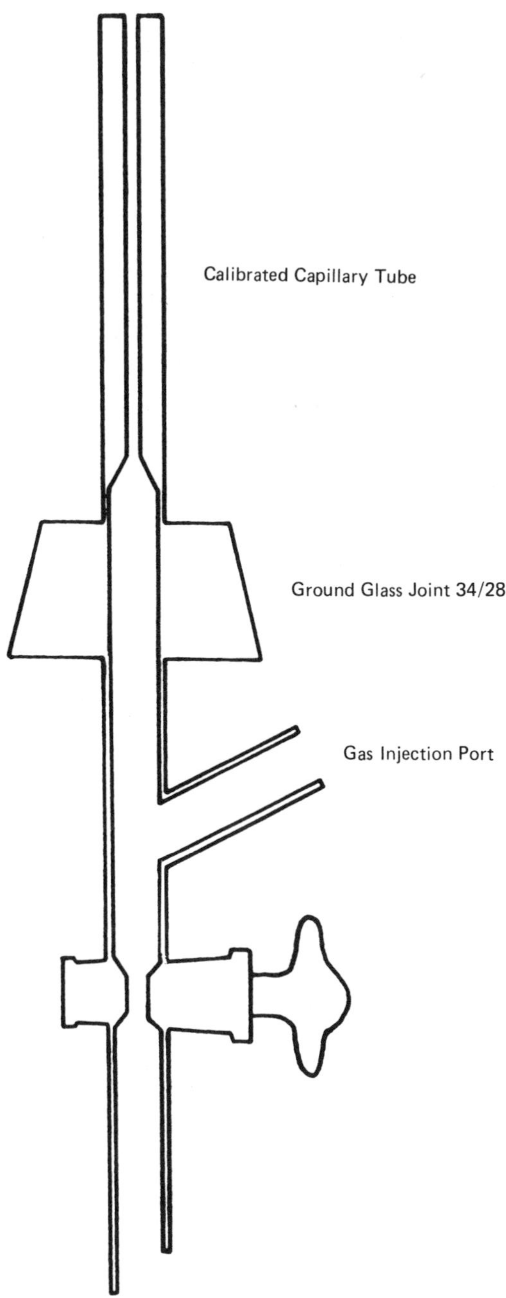

FIG. 4 Schematic diagram of part of the apparatus for measuring bubble size and rate of bubble ascent at a controlled temperature. (From Ref. [39] by permission.)

cylinder 70 cm high and 26 mm in internal diameter. The volume of air in each bubble is measured before its release, while the air is still contained inside a capillary tube of known diameter (0.397 mm), by measuring the length of the air slug with a cathetometer with a precision of ± 0.01 mm. The Stokes velocity of rise is calculated by use of Eq. (19). The time for each released bubble to rise through a specified distance (42.0 cm) is measured with a stopwatch (± 0.1 sec). The measured velocities are corrected for the effect of the presence of the confining walls of the cylinder by the Ladenburg-Faxen formula [38]:

$$V_c = V_m \left(\frac{1 + 2.1\ d}{D}\right) \tag{23}$$

where V_c is the corrected rate of rise; V_m the measured rate of rise; d the bubble diameter; and D the diameter of cylinder. For bubbles with an average diameter of 1.1 mm, rising in a tube of diameter 26 mm, this equation gives a correction of 8.9%. The temperature is controlled by pumping thermostated water through the cylinder's jacket. To minimize temperature gradients, the cylinder is enclosed in thermal insulation.

2. Experimental Results: Behavioral Limits

To obtain the rate of bubble rise in any solution at any temperature, a number of separate measurements are made. Since it is impossible to replicate the bubble size each time, these data give the variation of bubble velocity with diameter. The Rybczynski-Hadamard theory embodies the implicit supposition that the factor κ in Eq. (16) is independent of bubble size, within certain well-understood limits. If bubbles are too small, $2a < 0.03$ cm, according to Garner and Hammerton [33], the surface tension gradient has too limited a scope to set up a vigorous toroidal circulation of the internal gas, and the theory is inapplicable; and if bubbles are too large they depart too far from the spherical shape for either Stokes' law or its Rybczynski-Hadamard modification to apply. Within these limits of size, the velocity of the bubble should vary with the square of its diameter, so that the factor κ, or the product $C_D Re$, is independent of bubble size. The first treatment of the data, therefore, is to plot the corrected velocity V_c against the square of the bubble diameter $4a^2$. Typical plots are shown in Fig. 5, measured on a synthetic-ester lubricant, Mobil Ester P-41, at different temperatures [39]. Such plots demonstrate that the implicit assumption of the Rybczynski-Hadamard theory is in accord with observation.

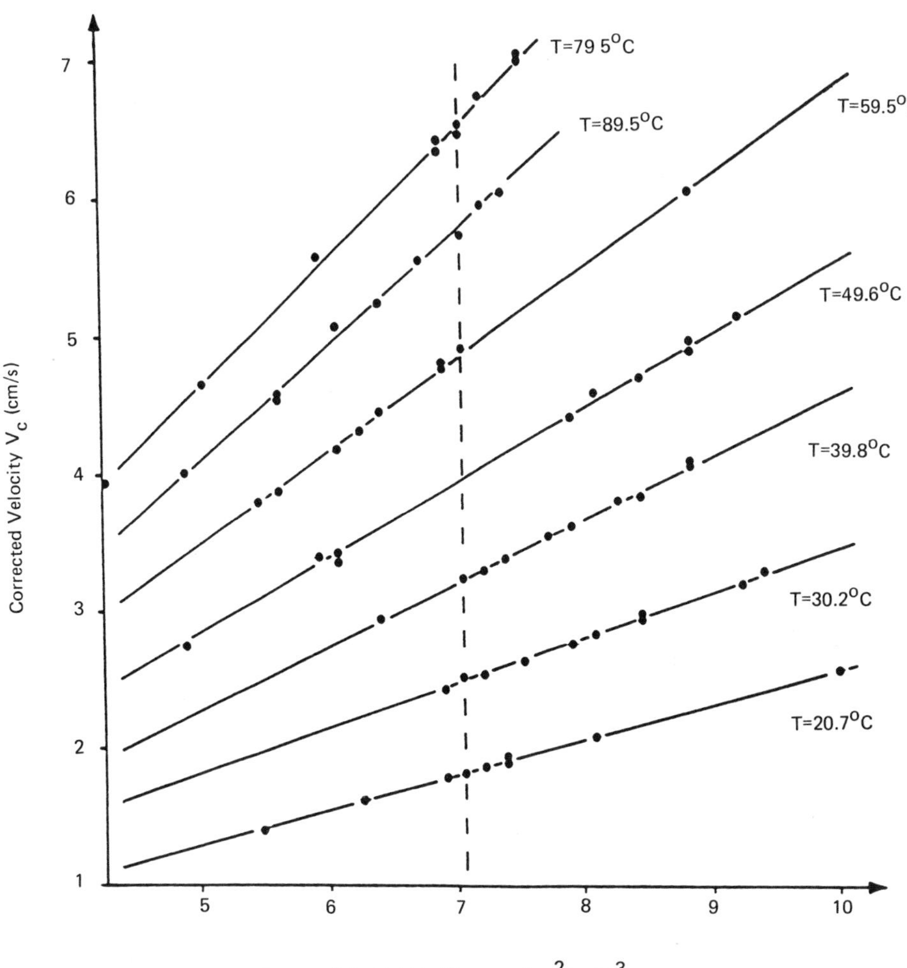

FIG. 5 Linear relations between the corrected velocity of bubble ascent and the square of the diameter of the bubble for a series of isotherms measured on Mobil Ester P-41. The dashed vertical line indicates an interpolation at 2a = 0.084 cm. (From Ref. [39] by permission.)

The next treatment of the data is to reduce the variety of measured bubble diameters to a single representative value, which can readily be done by using the linear relation between the square of the bubble diameter and the velocity of rise, then to interpolate for the value of the velocity corresponding to any desired diameter, from which drag coefficients can be calculated by Eq. (17). For the comparison on the theoretical side, use is made of Ryskin and Leal's tabulation of drag coefficients C_D at various values of Re and W, taking W = 0 for near-spherical bubbles. The relation between C_D and Re (W = 0) for $0.75 < \text{Re} < 20$ is given by Eq. (21), and for Re (W = 0) < 0.75 by Eq. (20). In Fig. 6, Eqs. (20) and (21) are plotted on a log-log scale to show the variation of C_D with Re for the two cases of bubbles with a fluid and with a rigid interface. Experimental data, represented in the diagram by discrete points, agree with these theoretical descriptions. Solutions at various temperatures in the range 20 to 80°C and bubbles of various sizes were used in obtaining these measurements, which, because of different bubble sizes and of changes of kinematic viscosity with temperature, give a series of different Reynolds numbers for the same solution.

The agreement between theory and observation is seen to be excellent. Most lubricants evidently show no dynamic surface tension; that is, their surface tension is not affected by any solute that may be present. All oils contain solutes; even those without additives are not chemically pure; but whatever solutes are present have little or no surface activity. Systems that conform to the theoretical description of a rising bubble with a constant surface tension are:

1. Mobil Ester P-41
2. Emolein 2917
3. Emolein 2917 + 5% w/w trimethyl-4-nonanol
4. Emolein 2917 + 5% w/w poly(propylene glycol) (molecular weight 4000)

The observations for the oil designated Mobil Ester P-41 agree with the Ryskin-Leal theory for a fluid interface, but the observations for the same oil containing 5% w/w *N*-phenyl-1-naphthylamine, lot A12A, do not conform to this description. They agree instead with an empirical equation suggested by Kuerten et al. based on numerous experimental observations of the drag coefficients of *solid* spheres in fluid media at $0.75 \leqslant \text{Re} \leqslant 10$, namely Eq. (22), and at $\text{Re} < 0.75$ with Stokes' equation, Eq. (19).

Comparisons of the results for the solvent Mobil Ester P-41 and the solution Mobil Ester P-41 + 5% w/w *N*-phenyl-1-naphthylamine, lot A12A, are shown in Figs. 7 and 8, where the solid lines represent Eqs. (19), (20), (21), and (22) while the observational data are represented by the positions of points. Figure 7 shows the

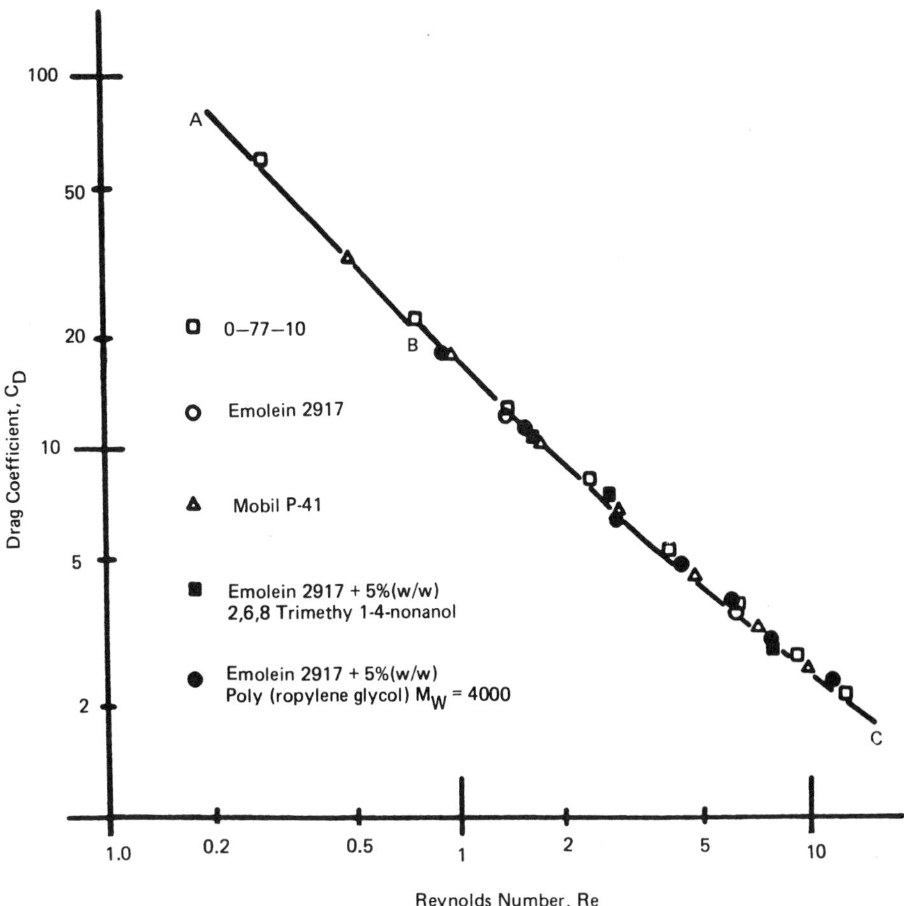

FIG. 6 Variation of the drag coefficient C_D with Reynolds number Re. Solid lines are theoretical equations; points are observational data. AB is the Rybczynski-Hadamard law, Eq. (20); BC is the Ryskin-Leal theory, Eq. (21). The data are for various synthetic lubricants, with 2a = 0.084 cm by interpolation. (From Ref. [39] by permission.)

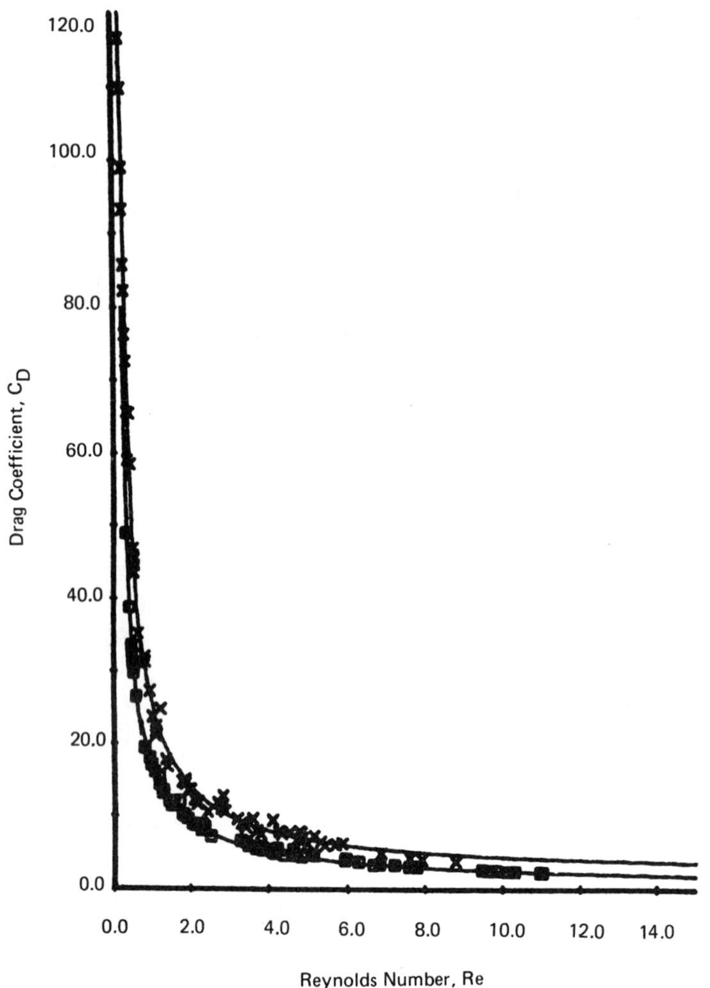

FIG. 7 Variation of the drag coefficient C_D with Reynolds number Re. Solid lines are theoretical equations; points are observational data. The lower curve refers to Mobil Ester P-41 and the upper curve refers to Mobil Ester P-41 + 5% (w/w) *N*-phenyl-1-naphthylamine, lot A12A. (From Ref. [39] by permission.)

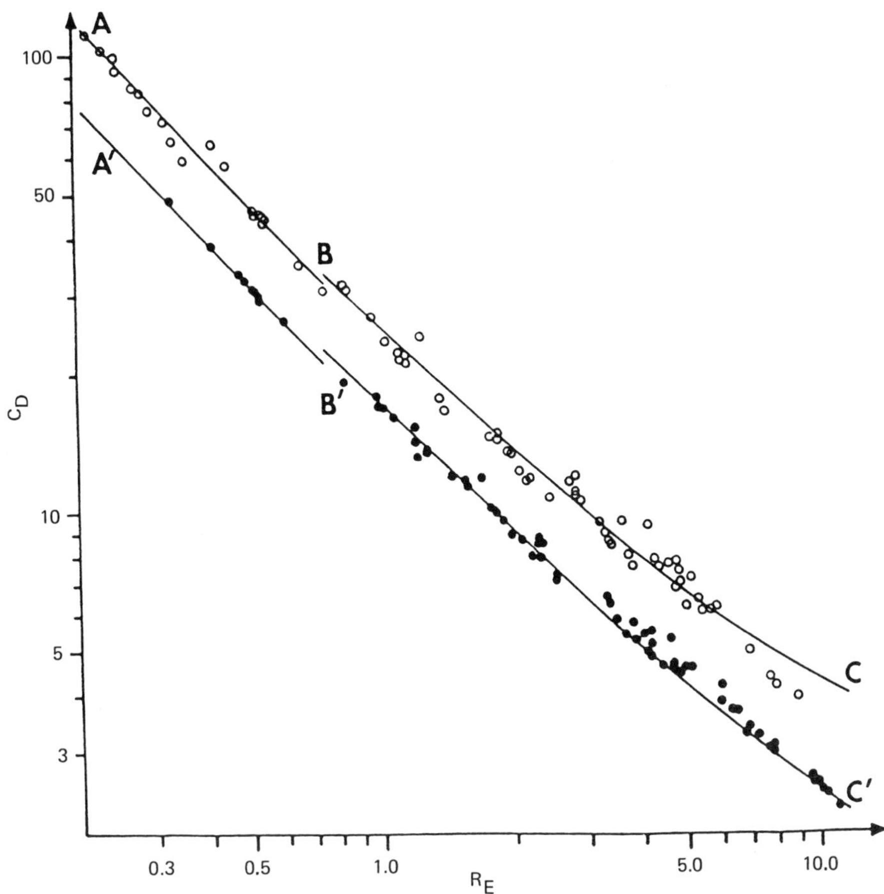

FIG. 8 Behavioral limits of the variation of the drag coefficient C_D with Reynolds number Re, on logarithmic scales. Solid lines are theoretical equations; points are observational data. AB is Stokes' law, Eq. (19); BC is Eq. (22); A'B' is the Rybczynski-Hadamard law, Eq. (20); B'C' is the Ryskin-Leal theory, Eq. (21). (From Ref. [39] by permission.)

nature of the variation of C_D with Re for the two cases of bubbles with a rigid and with a fluid interface; Fig. 8 shows the same data on a log-log plot, in which the two cases are better distinguished visually. As can be seen in Fig. 8, the experimental data conform to the theoretical description of a fluid or a rigid interface, whichever happens to occur in the system observed.

Mobil Ester P-41 is typical of oils in which the air-liquid interface is fluid, and the data for the solution of *N*-phenyl-1-naphthylamine in Mobil Ester P-41 show that the solute has caused the interface to become effectively rigid under the dynamic conditions of the experiment.

Subsequent tests with a different batch of *N*-phenyl-1-naphthylamine, namely lot A13B, from the same supply house (Eastman Kodak Co.) and used at the same concentration of 5% w/w disclosed it to have no effect on the oil Mobil Ester P-41. The surface activity observed in Figs. 7 and 8 for this solute is, therefore, probably due to an impurity in the solute, perhaps a trace of a silicone antifoam.

None of the above oil solutions in which additives had been dissolved showed any measurable reduction of the equilibrium surface tension of the solvent by the additive. This observation is true even of the solution in which a rigid interface was observed. It seems, therefore, that the apparent inability of a solute to reduce the equilibrium surface tension of an oil solution below that of the oil does not necessarily indicate the absence of surface activity. Even the present test, that is, the rate of ascent of a single bubble, is not a definitive indicator of the absence of surface activity in an oil solution. Foam tests showed that some of these solutions for which a fluid interface was demonstrated were able to stabilize bubbles, albeit briefly, and also that the system for which a rigid interface was observed produced no foam by a standard foam test [40].

Dynamic surface tension is manifested throughout a wide range of time scales, depending on various mechanisms of relaxation after the surface has been extended or contracted. A test for the effect, such as the present one, may well fail to show its existence if the time required for the operation of the test is much longer than the time required for the system to equilibrate. That appears to have been the case with some of the oil solutions reported above for which fluid interfaces were demonstrated.

3. Experimental Results: Transitional Behavior

The limits of the behavior, fluid to rigid, of the liquid-gas surface were first described by a theoretical model, which was then verified experimentally. It becomes of interest to follow the course of the transition between those limits, either as a function of concentration at a fixed temperature, or as a function of temperature at a fixed concentration. The former of these has been reported [41]. The isothermal transitions with concentration of solute are continuous; the data, best expressed as the concentration required to bring the transition to the halfway point, provide quantitative inverse measures of the surface activity of a solute in an apolar solvent. The technique is then of particular value in detecting either synergistic or antagonistic effects of combinations of solutes on surface activity.

The velocity of ascent of bubbles was measured by the technique described above in four series of solutions:

1. Sorbitan monolaurate (Span 20) in white mineral oil
2. Poly(dimethylsiloxane) (pdms) in white mineral oil
3. Poly(dimethylsiloxane) (pdms) in trimethylolpropane (tmp)-heptanoate
4. *N*-phenyl-1-naphthylamine in trimethylolpropane (tmp)-heptanoate

For each bubble, the ratio of observed velocity to Stokes terminal velocity was obtained from the relation:

$$K = \frac{V_c}{V_s} \tag{24}$$

For the mineral oil solvent the value of K obtained was 1.53, which is within the experimental error of the theoretical value of K = 1.50; this finding and the invariability of the velocity of ascent with height indicate that the solvent was free of surface-active contaminants. The dependence of the value of K on concentration of solute is shown in Figs. 9, 10, and 11. In Fig. 9, K is reported (A) for solutions of poly(dimethylsiloxane) (1000 cSt) in mineral oil in the range of concentration 0.1 to 60 ppm and (B) for solutions of Span 20 in mineral oil in the range of concentrations 50 to 7200 ppm. In Fig. 10, K is reported for solutions of poly(dimethylsiloxane) (1000 cSt) in tmp-heptanoate in the range of concentrations 0.1 to 30 ppm. Figure 11 records the values of K for solutions of *N*-phenyl-1-naphthylamine in tmp-heptanoate in the range of concentrations 0.01 to 5% w/w.

Figures 9, 10, and 11 show that the ratio of the observed velocity of ascent of a bubble to the calculated Stokes velocity varies between the limits $0.99 < K < 1.52$ for the solutions in mineral oil and the limits $0.95 < K < 1.47$ for the solutions in tmp-heptanoate. The values K = 1.52 for the mineral oil and 1.47 for the tmp-heptanoate demonstrate the virtual absence of any surface-active contaminant in these solvents. No significance is attached to the difference of 0.05 unit from the theoretical value of the lower limit, K = 1.00, as fluctuations of that magnitude at that limit are within the experimental error. Figures 9 and 10 also show that the range in which K is

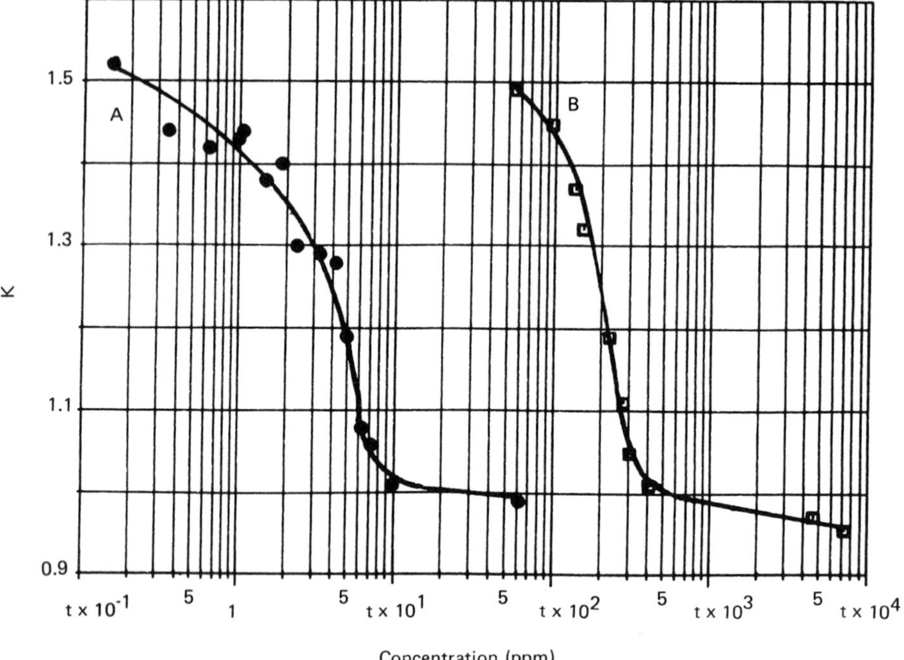

FIG. 9 Comparative rates of ascent (corrected for the effect of the wall) of air bubbles in mineral oil at 22°C containing (A) various concentrations of poly(dimethylsiloxane) (1000 cSt), and (B) various concentrations of Span 20 (sorbitan monolaurate). (From Ref. [41] by permission.)

concentration dependent is $0.1 < c < 20$ ppm for solutions of poly-(dimethylsiloxane) in mineral oil and in tmp-heptanoate. For solutions of Span 20 in mineral oil, the lowest concentration for any measurable variation in the value of K is 500 times greater; but the range of concentration through which the variation of K takes place is relatively narrower, i.e., $50 < c < 500$ ppm. The solute *N*-phenyl-1-naphthylamine in tmp-heptanoate requires a still greater concentration before its effect is found (see Fig. 11), showing that it is even less surface active in this solvent.

A simple model for the effect of surface-active solutes on the terminal velocity of drops and bubbles at small Reynolds numbers was proposed by Griffith [42]. He assumed that the solute is so slowly adsorbed or desorbed as to be essentially insoluble in the bulk liquid phase during the time of motion of the drop or bubble.

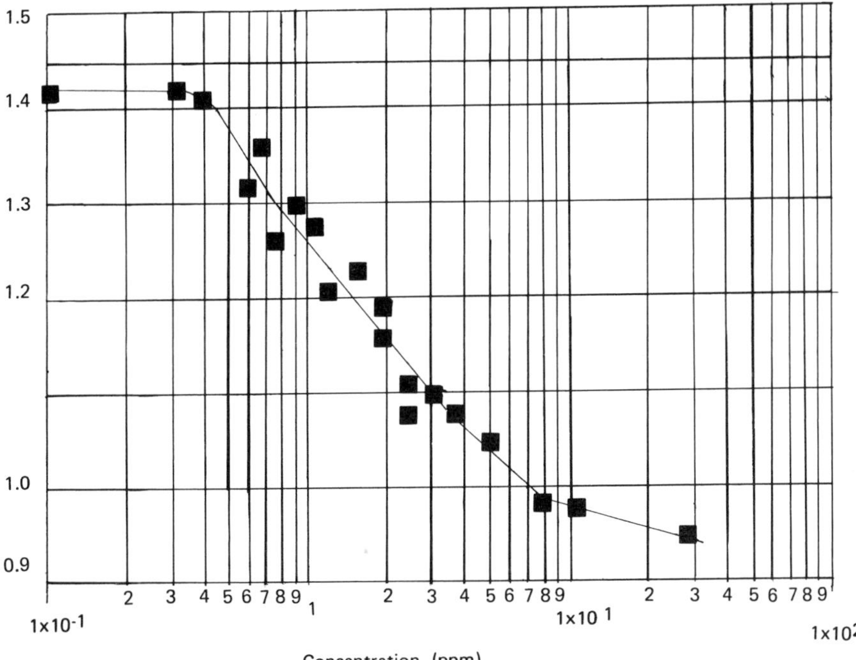

FIG. 10 Comparative rates of ascent (corrected for the effect of the wall) of air bubbles in tmp-heptanoate containing various concentrations of poly(dimethylsiloxane) (1000 cSt) at 22°C. (From Ref. [41] by permission.)

He further assumed that at the low surface concentrations of solute at which the transition from Hadamard to Stokes velocity takes place, the adsorbed solute behaves as an ideal two-dimensional gas. For these reasons he named his theory that of the *insoluble gaseous monolayer*. For the systems investigated these assumptions are indeed reasonable. A later extension of the theory in which the supply of surface-active solute is not limited to what is introduced initially [43] is a less likely model for the very dilute solutions of a strongly adsorbed solute or the more concentrated solutions of a weakly adsorbed solute that are the subjects of the work under discussion. The Griffith model is taken, therefore, as preferable. Griffith had already compared the results of his model with measured terminal velocities of ethylene glycol drops in mineral oil and carbon tetrachloride drops in glycerol. These observations were extended [41] to the case of air bubbles in solvents containing surface-active solutes.

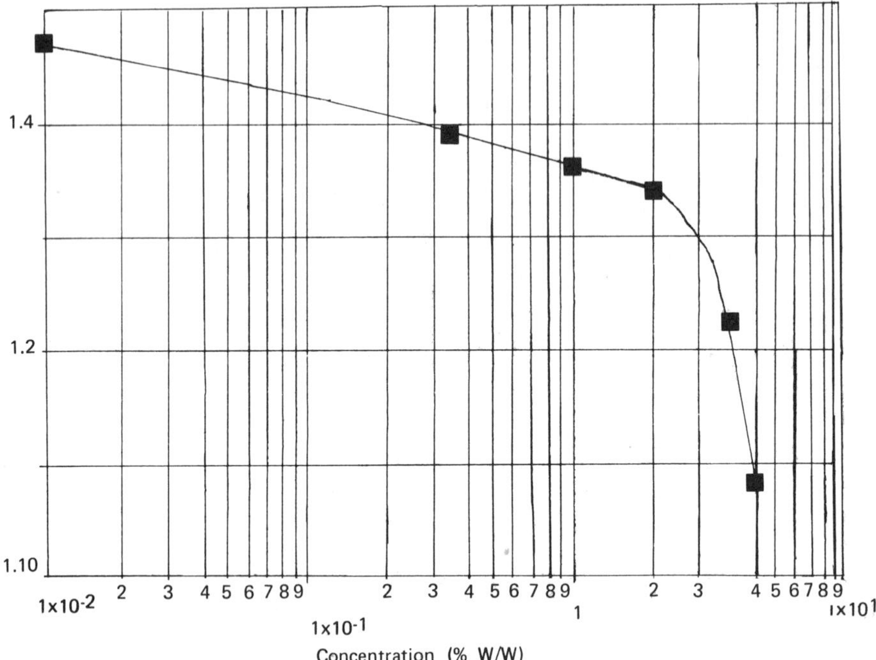

FIG. 11 Comparative rates of ascent (corrected for the effect of the wall) of air bubbles in tmp-heptanoate containing various concentrations of *N*-phenyl-1-naphthylamine at 26°C. (From Ref. [41] by permission.)

In the development of his theory, Griffith set the limiting surface tensions at the top ($\theta = 0$) and the bottom ($\theta = \pi$) as σ_i, the interfacial tension between the solute-free phases, and σ_e, the interfacial tension between the equilibrated phases, respectively. The compression of the insoluble monolayer at the rear of an ascending bubble would, however, cause the surface tension to fall below its equilibrium value, σ_e. The value of $\Delta\sigma$, the difference between σ_{max} and σ_{min}, the surface tensions at the top and bottom of the bubble, respectively, is therefore taken as the quantity to be evaluated from Griffith's plot of the dimensionless quantity $[(\rho_2 - \rho_1)ga^2/(\sigma_{max} - \sigma_{min})]^{\frac{1}{2}}$. The values of $\Delta\sigma$ so obtained are reported in Table 1. They show that relatively small differences of surface tension between the top and bottom of a bubble of 0.11 cm diameter establish a gradient sufficient to inhibit the shear stress in the gas phase inside the bubble.

TABLE 1 $\Delta\sigma$ Calculated by Griffith's Theory for Bubbles of 0.11 cm Diameter

Solution	Concentration	K	$\Delta\sigma$ (mN/m)
PDMS (1000 cSt) in mineral oil	0.5 ppm	1.46	0.09
	3.2	1.29	0.20
	10	1.01	0.72
Span 20 in mineral oil	55. ppm	1.49	0.09
	150	1.32	0.18
	400	1.02	0.65
PDMS (1000 cSt) in tmp-heptanoate	0.3 ppm	1.42	0.13
	0.9	1.28	0.23
	5	1.03	0.62
N-Phenyl-1-naphthylamine in tmp-heptanoate	0.05% w/w	1.44	0.11
	3.0	1.30	0.21
	4.8	1.09	0.44

In these studies, measurements of surface shear viscosity do not provide any evidence for higher viscosity in any of our solutions; therefore, Boussinesq's supposition [37] that a static plastic skin accounts for the rigidity of the bubble surface is ruled out.

The inhibition of movement of the surface and of air circulation inside the rising bubble is accepted as the explanation of the remarkable slowing of the bubble's rate of ascent. Okazaki et al. [44] found that a concentration of sodium dodecyl sulfate (SDS) in water as low at 10^{-5} M retards the rate of ascent of a bubble in that solution, and also confers stability to a single bubble at its surface, even though the static properties of the liquid, i.e., the density, bulk compressibility, viscosity, and surface tension, all remain unchanged from those of pure water. A time dependence of the surface tension is also absent in 10^{-5} M SDS in water, and no measurable increase of the shear viscosity of the surface of a solution so dilute was observed.

Oil solutions behave in many respects as do these very dilute aqueous solutions of SDS in water at concentrations of 10^{-6} to 10^{-4} M investigated by Okazaki et al. In oil systems, as in their aqueous systems, certain static and dynamic properties of the solution that might appear to be pertinent to surface activity show no change from those of the solvent, yet the rate of rise is greatly affected by the presence of the solute.

Foaminess of oil solutions, as measured by the standard test method for foaming stability of lubricating oils [40], shows that no necessary correlation exists between the retardation of the rate of bubble rise and the stability of a foam. Foams were not found to be stable at the concentrations of solute at which the effect described here were observed, although a single bubble floating on the surface of the solution is stabilized. From 10 to 100 times greater concentrations are required for production of foam by the test method used.

D. Effect of Colloidal Micelles—Robinson Regime

Robinson [45] reported experiments on the rate of rise of air bubbles in hydrocarbon lubricating oils. His experiments were designed to reveal the mechanism by which additives in a lubricating oil stabilize the "emulsified" air brought into the oil when it is circulated through a high-speed gear pump. His results established the existence of a third regime of concentration of a solute, a regime in which the velocity of rise of an air bubble in the solution is *less* than the computed Stokes law velocity, computed on the observed diameter of the air bubble.

The experimental method used by Robinson was similar to that reported above. Small air bubbles, up to 2 mm in diameter, were released from an extended syringe pipet into oil contained in a 100-ml graduated cylinder. The time was measured for the passage of the bubble between each pair of 10-ml graduations (equal to 1.90 cm). The diameter of the bubble was measured with a calibrated ocular micrometer set in a traveling microscope, at a magnification of about 10 times. The bubbles were formed in a capillary tube, with a U-turn at the end, inserted at the bottom of the cylindrical container.

A comparison was drawn between the behavior of a lubricating oil (A) containing no additives and one in which lubricating additives were present (B). The two oils were further differentiated inasmuch as oil B produced a stabilized foam from the air bubbles passing through it, whereas oil A did not. In terms of the rate of rise of air bubbles through these oils, the bubbles in oil A followed Stokes' law, the rate of rise being proportional to the square of the observed diameter and inversely proportional to the viscosity of the oil; the bubbles in B, the oil containing additives, rose more slowly than predicted by Stokes' law from their observed diameter, and the rate of rise decreased as the length of path the bubbles traveled increased.

A rate of rise less than that predicted by Stokes' law signifies a regime of behavior of a surface-active solute other than the Hadamard and Stokes regimes; and just as these regimes occur at different ranges of concentration, so the Robinson regime occurs at concentrations larger than theirs and well separated from them. The low concentration of solute at which the transition from Hadamard to Stokes regime takes place has already been remarked; the concentrations at which lubricant additives are used are many times greater than that (0.1% and up). Furthermore, the dynamic surface tension that determines Stokes-type behavior is not sufficiently pronounced to stabilize bubbles in a foam, although it does serve to stabilize for a short time the life of a single bubble, whereas Robinson's measurements of rates of rise less than that given by Stokes' law were taken with an oil able to stabilize foam.

The abnormally slow rate of rise of air bubbles in oils containing additives is accounted for by assuming that the rising air bubble is accompanied on its way by an ever growing stationary shell of oil, or additive, or both, which increases the resistance to passage of the bubble without a compensating increase in the oil displacement. Robinson calculated the factor by which the observed diameter of the bubble must be multiplied to obtain the outside diameter of the rigid shell of liquid carried by the bubble. This mode of description indicates the proportion of the shell for any size of bubble. He found that as the bubble rose through oil B, the factor increased from about 1 to nearly 2; this effect was completely absent in oil A, which contained no additive.

The immobilized shell was postulated by McBain [46] to have a structure similar to that of a liquid crystal, in which chains of oriented molecules of additive extend outward from a primary sorbed layer on the surface of the bubble, supplemented by cybotactic arrangement of the hydrocarbon molecules in the same region. Indeed, one of Robinson's oil solutions that showed abnormal retardation of the rate of rise of an air bubble, namely a solution consisting of oil A containing as additives 0.075% glycerol and 0.025% Aerosol OT, i.e., the sodium salt of bis-(2-ethylhexyl)sulfosuccinate, resembles in composition some of the solutions in isooctane investigated by Eicke and Kvita [47], who also used Aerosol OT but with water rather than with glycerol. In these solutions the presence of swollen reverse micelles was demonstrated. If, as is likely, Robinson's solution also contained reverse micelles, one may suppose that the air bubble by the time it reached the top of the column was weighted with a gathered burden of colloidal micelles. Similarly, in oil B the additives present were found to be concentrated in the foam, so they too were surface active. Whatever the fine structure of the immobilized shell might be in these oils, Robinson's observations account for the slow separation of finely divided air bubbles in lubricating oils containing certain additives.

ACKNOWLEDGMENT

Support of this effort was provided under Contract F33615-80-C-2017 by AFWAL/POSL, Air Force Wright Aeronautical Laboratories, Wright-Patterson Air Force Base, Ohio.

REFERENCES

1. J. W. Gibbs, *Scientific Papers*, Vol. 1, Longmans, Green, London, 1906, pp. 300–314.
2. J. W. Strutt (Lord Rayleigh), *Scientific Papers*, Vol. 3, Cambridge Univ. Press, Cambridge, 1902, pp. 351–362.
3. T. B. Thomas and J. T. Davies, *J. Colloid Interface Sci. 48*: 427 (1974).
4. A. I. Rusanov and V. V. Krotov, *Prog. Surf. Membr. Sci. 13*: 418 (1979).
5. M. van den Tempel, J. Lucassen, and E. H. Lucassen-Reynders, *J. Phys. Chem. 69*: 1798 (1965).
6. K. Malysa, K. Lunkenheimer, R. Miller, and C. Hartenstein, *Colloids Surfaces 3*: 329 (1981).
7. R. Cini and P. P. Lombardini, *J. Colloid Interface Sci. 81*: 125 (1981).
8. E. H. Lucassen-Reynders and J. Lucassen, *Adv. Colloid Interface Sci. 2*: 347 (1969), especially equations (84) and (85).
9. R. P. Borwankar and D. T. Wasan, *Chem. Eng. Sci. 38*: 1637 (1983).
10. A. Sheludko, *C. R. Acad. Bulg. Sci. 9*(1): 11 (1956).
11. R. J. Mannheimer and R. S. Schechter, *J. Colloid Interface Sci. 32*: 212 (1970).
12. T. Yasukatsu, K. Saburu, and T. Nobuyuki, *ASLE Trans. 21*: 351 (1978).
13. N. L. Jarvis, *J. Phys. Chem. 70*: 3027 (1966).
14. J. J. Bikerman, *Foams*, Springer-Verlag, New York, 1973, p. 106.
15. H. C. Maru, V. Mohan, and D. T. Wasan, *Chem. Eng. Sci. 34*: 1283 (1979).
16. B. V. Derjaguin and A. S. Titijevskaya, *Proc. 2nd Int. Congr. Surface Activity 1*: 210 (1957).
17. S. Ross and G. Nishioka, *J. Colloid Interface Sci. 65*: 216 (1978).
18. I. Langmuir, *Colloid Symp. Monogr. 3*: 62 (1925).
19. E. W. Becker, *Heavy Water Production*, International Atomic Energy Agency, Vienna, 1962, pp. 33–43.
20. L. R. Haywood and P. B. Lumb, *Chem. Can. 27*(3): 19 (1975).
21. N. H. Sagert and M. J. Quinn, *J. Colloid Interface Sci. 61*: 279 (1977).
22. F. J. Zuiderweg and A. Harmens, *Chem. Eng. Sci. 9*: 89 (1958).
23. E. F. Lundelius, *Kolloid Z. 26*: 145 (1920).
24. S. Ross and G. M. Nishioka, *Colloid Polym. Sci. 255*: 560 (1977).
25. S. Ross and G. M. Nishioka, *J. Phys. Chem. 79*: 1561 (1975).

26. G. M. Nishioka, L. L. Lacey, and B. R. Facemire, *J. Colloid Interface Sci. 80*: 197 (1981).
27. S. Ross and D. F. Townsend, *Chem. Eng. Commun. 11*: 347 (1981).
28. R. Defay and I. Prigogine, *Trans. Faraday Soc. 46*: 199 (1950).
29. R. M. Wellek, A. K. Agrawal, and A. H. P. Skelland, *AIChE J. 12*: 854 (1966).
30. W. Rybczynski, *Bull. Int. Acad. Sci. Cracovie Ser. A* (1911): 40.
31. J. Hadamard, *C. R. Acad. Sci. 152*: 1735 (1911).
32. F. H. Garner and A. R. Hale, *Chem. Eng. Sci. 2*: 157 (1953).
33. F. H. Garner and D. Hammerton, *Chem. Eng. Sci. 3*: 1 (1954).
34. G. Ryskin and L. G. Leal, *J. Fluid Mech. 148*: 1 (1984).
35. A. N. Frumkin and V. G. Levich, *Zh. Fiz. Khim. 21*: 1183 (1947); V. G. Levich, *Physicochemical Hydrodynamics*, Prentice-Hall, New York, 1962, Chap. 8.
36. H. Kuerten, J. Raasch, and H. Rumpf, *Chem.-Ing. Tech. 38*: 941 (1966).
37. J. Boussinesq, *Ann. Chim. Phys. [8] 29*: 349 (1913).
38. R. B. Bird, W. E. Stewart, and E. N. Lightfoot, *Transport Phenomena*, Wiley, New York, 1960, p. 206.
39. G. Furler and S. Ross, *Langmuir 2*: 68 (1986).
40. Foaming Characteristics of MIL-L-7808 Turbine Lubricants, AFAPL-TR-75-91 (1975), pp. 36–38. Air Force Wright Aeronautical Laboratories, Air Force Systems Command, Wright-Patterson Air Force Base, Ohio.
41. S. Ross and Y. Suzin, *J. Colloid Interface Sci. 103*: 578 (1985).
42. R. M. Griffith, *Chem. Eng. Sic. 17*: 1057 (1962).
43. R. E. Davis and A. Acrivos, *Chem. Eng. Sci. 21*: 681 (1966).
44. S. Okazaki, Y. Miyazaki, and T. Sasaki, *Proc. 3rd Int. Congr. Surface Activity, Mainz, Germany 2*: 549 (1961).
45. J. V. Robinson, *J. Phys. Colloid Chem. 51*: 431 (1947).
46. J. W. McBain, Study of the Foaming of Oil in Altitude Flying as a Problem in Colloid Chemistry, NACA ARR No. 4105 (1943), pp. 8–9. Also *idem, Nature 120*: 362 (1927).
47. H.-F. Eicke and P. Kvita, in *Reverse Micelles*, P. L. Luisi and B. E. Straub, eds., Plenum, New York, 1984, pp. 21–35.

2

Aqueous Nanophases in Liquid Hydrocarbons Stabilized by Ionic Surfactants

HANS-FRIEDRICH EICKE Institut für Physikalische Chemie der Universität Basel, Basel, Switzerland

I. INTRODUCTION

Although we consider in this chapter only one particular class of stabilized (aqueous) nanophases (i.e., water domains of colloidal dimensions, 10^{-8} to 10^{-7} m), in our opinion this class deserves, from an experimental and theoretical point of view, a certain distinction from other systems stabilized by nonionic or polymeric (block copolymer) surfactants.

Since the introduction of the term *microemulsion* (ME) by Hoar and Schulman in 1943, none of the many definitions of microemulsions have apparently been thought to be satisfactory, since many attempts have been made to find more suitable descriptions (e.g., [1–4]). The main reason for this has been the lack of an operational definition of the microemulsion phenomenon which could be used to analyze the steadily increasing experimental data on such multicomponent systems.

We consider this situation to be a stimulus for reconsidering the basic assumptions which must be fulfilled, in our opinion, in order to offer an operational definition of microemulsions. In particular, we wish to consider whether a microemulsion* can be experimentally distinguished within the multicomponent systems from other lyotropic structures, and also which statements such a definition should include in order to be meaningful.

It is noteworthy that most of the previous definitions of microemulsions emphasized the structural aspect (e.g., [1–4]). This is consistent with the fact that the microemulsion concept was originally derived from that of the generally known (technical) emulsion. The microemulsion is primarily distinguished from the emulsion not by being composed of smaller droplets (nanophases) but by being subject to a very restrictive condition, that it is thermodynamically stable. The latter condition is by no means sufficient but only necessary, since many multicomponent molecularly dispersed liquid mixtures are also thermodynamically stable. Such a thermodynamically stable emulsion (ME) is expected to show a pronounced similarity to binary liquid mixtures close to the critical point. This was pointed out by Tolman [5] as early as 1913; the mean specific free energy of the dispersed material is composed of (1) a contribution the material would have in a large quantity of undivided phase having the same composition as the particle in question and (2) a contribution of the dispersed state. The latter includes the interfacial free energy, i.e., the work to be done per gram of material on the respective component(s) to increase its (their) interface(s). If this energy is very small (compared with kT), as is the case with thermodynamically stable microemulsions, the difference between the microemulsion and a binary critical mixture becomes experimentally (in effect) indistinguishable. Accordingly, it is not surprising that phenomena typical of molecularly dispersed mixtures are also observed with microemulsions.

Hence, what is needed, after all, is unambiguous experimental evidence of long-lasting (permanent) nanophases consisting of a water core (in the present context) surrounded by surfactants. It appears from more recent Fourier transform (FT) spin-echo nuclear magnetic resonance (NMR) self-diffusion measurements [6] that this technique provides model-free† information on self-diffusion coefficients D of all components of the multicomponent mixtures. The

---

*The term microemulsion is retained in this chapter whenever it is used in the literature. However, if a system has been proved to obey the stricter conditions to be defined below, it will be called a *nanophase solution*.

†Model-free in this context does not refer to the NMR theory in general, which uses models, but stresses the possibility of directly obtaining information on diffusion coefficients without reference to structural details of the multicomponent system.

interpretation of individual self-diffusion coefficients assumes that the components constituting the continuous phase possess almost the same self-diffusion coefficients as they would in the pure state. In well-defined nanophase ensembles (i.e., microphase solution), such as those to be described below, the difference between the self-diffusion coefficients of dispersed material and components of the dispersed phase amounts to about one or two orders of magnitude [6,6a].

In following this idea we prefer to use the microemulsion concept under the restrictive condition that the self-diffusion coefficients of all colloidally dispersed components are small compared to the values of those components which are molecularly dispersed within the continuous phase, i.e., $D_{(\text{dispersed material})}/D_{(\text{continuous phase})} \ll 1$. This condition also implies (more indirectly) a statement regarding the lifetime of the aggregates. In order to see this, one must recall that the lifetimes of surfactant molecules in micelles or vesicles depend on the critical micelle concentration (CMC) [7], i.e, the concentration of surfactant molecules of the continuous phase in equilibrium with those in the aggregated state. (It must be assumed, however, that such an exchange is not prevented by a large energy of activation.) If this concept is applied analogously to nanophases, the mean lifetime of surfactants in the adsorbed state would simultaneously indicate the lifetime of the particular nanophase. Thus, an increased amount of molecularly dispersed surfactant molecules in equilibrium with the nanophases would necessarily reduce the lifetime of the aggregates. This increase of surfactant molecules in the continuous phase, on the other hand, invalidates the above condition $D_{(\text{dispersed material})}/D_{(\text{continuous phase})} \ll 1$ since the proportion of surfactants in the adsorbed state is diminished and hence the average self-diffusion coefficient of the surfactants (i.e., surfactants and cosurfactants) is increased toward their values in the molecularly dispersed state.

This situation is actually encountered with systems of four (and more) components composed of water, alcohol (cosurfactant), surfactant, and oil. Frequently, propanol, butanol, and pentanol are used as cosurfactants, components which are known [8] to increase the mutual solubility of water and oil. The obvious possibility of observing phenomena produced by a superposition of molecularly and colloidally dispersed components necessitates a particularly careful consideration of the composition of such systems. Fulfillment of the conditions stated above (regarding the ratio of the diffusion coefficients) is in every case sufficient reason to consider the particular system under consideration a microemulsion.

The emphasis placed here on the possibility of having an operational definition is due to the fact that in multicomponent systems containing surface-active material pronounced formation of transient hydrophilic or lipophilic structures is possible in spite of their similar self-diffusion coefficients. The information on particular structures in such systems has mostly been derived from classical light-scattering, quasi-elastic light-scattering (QELS), and neutron-scattering experiments and, hence, depends on the particular model chosen (e.g., [9–11]). These difficulties caused some authors (e.g., [12]) to question the existence of microemulsions in general. Another consequence of this situation was the suggestion of various "bicontinuous" structures, which are thought to comprise the whole region between the molecular dispersed state of multicomponent solutions and the well-defined permanent structural type of nanophase solution obeying the above operational definition.

II. THREE-COMPONENT MICROEMULSIONS

Starting with three-component systems to introduce the subject of multicomponent systems is not only appropriate from a scientific point of view; also, more specifically, the oil, water, and Aerosol OT (AOT) system represents a kind of reference system for a thermodynamically stable, incoherent dispersion of water in oil. From a theoretical standpoint it is advantageous to consider such a "simple" system (see also [13]). Insight into such a system is a prerequisite for predicting properties of microemulsions containing cosurfactants (i.e., amphiphilic molecules, mostly alcohols, used to titrate a multicomponent water-oil mixture to transparency) and/or brine. Hence, some thermodynamic relations applied to a microemulsion of the H_2O, AOT, oil type will be considered first.

A. Theoretical Considerations Regarding Three-Component Water-in-Oil Microemulsions

The spherical-droplet model of the polar nanophase (in general, an aqueous or aqueous electrolyte spherical core covered by a surfactant monolayer), on which the following considerations are based, has been experimentally confirmed in various ways (discussed later).

The Gibbs free energy of this microemulsion (G_{me}) with ionic surfactant (amphiphilic electrolyte) is composed of several contributions [14–16]: (1) the free energy of adsorption of the surfactant at the oil/water interface (G_s), (2) the free energy of the electrostatic interactions within the semidiffuse electric double layer for an isothermal process (G_f), (3) a free energy due to the (inhomogeneous) distribution of the counterions representing the semidiffuse part of the double layer (G_{mix}^{ion}), and (4) a small free energy term (compared to the other contributions, (1 to 3), mostly entropic, arising from the translational motion of the nanophases in the continuous oil phase (G_{mix}). Hence, one obtains

$$G_{me} = G^{\ominus} + G_s + G_f + G_{mix}^{ion} + G_{mix} \tag{1}$$

where $G_f + G_{mix}^{ion} = G_e$, which is the total electrostatic contribution to the Gibbs free energy of the system; G^0 comprises the (arbitrary) reference states of all free energy terms of Eq. (1) which are independent of the water content of the nanophases; $G_s = \gamma_{un}A_{total}$, where γ_{un} is the interfacial free energy that would obtain if no electric double layer (due to dissociation of the electrolyte [17]) had been formed; and $A_t = NA$, where A is the interface of a single nanophase. The term G_f is theoretically accessible via a numerical solution of the Poisson-Boltzmann equation [15], i.e.,

$$G_f = N\frac{\varepsilon_r\varepsilon_0}{2} \int_{\substack{\text{spherical} \\ \text{pol. core}}} (\nabla\Phi)^2 \, dV_{pol.core} \tag{2}$$

where N, ε_r, ε_0, and $\nabla\Phi(r)$ are the number of nanophases in the system, the relative dielectric constant, the permittivity of the vacuum, and the local electric field of the semidiffuse double layer, respectively, and $V_{pol.core}$ is the volume of the spherical polar core of a single nanophase (see Fig. 1).

The quantity H_e, the total electric energy contribution to the enthalpy of the system, is significant in calorimetric studies. It may be obtained from G_e via the Gibbs-Helmholtz relation [18], i.e.,

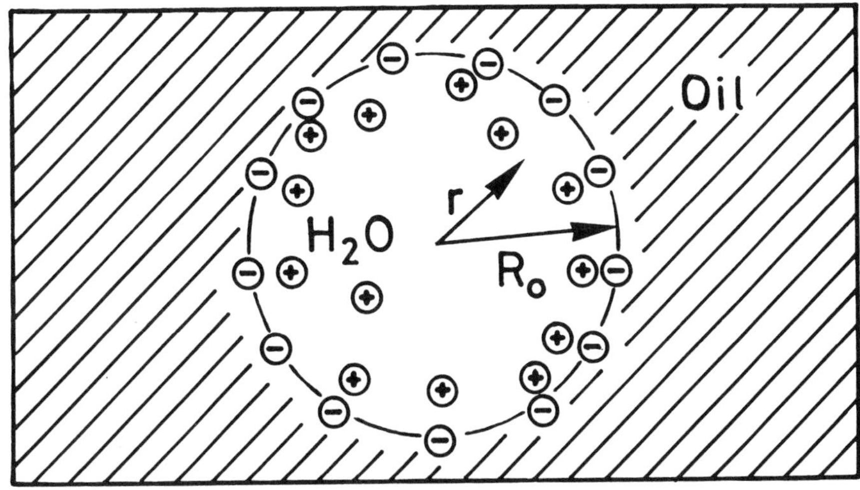

FIG. 1 Schematic representation of an aqueous nanophase [23a].

$$H_e = G_f\left(1 + \frac{\partial \ln \varepsilon_r}{\partial \ln T}\right) \tag{3}$$

and is also (in the same approximation as G_f) calculable.

Within the approximation of the Poisson-Boltzmann equation, the equilibrium condition for the nanophase ensemble (microemulsion) is obtained by forming the first derivative of G_{me} with respect to the total interface of the system (A_t) [15], i.e.,

$$\left(\frac{\partial G_{me}}{\partial A_t}\right)_{p,T} = 0 = \gamma_{un} - 2\frac{G_f}{AN} + 3\frac{kT\zeta}{A} \tag{4}$$

Here A is the interface of a single nanophase and $\zeta = \ln \phi - 1 + (4 - 3\phi)/(1 - \phi)^2 + \ln(V_{oil}/V_{me})$, where ϕ is the volume fraction of the dispersed phase and V_{oil} and V_{me} the molar volumes of oil and nanophases. This relation has been derived according to Percus and Yevick [19] and Carnahan and Starling [20] for hard-sphere suspensions [21].

Equation (4) shows that the overall surface tension $\gamma = \gamma_{un} - 2\,G_f/AN$ (which can be formally considered to be the experimentally observable interfacial tension within the ionic monolayer) must be compensated by the entropy of mixing of the nanophases in the oil-continuous phase. This term is obviously small, hence thermodynamically stable emulsions demand rather small interfacial tensions: a theoretical estimate shows that the third term in Eq. (4) is about -0.2 mJ m^{-2} (this value varies with the degree of dispersion and volume fraction; here the aqueous nanophase radius is 3 nm in i-C_8H_{18}, $c_{AOT} = 0.2$ mol dm^{-3}, $\phi_{H_2O} = 0.5$), hence γ must be a small (positive) quantity. Experimentally one finds [22] $\gamma = 10^{-4}$ to 10^{-3} mN m^{-1}, the so-called ultralow interfacial tensions.

If the small entropy of mixing term is neglected, the equilibrium condition demands

$$\gamma_{un} = 2\frac{G_f}{AN} \tag{4a}$$

If the relative dielectric constant is independent of temperature, then it follows from Eq. (3) that $H_e = G_f$, which may be introduced into Eq. (4a). Under this condition the right-hand side of Eq. (4a) equals the total electrostatic contribution to the interfacial free enthalpy of the system at a constant number of charges, as Frenkel has pointed out [17].

An interesting aspect of Eq. (4a) should be mentioned which has been experimentally confirmed and which justifies the splitting of the interfacial free energy into two contributions, γ_{un} and G_e/NA [see Eq. (1)]. The term γ_{un} contains only short-range interaction energies on a molecular scale, hence G_f/AN should be independent of the interfacial curvature. Since G_f/AN can be calculated according to Eq. (5), this assertion can be tested for systems of known R_0 and f_{am} (see Fig. 2). This can be achieved via a numerical solution of the nonlinearized Poisson-Boltzmann equation [23]. A large number of numerical solutions are fitted to experimental data and are functionalized. The resulting functions, $Y(R_0,f_{am}(R_0))$, depend on the radius R_0 of the nanophase and on the interfacial area f_{am} covered by one amphiphilic molecule. The function Y varies between zero ($R_0 \to 0$) and one ($R_0 \to \infty$, i.e., a plane interface) and can be thought of as the ratio of $G_f(R_0)/AN$ and $G_f(R_0 = \infty)/AN$. Thus, the relation between G_f/AN and Y may be written as

$$\frac{G_f}{AN} = Y(R_0,f_{am}(R_0))\frac{kT}{f_{am}} \tag{5}$$

Since calorimetric studies represent a relatively frequently applied technique [18,23,24] for investigating microemulsions, it is

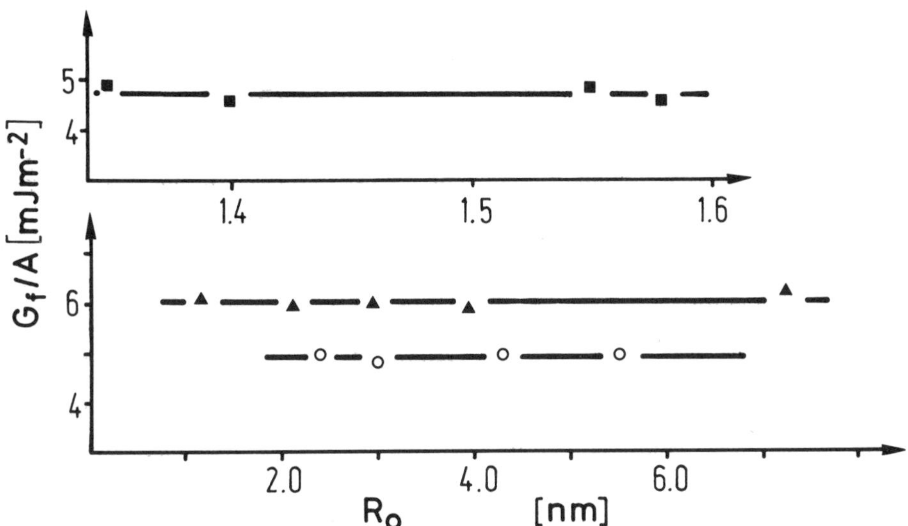

FIG. 2 Calculated parameter G_f/A for different water core radii: (■) H_2O/0.1 M AOT/toluene at T = 393 K [51]; (▲) H_2O/AOT/isooctane, T = 298 K [50]; (○) H_2O/0.05 M AOT/*n*-heptane, T = 298 K [52].

worth showing how Eqs. (3), (4a), and (5) can be utilized in handling experimental data. The experimentally determined enthalpy of a microemulsion, H_{me}, is obtained by mixing two sets of microemulsions with different volume fractions of water, i.e.,

$$H_{me} = H_{mix}(w_0) - H_{mix,ref}(\text{e.g., } w_0 = 0) \tag{6}$$

where the second term on the right-hand side is the enthalpy of a micellar solution and $w_0 = [H_2O]/[\text{ionic surfactant}]$. On the other hand, H_{me} is related to the interfacial and electric energy of the semidiffuse double layer by [23]

$$H_{me} = H^0 + H_{un}^s + H_e \tag{7}$$

where H^0 corresponds to a reference state analogous to that of G^0 in Eq. (1); H_{un}^s is obtained from $\gamma_{un}A_t$ by a relation analogous to the Gibbs-Helmholtz expression, i.e.,

$$H_{un}^s = \left[\gamma_{un} - T\left(\frac{\partial \gamma_{un}}{\partial T}\right)_p\right]A_t \tag{8}$$

and

$$H_e = \left[1 + \left(\frac{\partial \ln \varepsilon_r}{\partial \ln T}\right)_p\right]G_f \tag{8a}$$

Utilizing the equilibrium condition, Eq. (4a), H_{me} can be written as

$$H_{me} = H^{\ominus} + \xi G_f \tag{8b}$$

where $\xi = 3 + (\partial \ln \varepsilon_r/\partial \ln T)_p - 2(\partial \ln \gamma_{un}/\partial \ln T)_p$.

It is worth mentioning that the the total number of particles (nanophases) decreases during the mixing process (see Fig. 3). This causes a reduction of the translational entropy of the nanophase solution. The driving force for this process is to be found in the formation of the semidiffuse electric double layer (see Figs. 4 and 5).

Appendix

Equation (4) can be straightforwardly shown to be valid for a plane interface. Since no curvature effects are encountered with the microemulsions studied here, the considerations are significant for

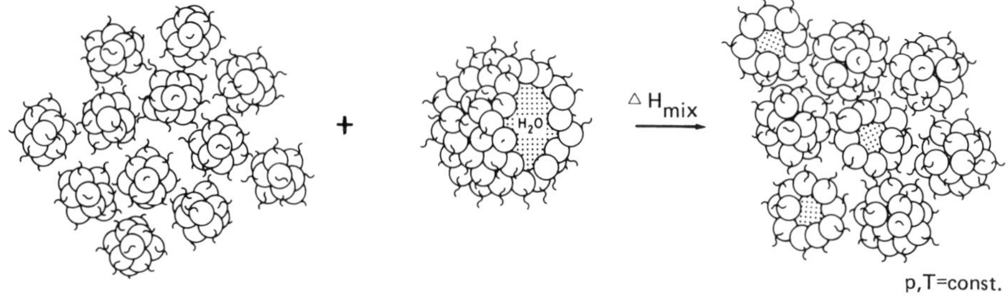

FIG. 3 Schematic representation of mixing process between nanophases containing different amounts of solubilized water [23a].

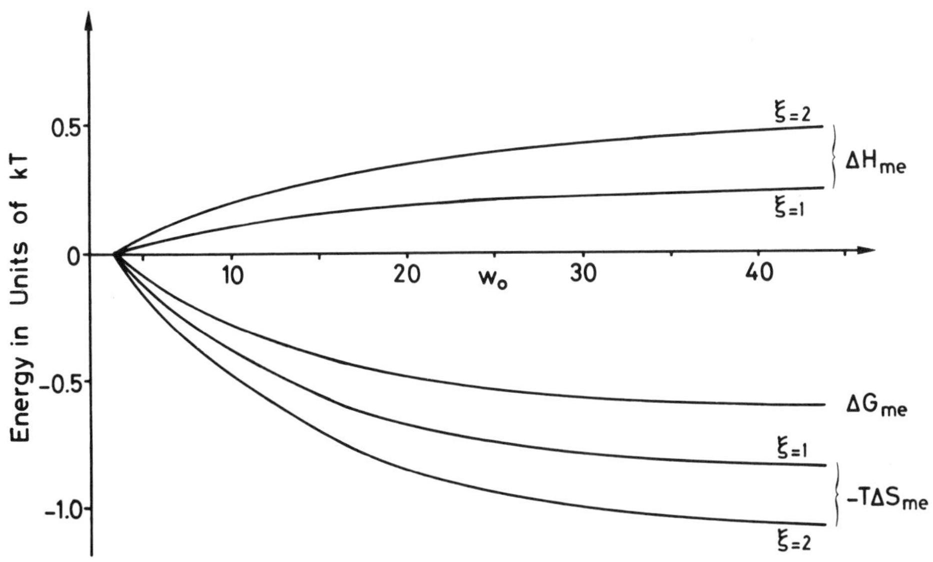

FIG. 4 ΔG_{me}, ΔH_{me}, and $T\ \Delta S_{me}$ as calculated for the mixing process in Fig. 3; $G_f/A = 6.0$ mJ m^{-2} and T = 298 K [23a].

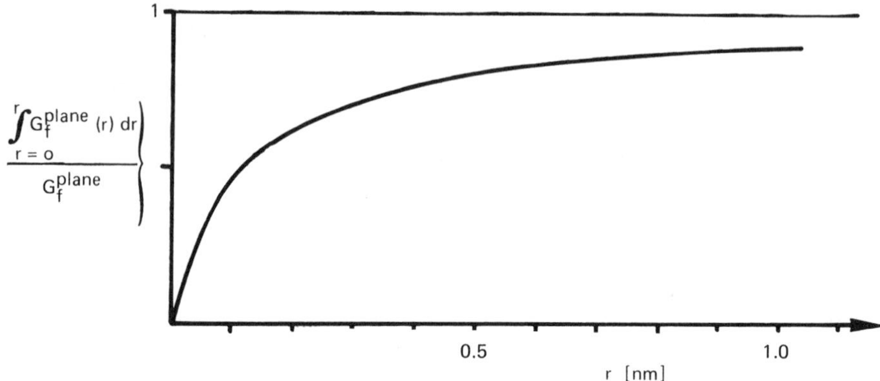

FIG. 5 Contribution of the free energy of the electric field within a layer of thickness r compared with the total energy G_f^* of a plane semidiffuse ionic double layer. The parameters are the same as in Fig. 4 (see appendix in Sec. II.A) [23a].

the present discussion. Following Marcus [25], one can express the free energy due to the inhomogeneous distribution of a counterion (i) within the plane (*) semidiffuse double layer, i.e.,

$$\frac{G_{mix}^{ion*}}{A^*kT} = \int_{r=0}^{\infty} c_i(\Phi)\left\{\ln\left[\frac{c_i(\Phi)}{\bar{c}}\right] - 1\right\} dr \tag{A1}$$

where A* is an element of the plane oil/water interface, $\bar{c}$ is a reference water concentration, and r is the distance of the volume element A* dr with ion concentration c_i from the interface. The solution of the Poisson-Boltzmann equation for a plane interface with the boundary conditions $\Phi(r = \infty) = 0$, $(d\Phi/dr)_{r=\infty} = 0$, and $(d\Phi/dr)_{r=0} = -\sigma/\varepsilon_0\varepsilon_r$ is [26]

$$\left(\frac{d\phi}{dr}\right)^2 = \frac{2kT}{\varepsilon_0\varepsilon_r} c_i(r,\sigma) \tag{A2}$$

where

$$c_i(r,\sigma) = \left(\sqrt{\frac{e^2}{2\varepsilon_0\varepsilon_r kT}}\, r - \sqrt{\frac{2\varepsilon_0\varepsilon_r kT}{\sigma}}\right)^{-2} \equiv (ar - b)^{-2} \tag{A3}$$

The symbols used in Eq. (A3) have been explained in the text, except σ, the surface charge density (= e/f_{am}), where f_{am} is the area of the oil/water interface covered by one surfacant molecule. If Eq. (A3) is inserted into Eq. (A1) the integral can be solved to yield

$$G_{mix}^{ion*} = \frac{kTA^*}{f_{am}} \left[\ln\left(\frac{\bar{c}}{c_{io}}\right) + 3\right] \tag{A4}$$

Here $c_{io} = b^2$ is the counterion concentration where $r = 0$ and $A^*/f_{am} = n_{am}$, the number of surfactant molecules covering the interfacial element A^*. Hence

$$\frac{1}{kTn_{am}} \left(\frac{\partial G_{mix}^{ion*}}{\partial f_{am}}\right) = -\left(\frac{\partial \ln c_{io}}{\partial c_{io}}\right)\left(\frac{\partial c_{io}}{\partial f_{am}}\right) = -\frac{2}{f_{am}} \tag{A5}$$

From analogous calculations for $G_f^*/A^* = (\varepsilon_0\varepsilon_r/2) \int_0^\infty (d\Phi/dr)^2/dr$ it follows that $G_f^* = kTA^*/f_{am} = kTn_{am}$. The derivative of G_e^* (electrostatic contribution to the interfacial free energy) with respect to f_{am} is, accordingly,

$$\frac{1}{n_{am}} \left(\frac{\partial G_e^*}{\partial f_{am}}\right) = \left[\left(\frac{\partial G_f^*}{\partial f_{am}}\right) + \left(\frac{\partial G_{mix}^{ion*}}{\partial f_{am}}\right)\right] \frac{1}{n_{am}} = -\frac{2G_f^*}{A^*} \tag{A6}$$

This result is generally valid since it can be shown [15] that Eq. (A6) is independent of curvature.

B. Experimental Characterization of the Three-Component Water, Aerosol OT, Oil System

1. Introduction

By far most of the investigations on microemulsions containing ionic surfactants have dealt with systems which require alcohol or other cosurfactants and/or salt (brine) for their formation or stability [27–31]. The enrichment of cosurfactants in the interface is thought to reduce the interfacial tension [32] and induce increased flexibility [9,13]. However, it is important to realize [33] that there is no theoretical requirement to use cosurfactants in general. On the contrary, systems of four and five (with salt) components pose formidable obstacles to experimental and theoretical progress. As already mentioned, these systems frequently do not fulfill the requirement $D/D_{(\text{continuous phase})} \ll 1$.

Hence, it is reasonable to emphasize the importance (and existence) of reference systems such as those formed with AOT. More recently, however, another cationic surfactant, di-dodecylammonium bromide [$(C_{12})_2$DAB] [13], has been recommended; it appears to form aqueous nanophases of spherical shape which are reported to show a pronounced solvent dependence [13]. Since the ternary water, Aerosol OT, oil system shows features which are unique in many respects, its separate presentation seems to be justified.

Aerosol OT (AOT; sodium di-2-ethylhexylsulfosuccinate), Fig. 6, has been known at least since the early 1940s [34] to form reversed micelles and to take up (solubilize) considerable amounts of water if this surfactant is dispersed in nonpolar, aprotic hydrocarbons (see Fig. 1). A typical phase diagram for the water, AOT, isooctane (2,2,4-trimethylpentane) system is shown in Fig. 7, displaying the characteristic large water-in-oil (W/O), i.e., L_2, phase at 288 K.

2. The Proper Solubilization Process as Studied by NMR and Vapor Pressure Measurements

The solubilization process itself may easily be followed with proton-, ^{2}H-, and ^{23}Na-NMR [35,23] and vapor pressure [36,37] measurements. The ^{1}H-NMR studies display several regions of solubilized

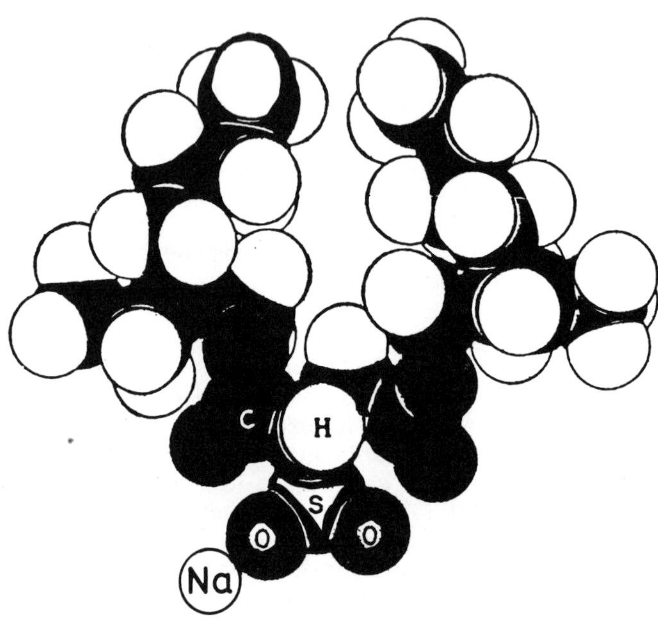

FIG. 6 Calotte model of Aerosol OT.

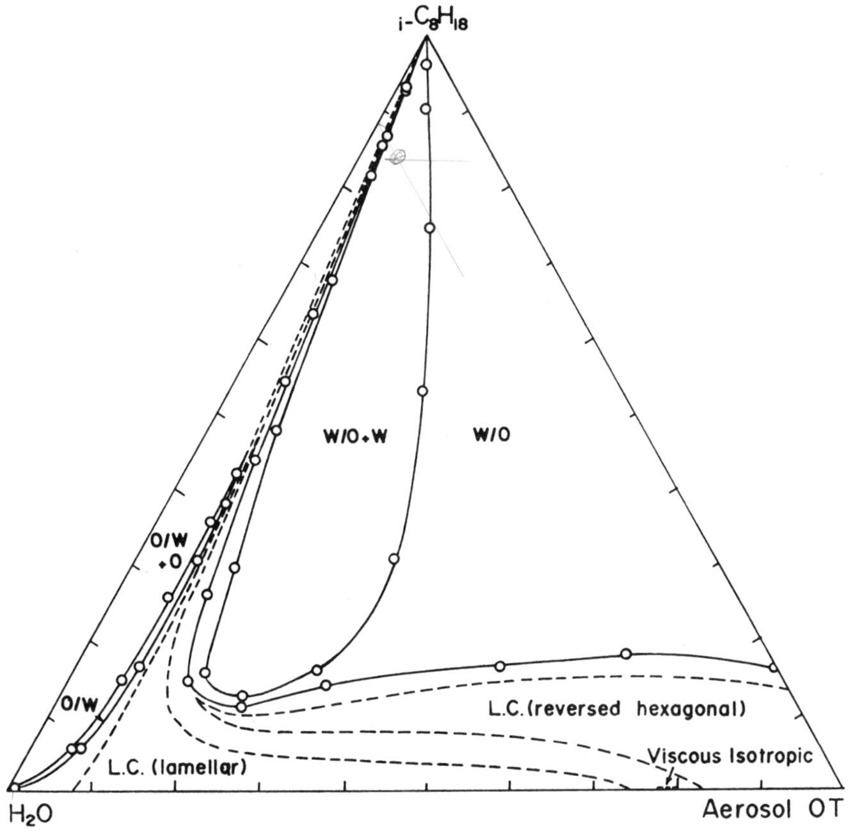

FIG. 7 Phase diagram of the three-component system H_2O/AOT/isooctane at T = 288 K. Scaling is in weight percent of the respective components [31a].

water which can be attributed to different binding states between surfactant and water (Fig. 8). Figure 9 shows ^{2}H-NMR measurements, again indicating that solubilized water approaches bulk properties with increasing weighed-in amounts of water. Figure 10 shows the proton-NMR line width dependence on the amount of solubilized water. This plot offers direct evidence for the restricted mobility of the water and surfactant molecules with small amounts of water in the polar core. The apolar tails, however, are almost unaffected, as is apparent from ^{13}C-NMR measurements [38,39].

A more detailed investigation was initiated [40] by considering the rotational isomerism of AOT and its possible effect on AOT's

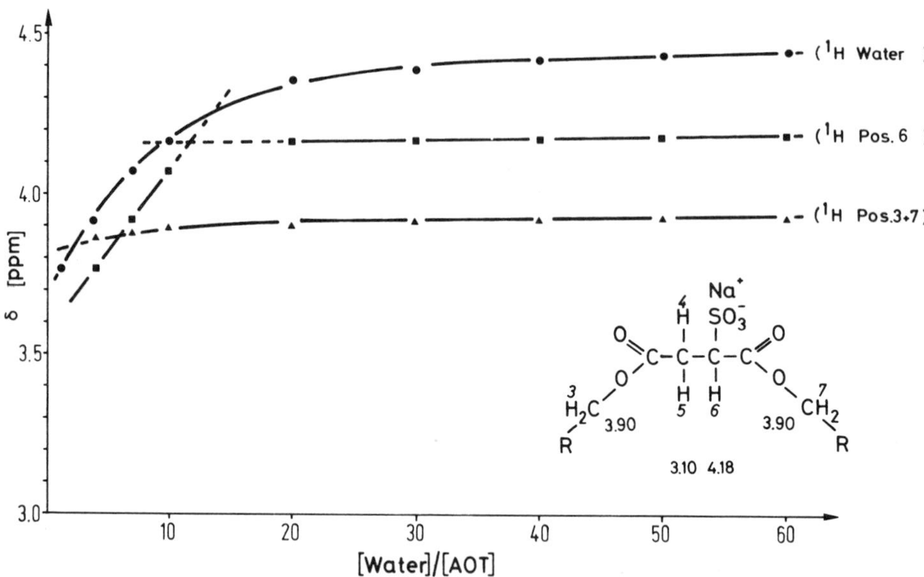

FIG. 8 Chemical shifts of three different protons (see insert) against the amount of solubilized water (w_0 = [water]/[AOT]). System water/AOT/isooctane, T = 298 K [35].

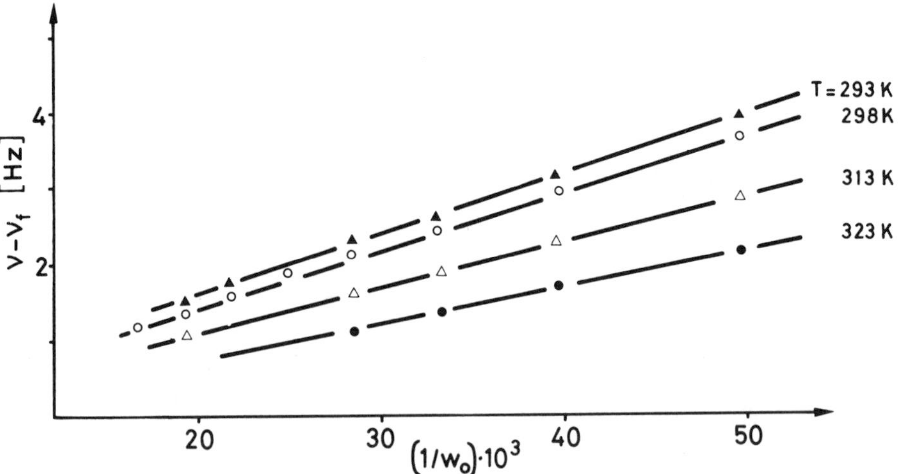

FIG. 9 Resonance frequency of chemical shift of the D nucleus in solubilized D_2O against $1/w_0$. System: D_2O/0.19 M AOT/cyclohexane [23a].

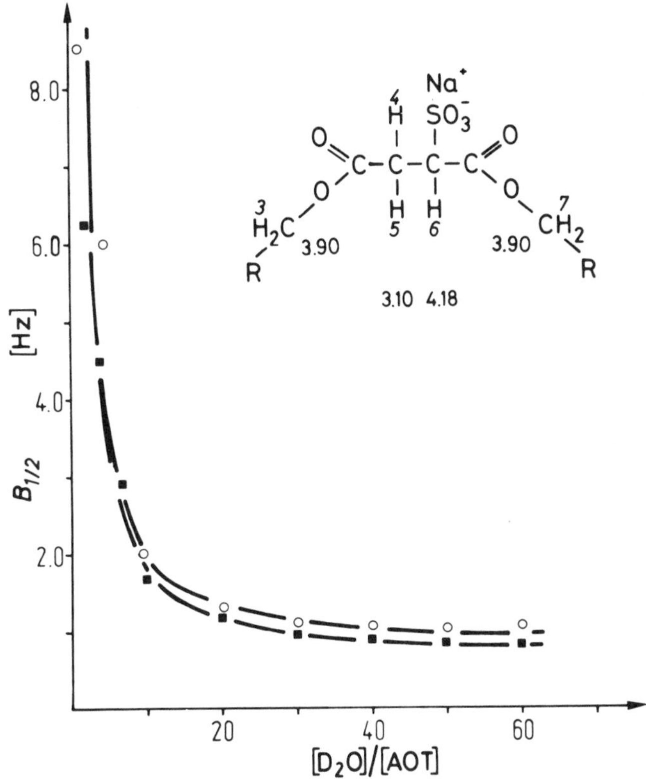

FIG. 10 Half-linewidth ($B_{1/2}$) of the proton in sixth position (■), water proton (○) (see insert) against the amount of solubilized water; T = 298 K [35].

solubilizing properties. The influence of different solvents was also studied in this context. In particular, the temperature dependence of various proton-proton spin coupling constants was investigated (Fig. 11) and the relative energies of the rotamers were determined (Fig. 12) by ^{1}H-NMR spectroscopy. With the help of Fig. 13 it is easily possible to interpret the temperature dependence of the coupling constants $J_{AX} + J_{BX}$ of AOT in different solvents. Relative populations of rotamers 1, 2, and 3 and their relative order of energies are derived from this temperature dependence by using typical vicinal coupling constants for gauche and trans with $J_{gauche} < J_{trans}$. Since a temperature increase would be expected to result in an increased contribution of rotamers with higher energies, a change in the coupling constant values with temperature would indicate directly whether the rotamers with gauche or trans orientation are

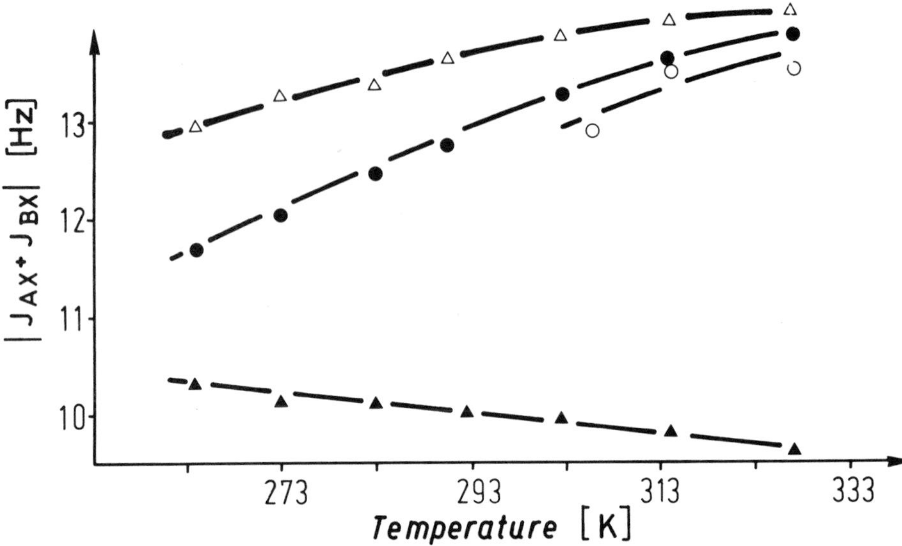

FIG. 11 Temperature dependence of coupling constants (J_{AX} + J_{BX}) of AOT in different solvents: (▲) CD_3OD with w_0 = 0.11; (●) $CDCl_3$ with w_0 = 0.11; (△) $CDCl_3$ with w_0 = 0.48; (○) isooctane with w_0 = 0.48 [40].

stable at lower temperature in a particular AOT-solvent system. Hence, the decreasing trend of J_{AX} + J_{BX} with increasing temperature indicates that the energetically least favored conformation is rotamer 1 in the AOT-methanol system. The same argument leads one to conclude that rotamer 1 has the lowest energy in the AOT-chloroform system. It thus appears that such an anisotropic interaction becomes highly favorable if all of the polar groups of the AOT molecule are spatially confined in such a way as to lead to a more pronounced amphiphilic character of the molecule. These investigations are in some respect supplemented by ^{13}C-NMR measurements [41] to determine spin-lattice relaxation times (T_1) and chemical shifts of water/AOT/organic solvents (c-C_6H_{12},CCl_4,C_6H_6). Upon addition of water to a micellar solution, gauche-to-trans conformational changes were reported to occur throughout the AOT molecule. Also, solvent penetration has been observed to decrease in the order: benzene > carbon tetrachloride > cyclohexane. Moreover, the authors found that the conformational changes resulted in a more open, extended structure of the AOT molecule. Benzene and carbon tetrachloride enhance this effect because of solvent penetration; i.e., the extended structure starts at smaller amounts of solubilized water.

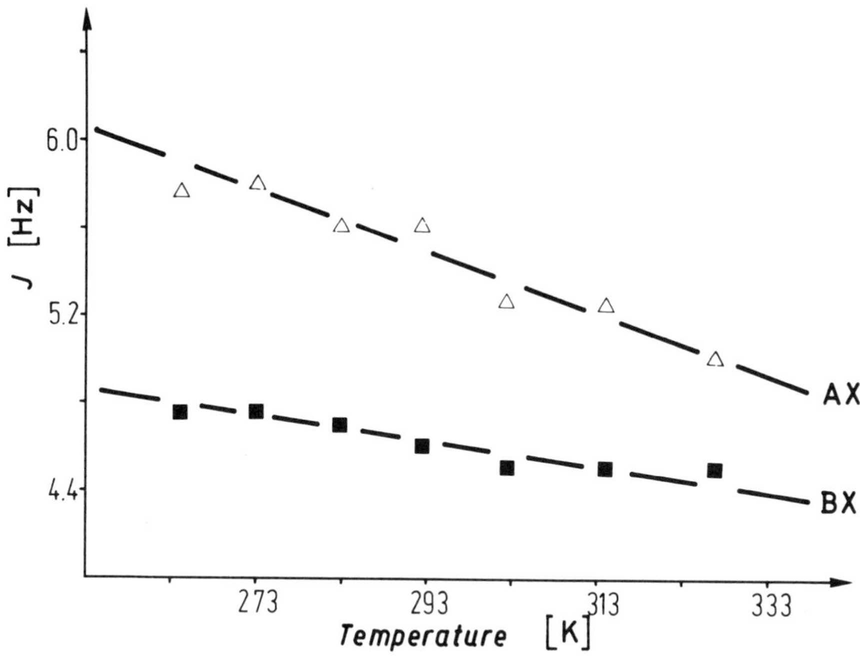

FIG. 12 Temperature dependence of coupling constants (J_{AX}, J_{BX}) of AOT in methanol [40].

Direct information on the activity of the solubilized water as a function of its concentration (w_0 = $[H_2O]/[AOT]$) can be obtained from measurements of the partial pressure of water above an equilibrated nanophase solution [37] (see Fig. 14). Neglecting the influence of the oil-continuous phase, it can be assumed for the present discussion that the systems consist exclusively of surfactant and water. This is the assumption of a "one-component macrofluid," [80]. Considering this model, the vapor pressure follows from Raoult's law. The mole fraction of water, x_{H_2O}, is $w_0/(1 + w_0)$. Figure 14 displays Raoult's law (solid line) and the curves calculated according to the Poisson-Boltzmann model [23]. For smaller amounts of solubilized water ($x_{H_2O} < 0.8$) the entropy of mixing of water within the nanophases plays a significant role with respect to a_{H_2O}. However, the uncertainty regarding the validity of the model (since the core consists of only a few H_2O molecules!) should be noted. The tendency of the deviation from Raoult's law is, however, correctly described. At $x_{H_2O} > 0.8$ the partial pressure is larger than described

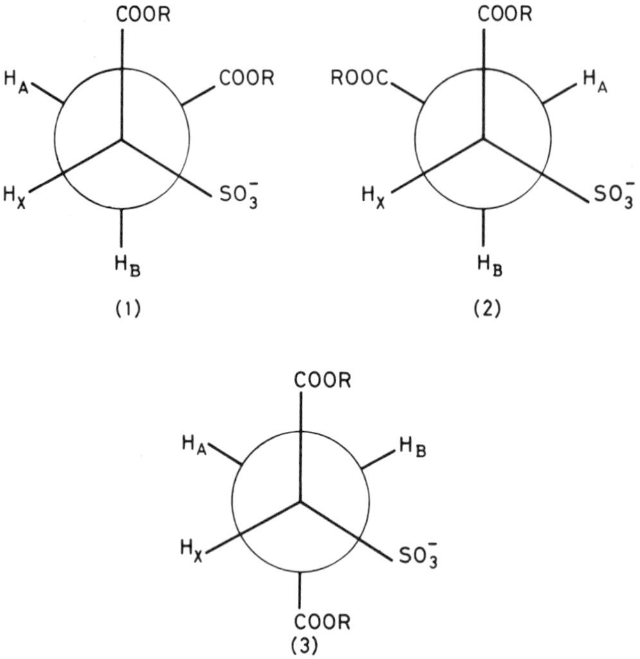

FIG. 13 Different conformations of monosubstituted succinic acid esters [40].

by Raoult's law. This is also in line with the Poisson-Boltzmann theory (dot-dashed line). Considering the nonideal mixing entropy improves the coincidence (dashed line). These considerations are confirmed by ^{23}Na relaxation rate measurements as a function of w_0 (Fig. 15). Such studies offer a direct approach for investigating the semidiffuse electric double layer. It is particularly suited for looking into the counterion distribution within the nanophase and obtaining information regarding the binding parameter β, i.e., the ratio of bound counterions to the total number of surfactant ions in solution.

3. Fluorescence and Polarization Decay Measurements

Fluorescence and polarization decay measurements with fluorescent (probe) molecules is a frequently applied technique for studying the interior (polar core and ionic interface) of Aerosol OT nanophases. The elegance of this technique, however, should not mislead the experimentalist regarding the possible disturbances of the nanophase by insertion of the probe molecule; principally, the probe molecule "sees" its disturbed environment. An incentive for applying this technique originates from catalysis and photochemical energy storage considerations.

Two methods were used to determine the Brownian motion (diffusion) of fluorescing molecules in order to obtain the microviscosity of their surroundings. The time-dependent polarization anisotropy yielded the rotational diffusion constant, while the fluorescence intensity decay (in the presence of a quenching process) was used to determine the translational diffusion of the probe molecule within the nanophase. Both techniques yield complementary information.

Predictions based on theory [42] regarding the decay of the fluorescence intensity indicate a highly nonexponential behavior if the ratio of the diffusion coefficient to the squared polar core radius of the nanophase is of the order of k, the rate constant describing the various intramolecular quenching processes including quenching at the interface. Figure 16 displays experimental results of fluorescence intensity against time for water plus 1-aminonaphthylene-4-sulfonic acid (i.e., 1-N)/AOT/i-C_8H_{18}. The evaluation of these data is based on the nanophase model with a solubilized dye molecule (see Fig. 16), where different physical states of the probe molecule

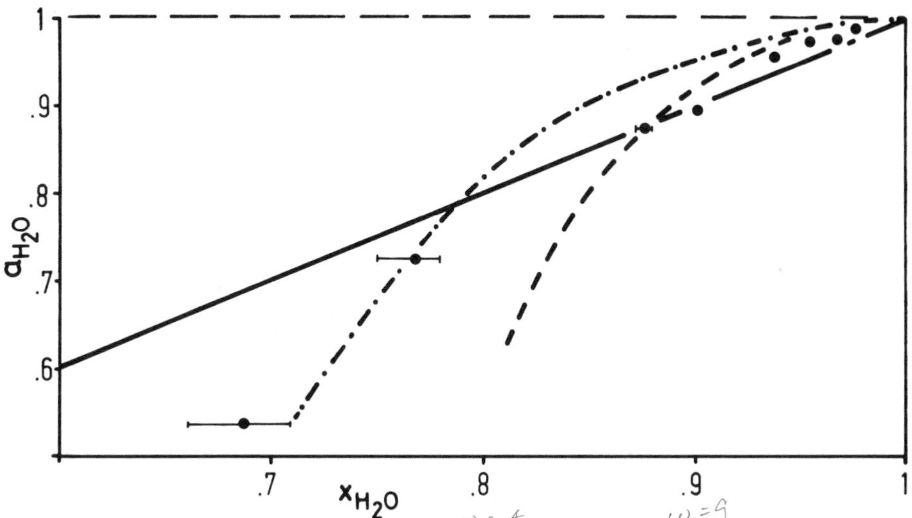

FIG. 14 Activity of water (a_{H_2O}) according to Raoult's law (—) and according to theoretical calculations (---- and -·-·-) (see [23a]) against mole fraction of water $x_{H_2O} = w_0/(1 + w_0)$; T = 298 K.

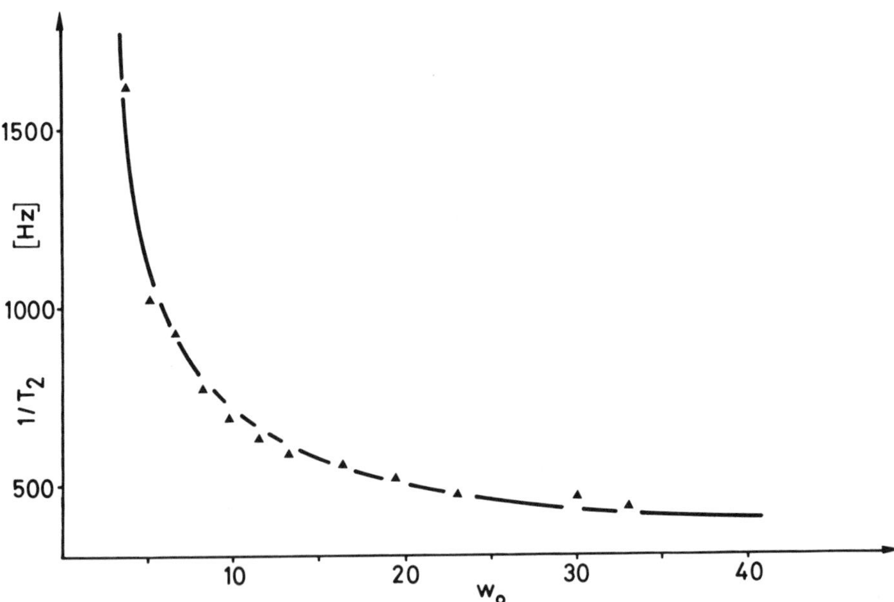

FIG. 15 Sodium-23 relaxation rate ($1/T_2$) against $[D_2O]/[AOT] = w_0$. System: D_2O/0.15 M AOT/cyclohexane, T = 298 K; theoretical calculations (—) [23a].

are sketched. Figure 18 shows the normalized intensity decay of the fluorescence (convolution of the sum of two exponentials with the time resolution function; see [42]). This evaluation procedure utilizes a function $\eta(r)$ (i.e., the anisotropic viscosity of the nanophase interior with the radius of the water core) and the corresponding radius-dependent diffusion coefficient D(r) (see Fig. 17). A series of measurements with different w_0 values were fitted, keeping the average rate constant k and a parameter (P) unchanged. The latter denotes the probability per unit time that an excited molecule which reaches the interface by diffusion will be quenched [42]. The results are in agreement with two possible models: (1) $D = D(w_0)$, homogeneous interior of the nanophase, average viscosity $\bar{\eta}$ (= A; see Fig. 17) or (2) $D = D(w_0,r)$, viscosity function C (see Fig. 17),

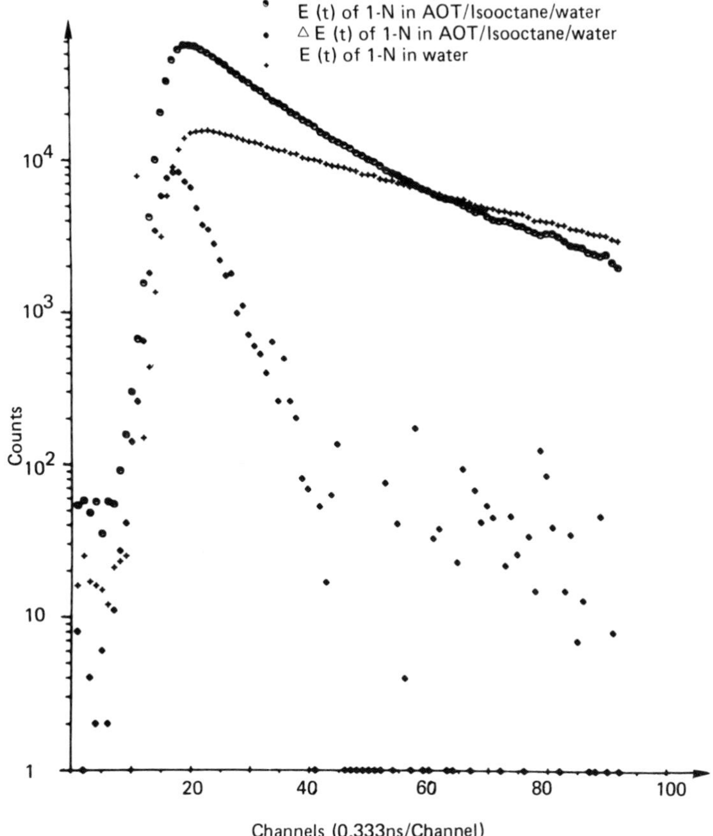

FIG. 16 Comparison of typical experimental decay plots. Fluorescence decay E of 1-N in H_2O/AOT/isooctane (w_0 = 3.7) and of 1-N in water at the same dye concentration (for abbreviations see text). Normalized difference ΔE of experimental decays with parallel and crossed polarizers [42].

i.e., low-viscosity inner region, analogous to bulk water, and high-viscosity interfacial boundary layer about 0.6 nm thick.

4. Measurements of Polarization Anisotropy Decay

The time decay of the polarization anisotropy is determined by the time-dependent ratio of the difference between the fluorescence intensities

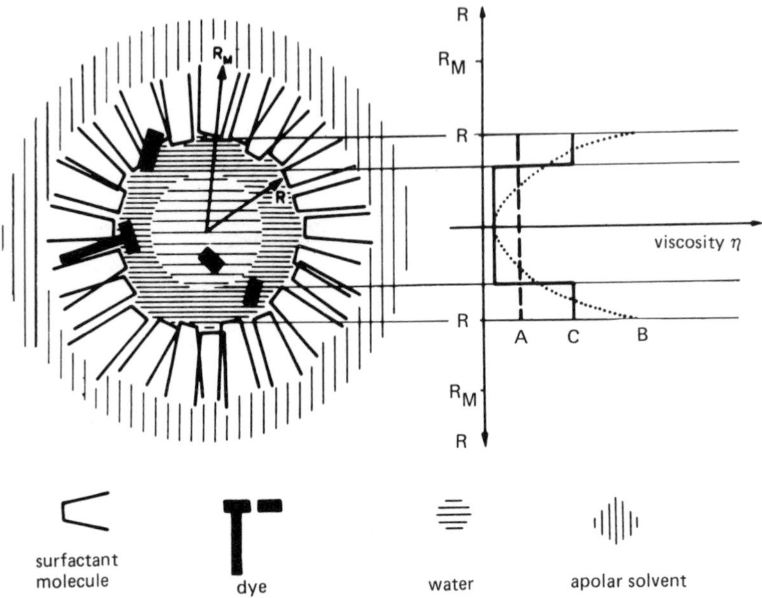

FIG. 17 Model of AOT micelle with solubilized water; viscosity of water core may be inhomogeneous as indicated by different hatchings. The effective viscosity $\eta(r)$ for the three cases A, B, and C is determined by measuring the diffusion constants of the fluorescing dye molecules. Probe molecules either move in the bulk of the water core or are bound in the micellar film [42].

measured with a polarization analyzer parallel and perpendicular with respect to the electrical vector of the exciting light and the total fluorescence intensity. The relationship between the polarization anisotropy and the rotational diffusion of a spherical nanophase is given [39,43] by the product of the anisotropy of the dye molecule in a nanophase-fixed coordinate system and that of the nanophase. In the case of rotating spheres (the dye molecule is also considered, hydrodynamically, as a sphere) an exponential relationship between the anisotropy and the sum of the diffusion coefficients of the nanophase and the label (dye) is found. The rotational correlation time is then given (apart from a numerical constant) according to Stokes-Einstein by the reciprocal of the sum of the rotational diffusion coefficients mentioned above. Hence, it is to be expected that the rotational correlation time is determined by the rotational diffusion coefficient of the nanophase at small amounts of solubilized water (i.e., the label is "fixed") and by the rotational diffusion of the label at large amounts of water.

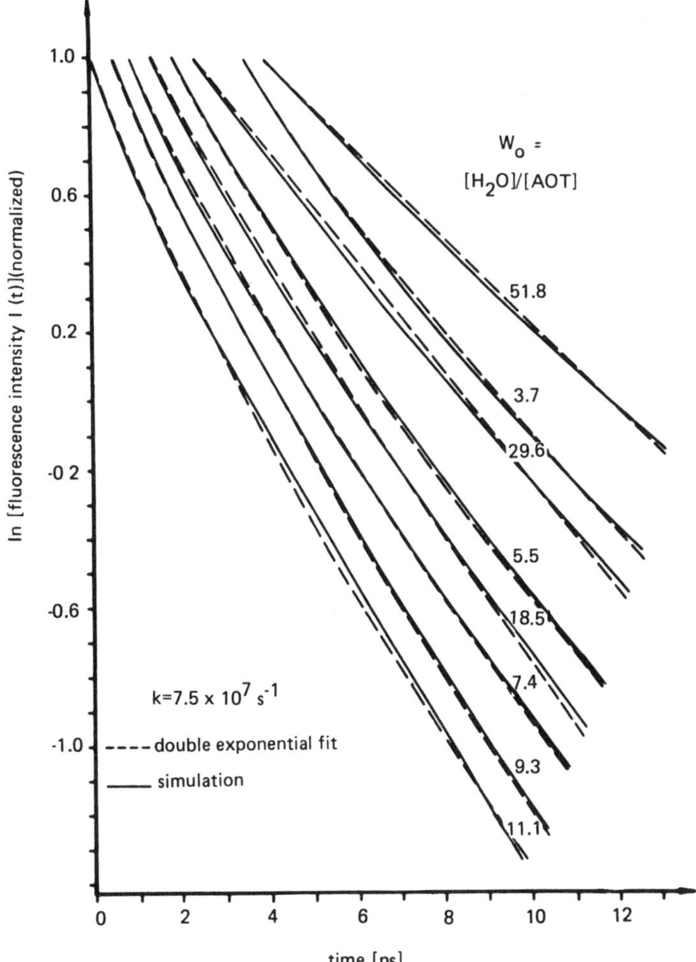

FIG. 18 Fluorescence intensity decay curves [I(t), dashed lines] obtained by fitting the sum of two exponentials to the experimental decays, and optimal numerically simulated decay curves (solid lines) obtained by fitting model II (viscosity function C of Fig. 17) to I(t); the curves are normalized and displaced from each other [42].

Results which display this behavior are shown in Fig. 19, where the rotational correlation times (τ_1) of the label are plotted against w_0 for low (upper curve) and high (lower curve) AOT concentrations. Also given in this diagram are "microviscosities," which were derived from the volume of the label and its correlation time. The remarkable drop of τ_1 starting with very small w_0 values at low AOT concentrations (where the label is tightly bound to the AOT aggregate) indicates the sensitivity of the mobility of the label to added water. Increasing amounts of the latter detach the label from the interfacial boundary layer. This seems to confirm an estimate [39, 43] that η_{micro} is proportional to $[H_2O]^{-1}w_0^{-1}$. This, however, implies that the process is confined to individual aggregates [micelles with negligible exchange of the substrate (label)], which does not appear unreasonable because of the low AOT concentration. The second branch, corresponding to a high AOT concentration, exhibits a considerably smaller sensitivity to the solubilized water. Apparently, an exchange process between aggregates is contributing increasingly to τ_1. Corresponding microviscosity values (Fig. 19) are similar to those reported by Shinitzky et al. [44].

Figure 20 exhibits the radius dependence on w_0. The plot consists of two straight lines, one corresponding to lower and the other (upper curve) to higher AOT concentrations. The necessity of fitting the experimental rotational correlation times at high AOT concentrations with larger aggregate radii indicates that the label occurs in "extended structures." (This conclusion has been confirmed by water diffusion processes and electrical conductivity studies [58,58a,58b].) Similar investigations were conducted by Keh and Valeur [45] to determine the radii of AOT nanophases in *n*-heptane and *n*-decane. A comparison with hydrodynamic radii obtained from quasi-elastic light-scattering data [46] for AOT nanophases in isooctane showed reasonable agreement between both techniques (see Fig. 21). These authors, however, used a slightly different calculation procedure for the hydrodynamic radii which considered a coupling between the two rotational processes of label and nanophase. This procedure does not seem unreasonable; however, this evaluation implies other restrictions [45] which might not be generally fulfillable. Hence, the actual improvement is difficult to estimate.

5. Time-Average (Static) and Dynamic Light-Scattering Studies

Light-scattering techniques (both static and dynamic) have proved to be most powerful and suitable for the investigation of microemulsions. Certainly, in scattering studies, multiple scattering effects must be absent or corrected for [47] to interpret the results in terms of particle-particle interactions. This means that light scattering is

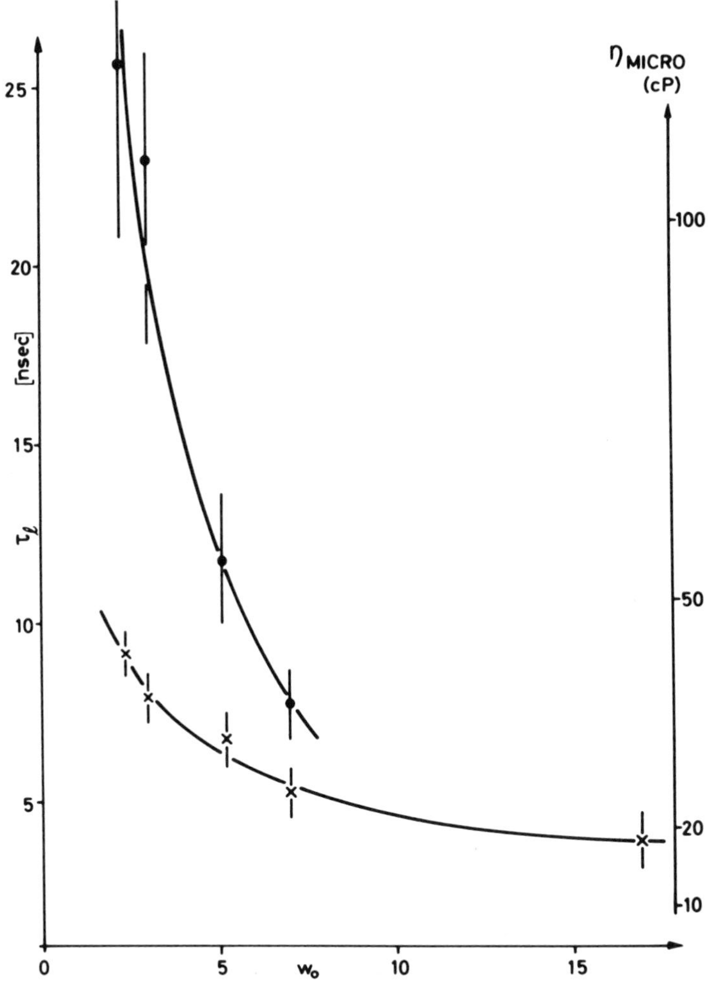

FIG. 19 Rotational correlation times of the label (τ_l) and "microviscosities" (η_{micro}) against $w_0 = [H_2O]/[AOT]$. Microviscosities were calculated according to $\eta_{micro} = (kT/V_l)\tau_l$; V_l = volume of dye label, radius $R_l = 0.59$ nm, T = 293 K [39]. (•) low, (×) high AOT concentrations.

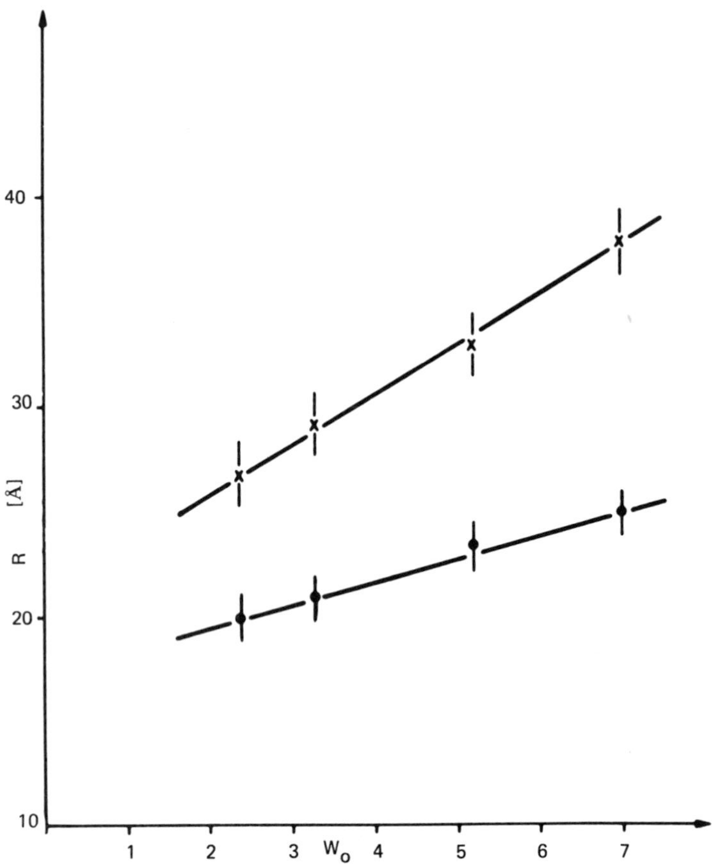

FIG. 20 Experimentally determined radii of swollen micelles against $w_0 = [H_2O]/[AOT]$ at low (•) and high (×) AOT concentrations in the range $0 < [AOT] < 1.0$ mol dm^{-3}. System: H_2O/AOT/isooctane, T=293 K [39].

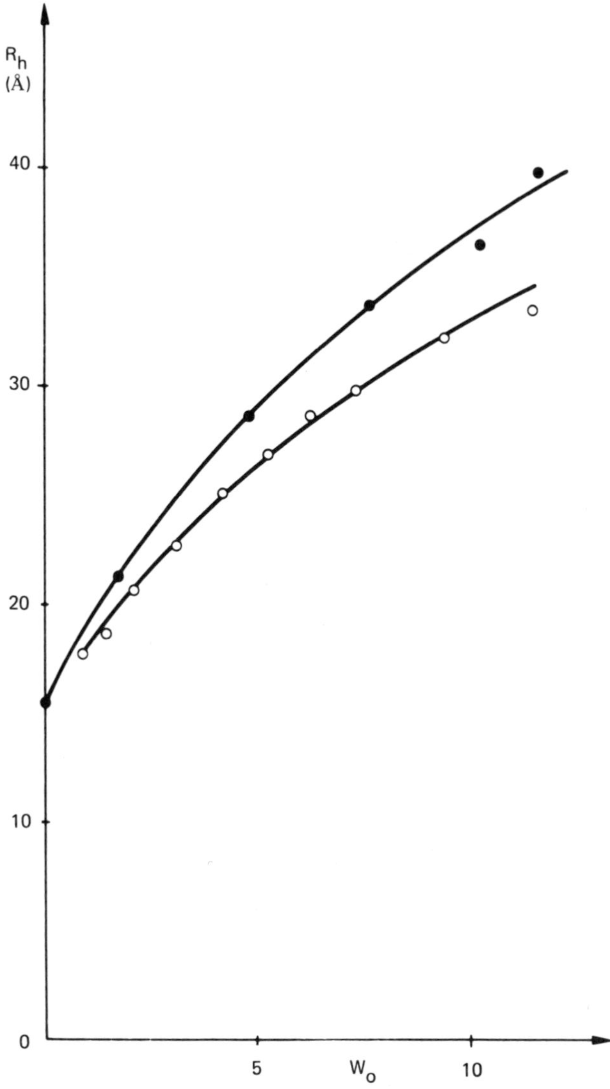

FIG. 21 Variation of hydrodynamic radii of aggregates against w_0: (○) present work, [45]; (•) data from [46].

(principally) restricted to rather low particle concentrations or particle sizes small compared to the wavelengths of the applied light.

a. Time-average light scattering. There apparently exists relatively meager information on static light scattering by microemulsions using Aerosol OT as surfactant.

Corkill et al. [48] applied this technique together with vapor pressure osmometry to an AOT/toluene solution with very small amounts of added water. Accordingly, this paper is at the boundary of our present topic of microemulsions. However, the paper reports on some generally interesting aspects. First, it demonstrates the importance of equilibration by comparing weight-average (by static light scattering) and number-average (by vapor pressure osmometry) molecular weights of AOT aggregates in toluene containing small amounts of water. The considerable deviations between both molecular weight figures indicate a pronounced polydispersity which was time dependent and disappeared as equilibrium was reached (see Fig. 22). The authors interpreted this observation as indicating that large aggregates were initially formed in the presence of water which subsequently dispersed to give a solution of uniform "nanodrops." The second observation is also noteworthy: small amounts of solubilized water, i.e., water which is bound within the hydration shell of the AOT, does not behave as a third component (Fig. 23). Similar conclusions have frequently been made (e.g., [49]).

Eicke and Rehak [50] applied static light-scattering measurements to the water/AOT/isooctane system by following up the mixture to the microemulsion domain, i.e., up to $w_0 \cong 44$. These authors compared results from static light scattering, ultracentrifugation, and vapor pressure osmometry to obtain Fig. 24. The latter displayed a satisfactory monodispersity of aggregate sizes. It also demonstrated the swelling region of the micellar aggregates up to a water-to-AOT ratio of about 9 to 10 (as indicated by the linear increase of the apparent molecular weight). Above $w_0 = 10$ the aggregates are expected to coalesce at constant temperature if the amount of the surfactant (AOT) is kept constant. From these experiments, which apparently supported the hypothesis of equipartition of water among the aggregates, the dependence of the interfacial area per AOT monomer on the amount of added water (assumed to be exclusively in the solubilized state) was calculated (see Fig. 25). The plot supports the idea of an equilibrium interfacial area of one AOT molecule which is approached for increasing amounts of solubilized water (for a discussion of this observation see [9]).

b. Dynamic light scattering (photon correlation spectroscopy, quasi-elastic light scattering). One of the first investigations to apply dynamic light scattering to microemulsions of the type water/AOT/organic solvent (c-C_6H_{12}, C_7H_8, C_6H_5Cl) was that of Day et al. [51].

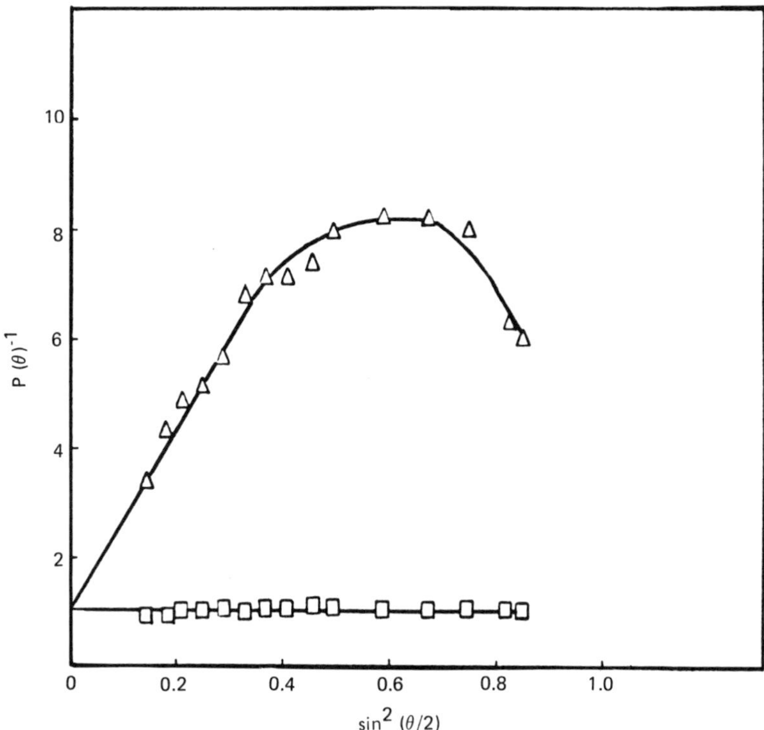

FIG. 22 Angular dependence ($\sin^2 \theta/2$) of the reciprocal scattering factor $P(\theta)^{-1}$: (△) freshly prepared 10^{-1} g/ml Aerosol OT (molar ratio H_2O/AOT = 2.46) toluene solution; (□) the same solution at equilibrium [48].

Single exponential correlation functions were obtained from which translational diffusion coefficients and, by application of the Stokes-Einstein relation, particle radii (hydrodynamic radii) could be derived. The droplet size was found to depend primarily on the ratio of the surfactant to water concentrations, but was essentially independent of solvent and concentration at a fixed surfactant/water ratio. These authors found reasonable agreement for droplet diameters obtained from translational diffusion, viscosity, and sedimentation ultracentrifugation measurements as seen in Fig. 26. There is, however, a difference between the values obtained from viscosity measurements and those from dynamic light scattering and ultracentrifugation. The latter techniques yielded almost the same droplet diameters and w_0 dependences for toluene and heptane. (The possibility cannot be excluded that this coincidence is a bit fortuitous.) The authors

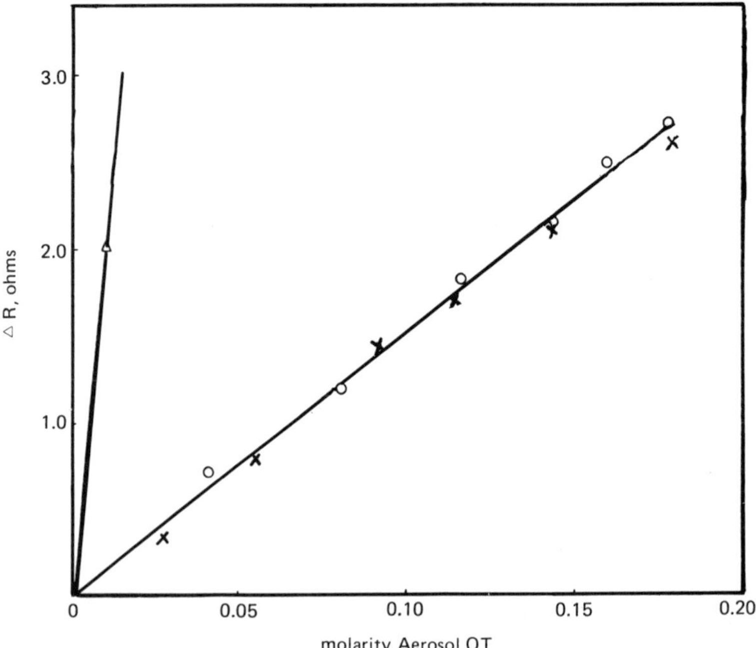

FIG. 23 Thermistor bridge readings (ΔR) as a function of concentration: (Δ) benzil in toluene; (○) AOT in toluene (H_2O/AOT = 0.73); (×) AOT in toluene (H_2O/AOT = 3.10) from [48].

found in particular some differences in behavior compared with other solvents: water solubilization was more difficult; maximum density of the water core was only 0.8 kg dm^{-3}, from which they inferred a more open structure of the aggregates. Finally, the larger diameters obtained from viscosity measurements might be indicative of some polydispersity.

Almost simultaneously, a paper by Zulauf and Eicke [46] appeared which was concerned with a rather similar study; water/AOT/isooctane was investigated starting with the binary (micellar) and extending to the ternary mixture, but adding much larger amounts of water ($w_0 \leqslant 60$) than reported in the former paper [51]. Also, for the first time, the temperature dependence of the system was followed up to 333 K. Experimental evidence was presented that a clear distinction is possible between micellar and swollen micellar solutions on the one hand and microemulsions on the other, according to the degree of solubilized water. The onset of the phenomenon characteristic of microemulsions (see Fig. 27) occurs when water becomes the

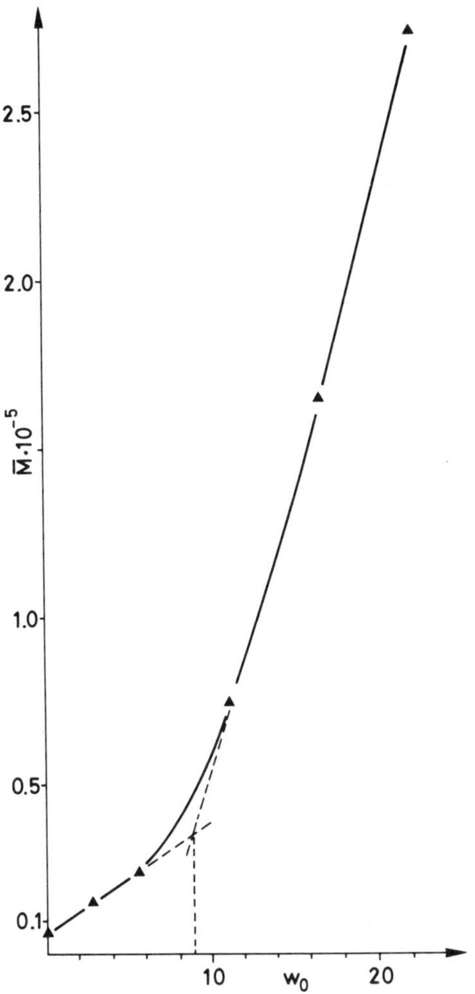

FIG. 24 Average apparent molecular weights ($\bar{M}$) of solubilizing AOT aggregates against $w_0 = [H_2O]/[AOT]$ [50].

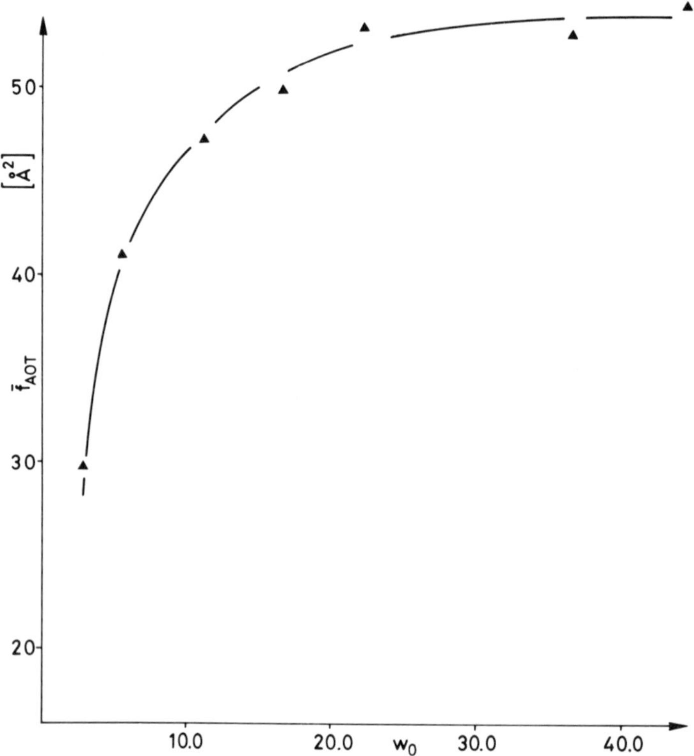

FIG. 25 Average surface area of one AOT molecule ($\bar{f}_{AOT}$) in the water/hydrocarbon interface against $w_0 = [H_2O]/[AOT]$ [50].

major component in the colloidal aggregates. From Fig. 28 it is apparent that some polydispersity of the droplet size must be considered; this follows from the residual scattering at about $w_0 = 30$, where optical matching is expected (for isooctane as solvent and T = 298 K), and the particular deviations of the experimental Stokes radii (hydrodynamic radii) from the (solid) curve corresponding to the equipartition model below and above the matching point. A comparison of the diffusion coefficients obtained from quasi-elastic light-scattering and sedimentation ultracentrifugation measurements showed that the latter technique yielded 10–20% higher values within the range $11 \leqslant w_0 \leqslant 56$. It should be noticed that this w_0 region is a straightforward continuation of the measurements reported in [51], where reasonable agreement between quasi-elastic light-scattering and sedimentation data was observed.

Still another investigation [53] applying dynamic light-scattering and viscosity measurements to water/AOT/CCl_4 in order to determine hydrodynamic radii from translational diffusion coefficients (D_T) yielded two notable results: (1) in the range of volume fractions of the dispersed phase between 0.05 and 0.12 and w_0 = 10.1 at 298 ± 0.1 K the hydrodynamic radii are reasonably constant, i.e., varying between 2.0 ± 0.1 and 2.1 ± 0.1 nm, which coincides with the observation reported in [51], and (2) the D_T dependence on the volume fraction of the dispersed phase (see Fig. 29). The linearly decreasing plot is noteworthy, as is its slope, which corresponds almost exactly to $-2.5D_0$, where D_0 is the extrapolated value of the diffusion coefficient at zero volume fraction of the dispersed phase and is equal to 1.21×10^{-10} m^2 sec^{-1}. This relation follows directly from the Stokes-Einstein equation and Einstein's viscosity relation for spheres. This is a strong hint as to the ideality of the system.

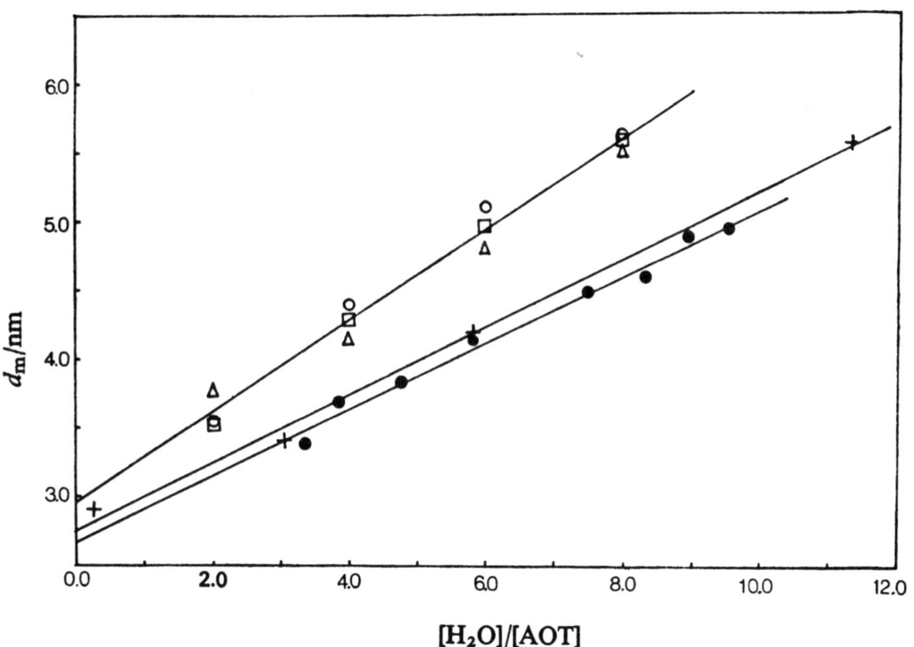

FIG. 26 Droplet diameter d_m as a function of [H_2O]/[AOT]. Open symbols are viscosity data for solvents (△) toluene, (○) chlorobenzene, and (□) cyclohexane; (+) ultracentrifuge data with *n*-heptane solvent; (•) dynamic light scattering with toluene solvent, [51].

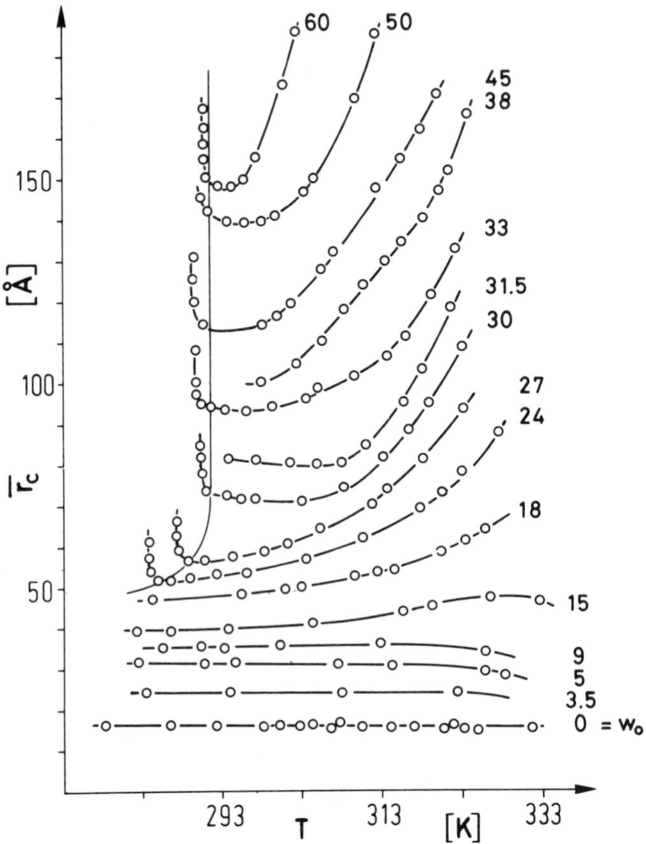

FIG. 27 Correlation radii ($\bar{r}_c$) of microemulsions (H_2O/AOT/isooctane) against temperature. The AOT concentrations were between 5.5 and 8.1 × 10^{-2} M; scattering angles varied between 8 and 90°; w_0 = [H_2O]/[AOT] (parameter). The vertical line at 291 K indicates a stability boundary; data points in this region were taken at 30-sec time intervals, [46].

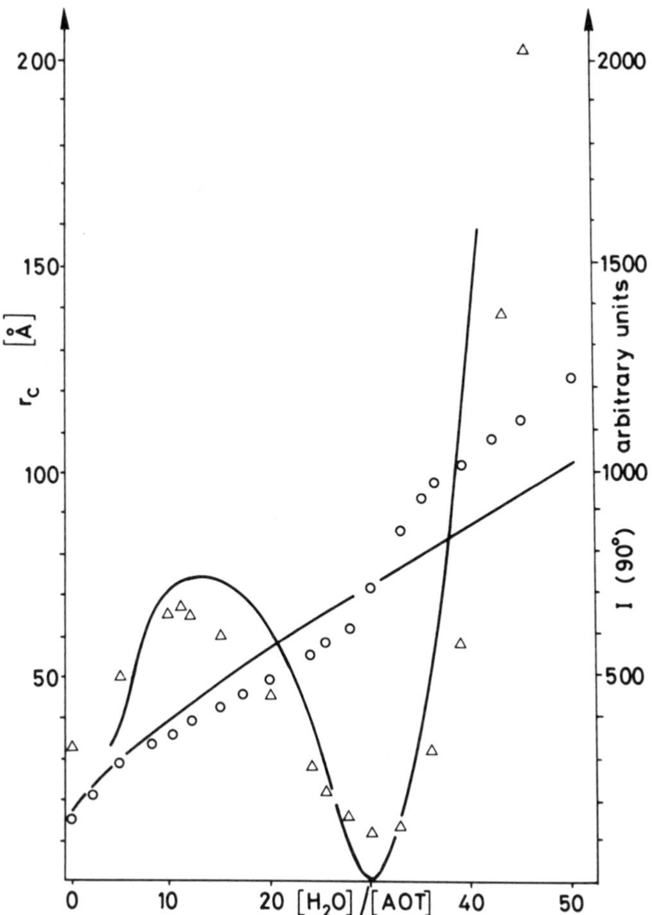

FIG. 28 Correlation radii ($\bar{r}_c$) (○) and scattered intensities (△) at 298 K as a function of $[H_2O]/[AOT]$. The solid lines show the predictions of the equipartition model [46].

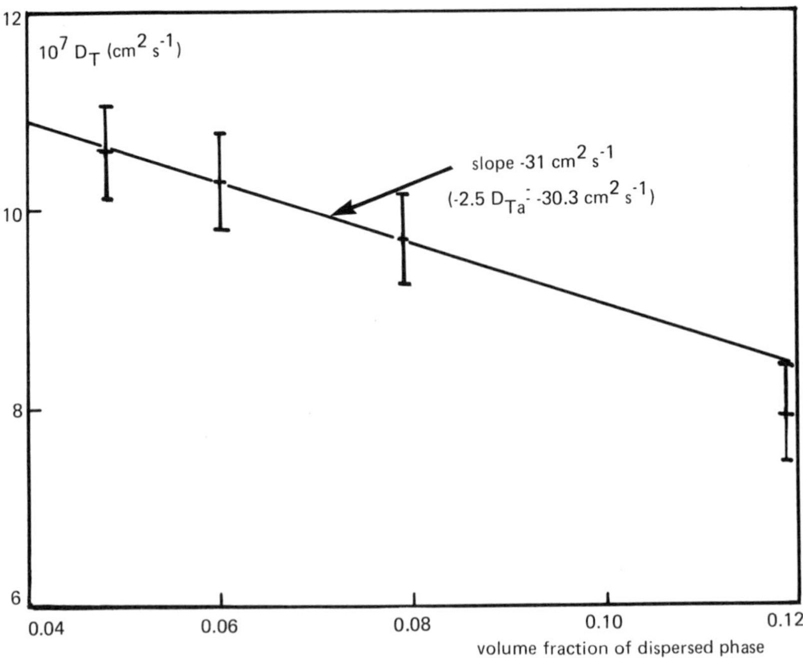

FIG. 29 Variations of translational diffusion coefficient (D_T) against volume fraction of dispersed phase, 298.3 ± 0.1 K [53].

More recently, a rather thorough investigation of three-component water/AOT/organic solvent systems with the help of dynamic light scattering, which was particularly devoted to the study of particle-particle interactions, was presented by Nicholson et al. [54]. Again, the sizes of nanophases and their dependences on the mole ratio w_0 (=$[H_2O]/[AOT]$), the nature of the solvent (from the dependence of the diffusion coefficient on concentration), and, finally, the characterization of polydispersity are considered. Figure 30 displays the variation of the apparent hydrodynamic radii (r_h) with w_0 for 0.1 mol dm^{-3} AOT in the respective solvents; for all solvents the effective hydrodynamic radii ($\bar{r}_h$) of the micelles or nanophases, respectively, increase with increasing w_0. This increase is approximately linear over the w_0 range studied. Since scattering from the cyclohexane solutions was weak, no firm conclusion could be drawn in this case. The radius $\bar{r}_h$ is not proportional to $w_0^{1/3} + 1$ (surfactant layer thickness), which appears reasonable, since it is known that the interfacial area covered by one AOT molecule expands with w_0 (up to a maximum value) (see also [50,9]). The effective hydrodynamic radius depends on the solvent used. For a fixed w_0 value these authors found $\bar{r}_h$ to be greatest in cyclohexane and smallest in toluene. A reasonable explanation for this difference is the different degree of solvation of the microphases by solvent molecules. This phenomenon has been observed frequently [55–57] and more indirectly in [58].

Similarly, as with light-scattering experiments in water/AOT/cyclohexane microemulsions, the induced birefrigence [59] was surprisingly weak, apparently due to a particularly strong solvation ($c\text{-}C_6H_{12}$ has a relatively small molar volume) which prevents any notable aggregation and hence anisotropic structuring.

Another important aspect is the interparticle interaction with respect to diffusion of the particles at constant w_0 ratio (11.2) in *n*-heptane. A typical result is shown in Fig. 31. It is seen that as the volume fraction (ϕ) of water increases, the diffusion coefficient

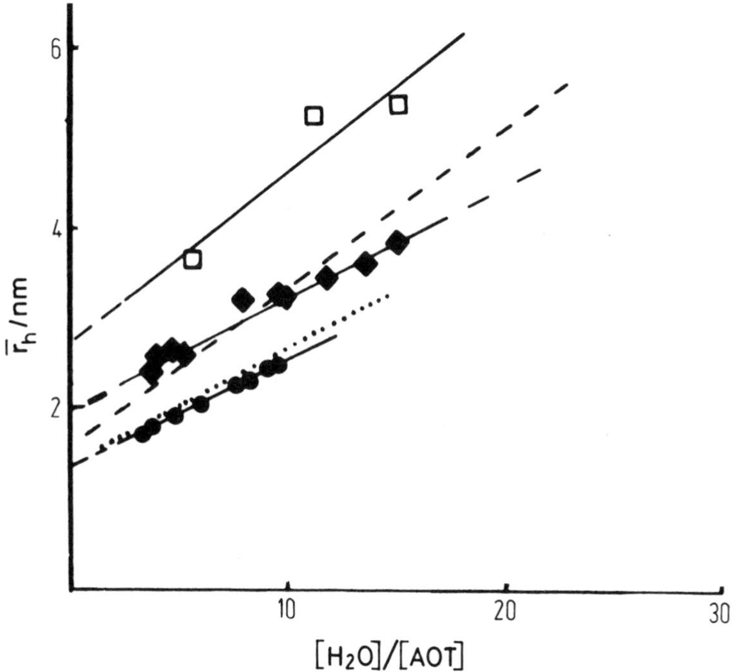

FIG. 30 Variation of apparent hydrodynamic radii ($\bar{r}_h$) with H_2O/AOT for solutions 0.1 mol dm^{-3} in AOT as determined from dynamic light scattering at 293 ± 1 K: (•) toluene, (♦) *n*-heptane, (□) cyclohexane, (dashed line) isooctane, (dotted line) *n*-heptane (ultracentrifuge data) [54].

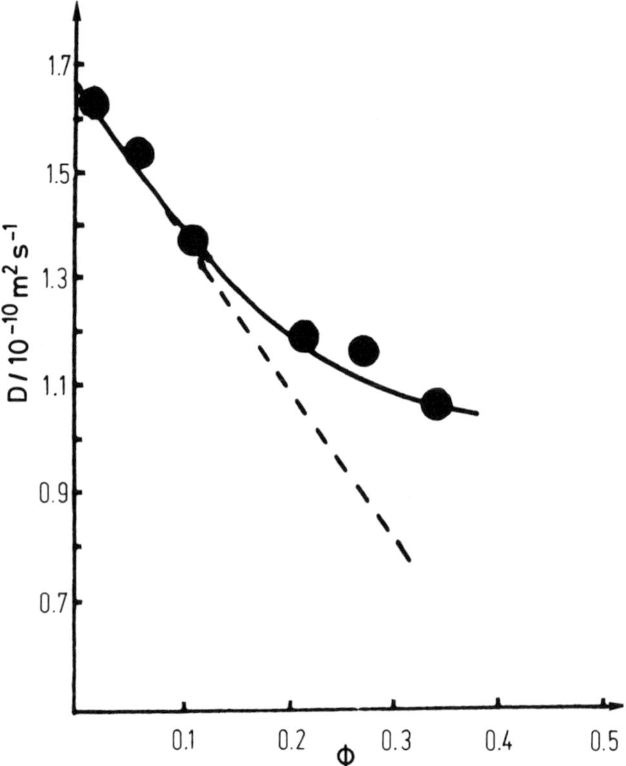

FIG. 31 Variation of mean diffusion coefficients (obtained using cumulants analysis) with volume fraction ϕ of water in n-heptane, $[H_2O]/[AOT] = 11.2$. The slope of the linear extrapolation to infinite dilution is −1.7 [54].

(D) decreases. Empirically, extrapolation to $\phi = 0$ should be used to obtain a diffusion coefficient (D_0) appropriate to the Stokes-Einstein relation. At $\phi = 0.11$ (=0.13 mol dm^{-3} surfactant/water solution) the error in using the observed value of D in place of D_0 is already 18.5%. According to the authors, there is little doubt that the decrease of D and increase of the viscosity (η) are a consequence of increasing particle interactions with an increase of ϕ. This decrease of D is linear at low volume fractions and can be represented by $D = D_0 \cdot (1 + B)$ with $B < 0$ (compare [53]). Similar behavior has also been observed for other microemulsions [60–62]. From the physical situation it appears natural that interactions in these systems are short-range. This conclusion, however, is probably only valid at small water contents, i.e., with so-called swollen micelles as used in this study, since for larger amounts of solubilized water charge exchange processes become relevant [63,64].

Several theoretical approaches toward predicting the dependence of D on volume fraction for spherical particles in viscous media were suggested (e.g., see [65–70]). All of the predictions apply only to very dilute solutions, where interparticle separations are large compared to particle sizes. The microemulsions studied by the present authors and others are in general too concentrated, with mean particle separations of three to four particle radii. Hence, substantial local structuring is to be expected, which is still a theoretically unresolved problem [115].

The observation of a negative B coefficient (see above) may be indicative of some (specific) attractive interactions between particles.

Regarding, finally, the effect of polydispersity, the authors [54] conclude that it results from particle association. This has been confirmed by recent investigations [71] with the help of small-angle light scattering at larger w_0 values and varying particle concentrations in water/AOT/isooctane. The excess Rayleigh factor as a function of the total weighed-in amount of water and surfactant at constant $w_0 = 46.3$ and temperature varying between 296 and 305 K could be very satisfactorily fitted with the hard-sphere approximation, where the structure factor had to be modified in view of some polydispersity [72] (see Fig. 32). The latter was described by an "open association" model [73]. Increasing deviations were observed with larger droplet sizes ($\propto w_0$) and increasing temperatures, as was to be expected for increasingly "soft" spheres (for details see [74]).

Again, two studies were devoted to the water/AOT/*n*-heptane system by Gulari et al. [75,76] using photon correlation spectroscopy. Translational diffusion coefficients (D) for various w_0 values were determined from which hydrodynamic radii were derived under the assumption of the validity of the Stokes-Einstein relation. However, since the applied concentrations of the dispersed phase were typically 0.12, the error in D is already about 20% according to the above [54]. In particular, the assumption of an increase in the nanophase sizes as concluded from a decreasing D value did not prove true.

The last investigation to be discussed in this section concerns mostly the critical state of the water/AOT/*n*-decane system (3% w/v AOT, 4.76% v/v H_2O, and the rest *n*-decane) [77]. Again, photon correlation spectroscopy was applied. The results can be summarized as follows: the scattering intensity obeyed the Ornstein-Zernicke [78] and Debye [79] relations. The extrapolated forward scattering, I(0), diverges at the critical point $T_c = 36.01°C$ (309.01 K). The forward scattering can be represented by $I(1) \propto (T_c - T)^{-\gamma}$ with $\gamma = 1.22 \pm 0.05$. The authors claim that this critical index γ is in good agreement with corresponding values for most critical binary mixtures. Multiple scattering was insignificant. Hydrodynamic radii

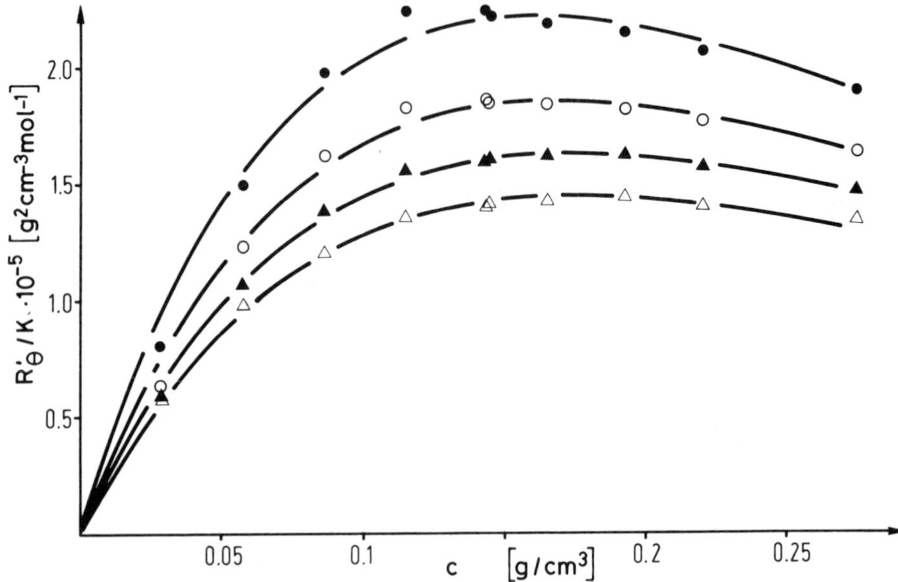

FIG. 32 Excess Rayleigh scattering in H_2O/AOT/isooctane against weighed-in amount of AOT and water at w_0 = 46.3. Temperature: (△) 296 K, (▲) 299 K, (○) 302 K, (●) 305 K [71,74].

were determined via the Stokes-Einstein relation and found to agree reasonably well with the radii of gyration. The authors believe the system to be polydisperse in the neighborhood of T_c with highly distorted droplets. The hydrodynamic radius diverged at T_c with a critical exponent ν of 0.75 ± 0.05 and ν' = 0.68 ± 0.08 in the upper phase (after phase separation). These data also agree with the value 0.64 ± 0.03 obtained for binary critical mixtures. This again might be indicative of one-component macrofluid-type behavior [80,80a] of water, AOT, organic solvent systems.

6. Small-Angle Neutron Scattering (SANS)

In the past several years a powerful, and now quite convenient and popular, scattering technique, small-angle neutron scattering, has become available which ideally supplements light-scattering techniques applied to water-in-oil microemulsions.

The particular method used is the so-called contrast-variation technique, fully described in [81–83]. The principle is rather simple: the water core size is obtained by mixing H (hydrocarbon), H (surfactant), and D_2O (aqueous core) and the overall size of the microemulsion droplet (nanophase) by mixing D (hydrocarbon), H (surfactant), and H_2O (aqueous core). From the difference the thickness of the surfactant coat (film) region can be determined and even finer details may be studied.

In this way Cabos and Delord [83] determined (in dilute dispersions of aqueous nanophases in *n*-heptane, using AOT as surfactant for $8 \leqslant w_0 \leqslant 50$) the particle volume $V_{tot} = V_{D_2O} + V_{pol.heads} + V_{HC\text{-}chain}$, V_{D_2O}, V_{sphere} (from radius of gyration), the aggregation number of surfactants, and the (mean) interfacial area per surfactant molecule (here $\bar{f}_{AOT}$). Of particular interest is $\bar{f}_{AOT}$ as a function of w_0, as shown in Fig. 33, which should be compared to the corresponding data of Eicke and Rehak [50] obtained from light-scattering studies with water, AOT, and isooctane. In an extension of their investigations, Cabos and Delord studied the four-component system water, AOT, *n*-heptane, and salt (NaCl) [84]. The authors obtained an increase of the droplet volume with

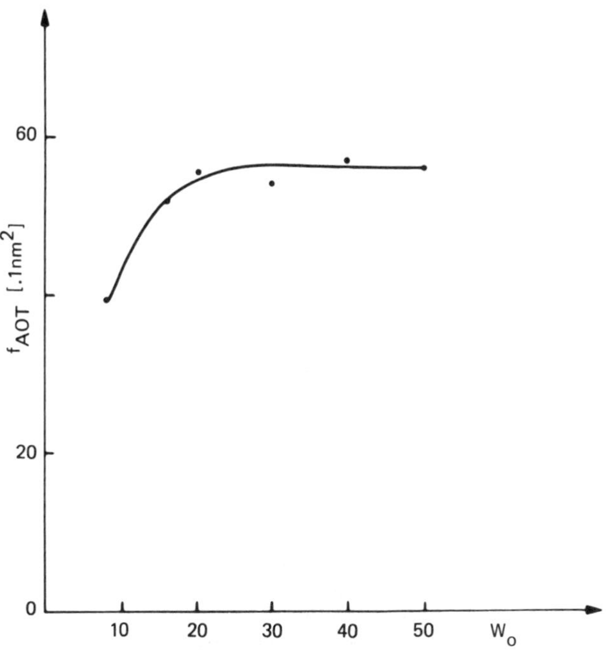

FIG. 33 Variation of average surface area of one AOT molecule in water/hydrocarbon interface against $[H_2O]/[AOT] = w_0$ [83].

increasing amount of salt for $w_0 = 8$ and 15. The simultaneous decrease of $\bar{f}_{AOT}$ (particularly pronounced at $w_0 = 8$) is reasonable and is also found from theoretical calculations if the screening effects produced by the added salt are considered. The authors again emphasized the spherical model of aqueous nanophases for dilute solutions of W/O microemulsions with AOT.

They also suggested nanophase dimerization as an interpretation of the observed anisotropy in electro-optical Kerr effect experiments [85] in favor of a droplet distortion.

Similar investigations were extended to larger w_0 variations (8 to 49) by Kotlarchyk et al. [86]. These authors were concerned with estimating the area of a single AOT molecule at the water/oil interface and its temperature dependence between 298 and 306.5 K. The values obtained were definitely larger than those derived by Cabos and Delord [83] and Eicke and Rehak [50]. This divergence might be due to different purification procedures for AOT, particularly with respect to the problem of salt elimination (Na_2SO_4). The latter could easily explain the larger interfacial areas found in [86]. Another point emphasized in this paper is the polydispersity of the samples. According to the authors' results, the D_2O core displays, within $33 \leqslant w_0 \leqslant 49$ and between 298 and 306.5 K, up to 31% polydispersity (defined by $\sigma/\bar{R}$, where σ is the width parameter and $\bar{R}$ the mean particle radius in a Gaussian distribution). Such large polydispersity is indeed surprising and contradictory to all previous QELS and SANS measurements [51,52,60,62,87]. The origin of the pronounced polydispersity is, according to the authors, the decreasing interfacial free energy (γ) on approaching the critical point. It is, however, still not convincing that polydispersity should increase to such an extent, since the interfacial free energy is already very small compared to kT far from the critical point, where relatively small γ values have been reported. Moreover, the fluctuation amplitude $(kT/\gamma)^{1/2}$ as cited by these authors is valid only for one-component systems and cannot be applied, schematically, to systems with more components. The inferred (dynamic) distortion of the droplets would inevitably result in a rise of the interfacial tension.

Finally, in line with several previous measurements [50,83], these authors also reported that the effective head group area per AOT molecule remains constant for $8 \leqslant w_0 \leqslant 49$, which, however, cannot be immediately seen from their results.

In a more recent paper [88] the same authors report again on small-angle neutron scattering close to a critical point of a water (D_2O), AOT, *n*-decane microemulsion with $w_0 = [D_2O]/[H_2O] = 40.8$. The Q range (momentum transfer, $Q = 2\pi/\lambda \sin \theta/2$) extended from 4×10^{-3} to 2.5×10^{-1} Å^{-1} and could be fitted by assuming that scattering was caused by density fluctuations of polydisperse spheres.

The reason for this pronounced polydispersity is still open to question in the view of these authors. On the other hand, the authors have to face the fact that in approaching a demixing curve pronounced clustering should be apparent and has actually been frequently reported in many of the references mentioned above.

The particular evaluation procedure applied by the authors mentioned above has been outlined [89] by two of them in considerable detail.

In closing this section, we note that all scattering techniques are model-dependent and hence, according to the above discussion, the physical phenomena close to the critical point are not yet fully understood, even for the comparatively simple three-component water, AOT, oil systems.

7. X-Ray and Small-Angle X-Ray Scattering

Relatively little work has been performed on x-ray scattering in water, AOT, oil systems. Early investigations were carried out by Mattoon and Mathews [34,92] and Philippoff [90]. These authors

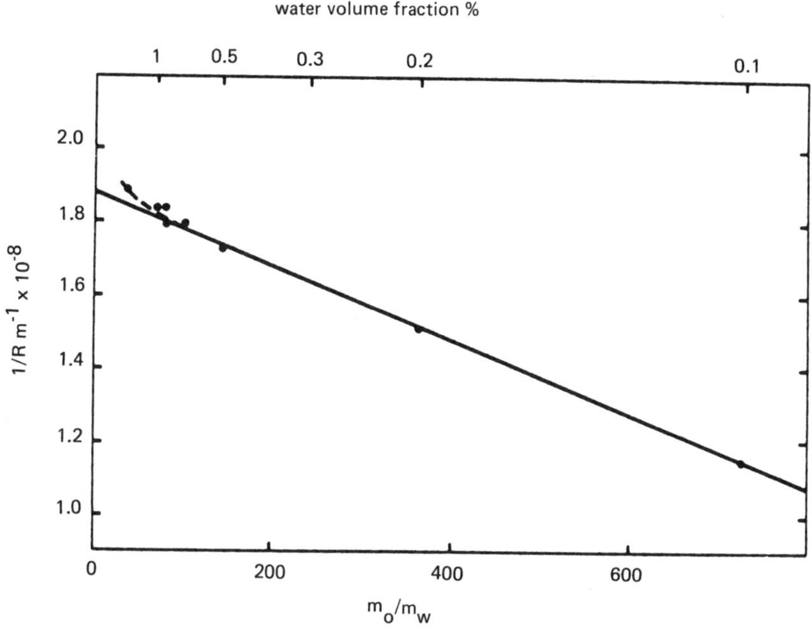

FIG. 34 Inverse of the radius of water core of inverse micelle against ratio of total mass of decane and total mass of water in solution. Straight line represents the high dilution regime [92].

worked with rather high AOT concentrations (about 1.6 mol dm^{-3}) and found that only the innermost diffraction rings were sensitive to the addition of water. A plot (not reproduced here) in [90] obtained by application of Bragg's equation on the innermost diffraction rings yielded a linear relationship between the long spacings and the amount of solubilized water. Geometric considerations which were put forward in order to explain such linear plots are not generally valid. Also, the solvent dependence observed seems to vary with the amount of dissolved AOT, which is difficult to understand; however, the high AOT concentrations might induce the formation of liquid crystal-like structures. Such phenomena are, however, outside our present discussion. Interested readers should consult [90] and the literature cited therein.

A more recent study with the small-angle x-ray scattering technique is that of Assih et al. [91] on water, AOT, and *n*-decane at $w_0 = 30$. This investigation was concerned with the high-dilution regime of this microemulsion. The essential result is an apparent slight increase of the nanophases with dilution as shown in Fig. 34.

In general, there seems to be no particular advantage in using x-ray scattering, since the scattering power of these systems is not pronounced. Neutron scattering, particularly the contrast variation method, is definitely superior.

III. MULTI(>3)-COMPONENT MICROEMULSIONS WITH IONIC SURFACTANTS

This chapter is not intended to give an extensive description of multicomponent microemulsions where ionic surfactants have been used. Such information can be found in a large number of existing reviews mentioned above. Some investigations on multicomponent microemulsions with ionic surfactants which are typical of either new or particular trends or the application of modern techniques are mentioned. Where appropriate, aspects are emphasized which have led to a more reserved attitude toward microemulsions with cosurfactants and/or other additives.

First, it is interesting to note that the relevant literature on microemulsions such as those to be discussed here is mainly concerned with a rather limited group of ionic surfactants, e.g., potassium oleate, sodium dodecyl sulfate (SDS), and sodium dodecyl benzenesulfonate.

Among investigations with scattering techniques the studies of Vrij and co-workers [93–95] should be mentioned. These authors studied the four-component system water, potassium oleate, benzene, and hexanol, a system which was extensively investigated by Schulman and co-workers [96] in the late 1940s. All the studies of the

Dutch group were essentially devoted to the application of "hard-sphere" fluid theories in order to describe the osmotic compressibility and its dependence on the simultaneous action of hard-sphere repulsions and weak attractive (van der Waals) forces. These authors also considered the effect of polydispersity and approaches toward concentrated dispersions.

A rather extensive, thorough, and exemplary investigation of a four-component microemulsion, i.e., water, sodium dodecyl benzenesulfonate, xylene, and hexanol, was carried out by Cebula et al. [62] with the help of light-scattering, SANS, and QELS techniques. Particular emphasis was put on the examination of concentrated dispersions up to a water volume fraction of 0.533. In spite of this large volume fraction, multiple scattering effects were not observed. Since extrapolation to zero concentration is not possible with microemulsions, the authors had to apply model calculations in order to obtain particle radii for their aggregates. Hence, they used the approved hard-sphere model in the Percus-Yevick approximation in order to calculate the structure factor which enters the Rayleigh ratio. Within the approximations which had to be made regarding, e.g., the size distribution of the particles, (i.e., assuming that the polydispersity is fairly small), all three techniques yielded satisfactorily consistent results.

A consequence of their study which was particularly emphasized by the authors was the necessity for assuming that a certain dilution procedure maintains a constant droplet size.

This is certainly an appropriate point, since dilution procedures were frequently applied in most of the papers to be discussed here. The first paper is the work of Finsy et al. [61], who used a microemulsion similar to that used by Vrij, but with butanol as a cosurfactant. These authors determined the volume fraction (ϕ)-dependent translational diffusion coefficient $D = D_0(1 + k_D\phi)$ between $\phi = 0.046$ and 0.151. The correction factor k_D turned out to be negative, indicating attractive forces between the particles. Nothing was said regarding the "softness" of the spheres as mentioned by Cazabat and Langevin, to be discussed later. Such softness effects could be related to the use of the cosurfactant butanol, which is known to considerably increase the mutual solubilities of water and oil, thus competing, from a phenomenological viewpoint, with the solubilizing properties of the surfactant aggregates.

Another study of the four-component system H_2O, potassium oleate, *n*-dodecane, and hexanol [97] used the QELS technique. Volume fractions of water up to about 0.5 were used. The authors were faced with several difficulties in interpreting their results and had to assume some kind of gel formation since they studied compositions close to the gel phase of this system.

Several investigations along similar lines were carried out by Cazabat, Langevin, and co-workers [98–100]. These authors studied quaternary mixtures of water, sodium dodecyl sulfate, alcohol (butanol, pentanol, or hexanol), and cyclohexane or toluene with the help of light scattering and QELS. From their results they concluded, in particular, that the strength of attractive interactions increased with increasing droplet (nanophase) size, with decreasing alcohol chain length, with decreasing solvent polarity, and with decreasing amount of salt (if any). It is interesting to note that the authors mentioned in a later paper [101] that the method applied in the dilution procedure is not very sensitive with respect to the dispersed phase, particularly in more concentrated regions ($\phi > 0.05$). According to them, it is necessary to suppose that the microemulsion structure is not changed by the dilution process. Such an assumption will inevitably affect the conclusions drawn from these experiments.

Another light-scattering investigation [102] of four-component microemulsions formed by water, SDS, dodecane, and alcohol (pentanol, hexanol, or heptanol) was again concerned with particle interactions and their dependence on alcohol chain length. Simultaneously, the corresponding variation of the second virial coefficient was studied and compared with theoretical data. The results demonstrate that the attraction between the droplets increases from heptanol to pentanol as cosurfactant. Model calculations are reported to be in satisfactory agreement with these observations. However, since again a dilution procedure was applied to vary the particle concentration, the results should be considered only approximately correct in view of the above remarks.

A number of papers were concerned with small-angle neutron scattering in four (or more) component microemulsions. Probably among the first groups to apply this technique extensively were Taupin and co-workers [10,103,104]. These investigations allow one to obtain more detailed insight into the structure of the interfacial region (film) of aggregates, i.e., regarding the penetration of oil (continuous phase) into the film and the distribution of alcohol between the three regions (water, film, oil) and its dependence on the amount of solubilized water [10]. It was also found [10] that the area per polar head group of the surfactant molecule is relatively little influenced by the degree of solubilization. This observation is in agreement with earlier findings [50] from light-scattering experiments with three-component microemulsions and AOT as surfactant.

Moreover, the same authors studied four (and five) component systems consisting of water, SDS, cyclohexane, and pentanol (and occasionally salt) with SANS, ultracentrifugation, viscosity, and conductivity measurements.

Apart from the determination of structural details of the aggregates, analyses of the particle-particle interactions were carried out in terms of the hard-sphere approximation model.

These authors also observed percolation phenomena in the range $0.06 < \phi_W < 0.09$, where ϕ_W is the volume fraction of water in the system. They claim that the polydispersity of the system is remarkably low, without, however, giving a numerical value. Again, it does not seem advisable and suitable to go into more details, since only more generalizable results are of prime interest in the present context.

Finally, a rather different group of microemulsions has been investigated by Ekwall, Danielsson, and co-workers (e.g., [105–109]) over many years and also by Friberg and co-workers [110,111]. The latter papers were concerned with ionic surfactants; these Scandinavian workers preferred microemulsions of three or more components consisting of water, surface-active compound, and alcohol (plus eventually hydrocarbons), where the alcohol played the role of cosurfactant and continuous oil phase. Depending on the alcohol chain length, the phase diagram exhibited features typical of three-component water, surfactant, and hydrocarbon systems. An apparently intrinsic problem in these systems is the bifunctional role of the alcohol in such microemulsions, which is expected to impede an analysis of the data.

In the papers mentioned above, Friberg and co-workers studied the enhancement of water solubilization by the dissolution of hydrocarbons in the continuous alcohol phase with the help of infrared, NMR, light-scattering, and electron microscopic techniques.

These investigations show once more the difficulty of data interpretation in view of the mutually dependent and interacting components.

Undoubtedly, the many investigations of so-called microemulsions, in the present context with ionic surfactants, produced much valuable information regarding such multicomponent colloidal solutions. However, in view of more recent objections, and considerations based on extensive thermodynamic [12] and particularly NMR self-diffusion measurements [6,112–114], a more thorough discussion of multicomponent microemulsions does not appear justified before more rigorous criteria are available (e.g., such as those suggested in the introductory section) to distinguish true nanophase solutions from so-called microemulsions.

In conclusion, it is worth noting from the preceding discussion that typical features of ionic surfactants (such as those due to the effect of the semidiffuse electric double layer) are hardly considered in the majority of the more recent papers on multicomponent microemulsions. Instead, interest seems to be centered around the application of modern fluid theories (since Vrij's first paper on the subject in 1976), the results of which are not yet fully accessible to an interpretation of details related to the specific composition of these systems or even to particular properties of the surfactants.

ACKNOWLEDGMENT

It is a pleasure to acknowledge the contributions of several co-workers: the many discussions with and the help received from Dr. R. Kubik (now with Ciba-Geigy SA Basel) concerning the first part of this chapter; in the second part I was very efficiently assisted by Dr. G. Furler (now at the Chemistry Department of Rensselaer Polytechnic Institute, Troy, New York) in many respects, particularly in collecting literature and proofreading. Finally, Mr. H. Hammerich was of great help in redrawing and improving figures and also in many organizational aspects.

REFERENCES

1. L. M. Prince, in *Micellization, Solubilization, and Microemulsions*, K. L. Mittal, ed., Plenum, New York, 1977, vol. 1, p. 45.
2. Healy and Reed, *J. Petrol. Technol. 14*: 491 (1974).
3. A. W. Adamson, *Physical Chemistry of Surfaces*, Interscience, New York, 1967.
4. J. Stauff, *Kolloidchemie*, Springer, Berlin, 1960.
5. R. C. Tolman, *J. Am. Chem. Soc. 35*: 317 (1913).
6. B. Lindman, P. Stilbs, and M. E. Moseley, *J. Colloid Interface Sci. 83*: 569 (1981).
6a. S. Geiger and H. -F. Eicke, *J. Colloid Interface Sci. 110*: 181 (1986).
7. J. Israelachvili, S. Marcelja, and R. G. Horn, *Q. Rev. Biophys. 13*: 121 (1980).
8. C. Wagner, *Z. Phys. Chem. 132*: 273 (1928).
9. P. G. DeGennes and C. Taupin, *J. Phys. Chem. 86*: 2294 (1982).
10. M. Dvolaitzky, M. Guyot, M. Lagües, J. P. Pesant, R. Ober, C. Sauterey, and C. Taupin, *J. Chem. Phys. 69*: 3279 (1978).
11. G. Fourche, A. M. Bellocq, and S. Brunetti, *J. Colloid Interface Sci. 88*: 302 (1982).
12. M. Kahlweit, E. Lessner, and R. Strey, *J. Phys. Chem. 88*: 1937 (1984); C. V. Hermann, G. Klar, and M. Kahlweit, *J. Colloid Interface Sci. 82*: 6 (1981).
13. L. R. Angel, D. F. Evans, and B. W. Ninham, *J. Phys. Chem. 87*: 538 (1983).
14. J. Th. G. Overbeek, *Faraday Discuss. Chem. Soc. 65*: 7 (1978).
15. B. Jönsson and H. Wennerström, *J. Colloid Interface Sci. 80*: 482 (1981).
16. E. Ruckenstein and R. Krishnan, *J. Colloid Interface Sci. 75*: 476 (1980).
17. J. Frenkel, *Kinetic Theory of Liquids*, Dover, New York, 1955.

18. G. Gunnarson, H. Wennerström, G. Olofsson, and A. Zacharov, *J. Chem. Soc. Faraday Trans. I 76*: 1287 (1980).
19. J. K. Percus and G. J. Yevick, *Phys. Rev. 110*: 1 (1958).
20. N. F. Carnahan and K. E. Starling, *J. Chem. Phys. 51*: 635 (1964).
21. A. Vrij, E. A. Nieuwhuis, H. M. Fijnaut, and W. G. M. Agterof, *Faraday Discuss. Chem. Soc. 65*: 101 (1978).
22. A. Pouchelon, J. Meunier, D. Langevin, and M. Cazabat, *J. Phys. Lett. 41*: L-239 (1980).
23. H. -F. Eicke and R. Kubik, *Faraday Discuss. Chem. Soc. 76*: 305 (1983).
23a. R. Kubik, Ph.D. thesis, Univ. of Basel (1983).
24. H. Christen, H. -F. Eicke, H. Hammerich, and U. Strahm, *Helv. Chim. Acta 59*:1297 (1976).
25. R. A. Marcus, *J. Chem. Phys. 23*: 1057 (1955).
26. G. M. Bell and S. Levine, in *Chemical Physics of Ionic Solutions*, B. E. Conway and R. G. Barradas, eds., Wiley, New York, 1966.
27. P. Sherman, ed., *Emulsion Science*, Academic Press, New York, 1968.
28. K. J. Lissant, ed., *Emulsions and Emulsion Technology*, Dekker, New York, 1974.
29. M. Rosoff, in *Progress in Surface and Membrane Science*, Vol. 12, Academic Press, New York, 1978.
30. J. H. and E. J. Fendler, *Catalysis in Micellar and Macromolecular Systems*, Chap. 10, Academic Press, New York, 1975.
31. K. Shinoda and S. Friberg, *Adv. Colloid Interface Sci. 4*: 281 (1975).
31a. H. Kunieda and K. Shinoda, *J. Colloid Interface Sci. 70*: 577 (1979).
32. C. Wagner, *Colloid Polym. Sci. 254*: 400 (1976).
33. D. J. Mitchell and B. W. Ninham, *J. Chem. Soc. Faraday Trans. II 77*: 609 (1981).
34. R. W. Mattoon and M. B. Mathews, *J. Chem. Phys. 17*: 496 (1949).
35. H. -F. Eicke, in *Microemulsions*, I. D. Robb, ed., Plenum, New York, 1982.
36. W. I. Higushi and J. Misra, *J. Pharm. Sci. 51*: 455 (1962).
37. R. Kubik, H. -F. Eicke, and B. Jönsson, *Helv. Chim. Acta 65*: 1970 (1982).
38. A. Denss, Ph.D. thesis, Univ. of Basel (1977).
39. H. -F. Eicke and P. Zinsli, *J. Colloid Interface Sci. 65*: 131 (1978).
40. A. Maitra and H. -F. Eicke, *J. Phys. Chem. 85*: 2687 (1981).
41. C. A. Martin and L. J. Magid, *J. Phys. Chem. 85*: 3938 (1981).

42. P. Zinsli, *J. Phys. Chem. 83*: 3223 (1979).
43. P. Zinsli and H. -F. Eicke, *Progr. Colloid Polym. Sci. 65*: 158 (1978).
44. M. Shinitzky, A. C. Dianoux, C. Gitler, and A. Weber, *J. Biochem. 10*: 2106 (1971).
45. E. Keh and B. Valeur, *J. Colloid Interface Sci. 79*: 465 (1981).
46. M. Zulauf and H. -F. Eicke, *J. Phys. Chem. 83*: 480 (1979).
47. A. Vrij, J. W. Jansen, J. K. G. Dhout, C. Pathmamanoharan, M. M. Kops-Werkhoven, and H. M. Fijnaut, *Faraday Discuss. Chem. Soc. 76*: 19 (1983).
48. J. M. Corkill, J. F. Goodman, and T. Walker, *Trans. Faraday Soc. 61*: 589 (1965).
49. M. B. Mathews and E. Hirschhorn, *J. Colloid Sci. 8*: 86 (1953).
50. H. F. Eicke and J. Rehak, *Helv. Chim. Acta 59*: 2883 (1976).
51. R. A. Day, B. H. Robinson, J. H. R. Clarke, and J. V. Doherty, *J. Chem. Soc. Faraday Trans. I 75*: 132 (1979).
52. P. D. I. Fletscher and B. H. Robinson, *Ber. Bunsenges. Phys. Chem. 85*: 863 (1981).
53. E. Sein, J. R. Lalanne, J. Buchert, and S. Kielich, *J. Colloid Interface Sci. 72*: 363 (1979).
54. J. D. Nicholson, J. R. Doherty, and J. H. R. Clarke, in *Microemulsions*, E. D. Robb, ed., Plenum, New York, 1982.
55. H. -F. Eicke, R. Kubik R. Hasse, and I. Zschokke, in *Surfactants in Solution*, Vol. 3, K. L. Mittal and B. Lindman, eds., Plenum, New York, 1984, p. 1533.
56. A. Maitra, G. Vasta, and H. -F. Eicke, *J. Colloid Interface Sci. 93*: 383 (1983).
57. D. J. Shaw, *Introduction to Colloid and Surface Chemistry*, Butterworths, London, 1978, p. 87.
58. H. -F. Eicke, R. Hilfiker, and M. Holz, *Helv. Chim. Acta 67*: 361 (1984).
58a. H. -F. Eicke, R. Hilfiker, and H. Thomas, *Chem. Phys. Lett. 120*: 272 (1985).
58b. R. Hilfiker, H. - F. Eicke, S. Geiger, and G. Furler, *J. Colloid Interface Sci. 105*: 378 (1985).
59. D. Dünnenberger, Ph.D. thesis, Univ. of Basel (1984).
60. A. M. Cazabat, D. Langevin, and A. Pouchelon, *J. Colloid Interface Sci. 73*: 1 (1980).
61. R. Finsy, A. Devriese, and H. Lekkerkerker, *J. Chem. Soc. Faraday Trans. II 76*: 767 (1980).
62. D. J. Cebula, R. H. Ottewill, J. Ralston, and P. N. Pusey, *J. Chem. Soc. Faraday Trans. I 77*: 2585 (1981).
63. H. F. Eicke and A. Denss, in *Solution Chemistry of Surfactants*, Vol. 2, K. L. Mittal, ed., Plenum, New York, 1979, p. 699.
64. Z. A. Randriamalala, Ph.D. thesis, Institut National Polytechnique, Grenoble (1983).
65. B. J. Berne and R. Pecora, *Dynamic Light Scattering*, Wiley, New York, 1976.

66. G. D. J. Phillies, *J. Chem. Phys. 60*: 976 (1974); *62*: 3925 (1975); *67*: 4690 (1977).
67. G. K. Batchelor, *J. Fluid Mech. 74*: 1 (1976).
68. J. L. Anderson and C. C. Reed, *J. Chem. Phys. 64*: 3240 (1986).
69. B. U. Felderhoff, *J. Phys. A 11*: 929 (1978).
70. B. J. Ackerson, *J. Chem. Phys. 69*: 684 (1978).
71. G. Furler, Ph.D. thesis, Univ. of Basel (1984).
72. P. N. Pusey, A. M. Fijnaut, and A. Vrij, *J. Chem. Phys. 77*: 4270 (1982).
73. H. G. Elias, in *Light Scattering from Polymer Solutions*, M. B. Huglin, ed., Academic Press, London, 1972.
74. R. Hilfiker, H. -F. Eicke, S. Geiger, and G. Furier, *J. Colloid Interface Sci. 105*: 378 (1985).
75. E. Gulari, B. Bedwell, and S. Alkhafaji, *J. Colloid Interface Sci. 77*: 202 (1980).
76. B. Bedwell and E. Gulari, in *Solution Behavior of Surfactants*, Vol. 2, K. L. Mittal, ed., Plenum, New York, 1982, p. 833.
77. J. S. Huang and M. W. Kim, *Phys. Rev. Lett. 47*: 1462 (1981).
78. L. S. Ornstein and F. Zernike, *Proc. K. Ned. Akad. Wet. 17*: 793 (1914).
79. P. J. W. Debye, *J. Chem. Phys. 31*: 380 (1959).
80. J. B. Hayter, *Faraday Discuss. Chem. Soc. 76*: 7 (1983).
80a. H. -F. Eicke, R. Hilfiker, and H. Thomas, *Chem. Phys. Lett.*, *125*: 295 (1986).
81. B. Jacrot, *Rep. Prog. Phys. 39*: 911 (1976).
82. B. Stuhrmann and A. Miller, *J. Appl. Crystallogr. 11*: 325 (1978).
83. C. Cabos and P. Delord, *J. Appl. Crystallogr. 12*: 502 (1979).
84. C. Cabos and P. Delord, *J. Phys. Lett. 41*: L-455 (1980).
85. J. Rouvière, J. M. Couret, A. Lindheimer, M. Lindheimer, and B. Brun, *J. Chim. Phys. 76*: 297 (1979).
86. M. Kotlarchyk, S. -H. Chen, and J. S. Huang, *J. Phys. Chem. 86*: 3273 (1982).
87. J. S. Huang and M. W. Kim, in *Scattering Techniques Applied to Supramolecular and Nonequilibrium Systems*, S. H. Chen, B. Chu, and R. Nossal, eds., Plenum, New York, 1982, p. 809.
88. M. Kotlarchyk, S. -H. Chen, and J. S. Huang, *Phys. Rev. A 28*: 508 (1983).
89. M. Kotlarchyk and S. -H. Chen, *J. Chem. Phys. 79*: 2461 (1983).
90. W. Philippoff, *J. Colloid Sci. 5*: 169 (1950).
91. T. Assih, F. Larché, and P. Delord, *J. Colloid Interface Sci. 89*: 35 (1982).
92. R. W. Mattoon, R. S. Stearns, and W. D. Harkins, *J. Chem. Phys. 16*: 644 (1948).
93. W. G. M. Agterof, J. A. J. van Zomeren, and A. Vrij, *Chem. Phys. Lett. 43*: 363 (1976).

94. A. A. Caljé W. G. M. Agterof, and A. Vrij, in *Micellization, Solubilization and Microemulsions*, K. L. Mittal, ed., Plenum, New York, 1977, Vol. 2, p. 779.
95. A. Vrij, E. A. Nieuwenhuis, H. M. Fijnaut, and W. G. M. Agterof, *Faraday Discuss. Chem. Soc. 65*: 101 (1978).
96. J. H. Schulman and J. A. Friend, *Kolloid-Z. 115*: 67 (1949).
97. A. M. Bellocq, G. Fourche, P. Chabrat, L. Letamendia, J. Rouch, and C. Vaucamps, *Opt. Acta 27*: 1629 (1980).
98. A. M. Cazabat and D. Langevin, *J. Chem. Phys. 74*: 3148 (1981).
99. A. M. Cazabat, D. Langevin, J. Meunier, and A. Pouchelon, *Adv. Colloid Interface Sci. 16*: 175 (1982).
100. A. M. Cazabat, D. Chatenay, D. Langevin, and J. Meunier, *Faraday Discuss. Chem. Soc. 76*: 291 (1983).
101. A. M. Cazabat, *J. Phys. Lett. 44*: L-593 (1983).
102. S. Brunetti, D. Roux, A. M. Bellocq, G. Fourche, and P. Bothorel, *J. Phys. Chem. 87*: 1028 (1983).
103. M. Dvolaitzky, M. Lagües, J. P. LePesant, R. Ober, C. Sauterey, and C. Taupin, *J. Phys. Chem. 84*: 1532 (1980).
104. R. Ober and C. Taupin, *J. Phys. Chem. 84*: 2418 (1980).
105. P. Ekwall, I. Danielsson, and L. Mandell, *Kolloid-Z. 169*: 113 (1960).
106. L. Mandell and P. Ekwall, *Acta Polytech. Scand. Chem. 74I*: 5 (1968).
107. L. Mandell and P. Ekwall, *5th Int. Congr. Surface Active Substances, Barcelona*, 1968, preprint B IV 183.
108. P. Ekwall and L. Mandell, *Acta Chem. Scand. 21*: 1630 (1967).
109. B. Lindman and P. Ekwall, *Mol. Crystallogr. 5*: 79 (1968).
110. G. Gillberg, H. Lehtinen, and S. Friberg, *J. Colloid Interface Sci. 33*: 40 (1970).
111. E. Sjöblom and S. Friberg, *J. Colloid Interface Sci. 67*: 16 (1978).
112. B. Lindman, N. Kamenka, Th. -M. Kathopoulis, B. Brun, and P. -G. Nilsson *J. Phys. Chem. 84*: 2485 (1980).
113. H. Fabre, N. Kamenka, and B. Lindman, *J. Phys. Chem. 85*: 3493 (1981).
114. B. Lindman, N. Kamenka, B. Brun, and P. -G. Nilsson, in *Microemulsions*, I. D. Robb, ed., Plenum, New York, 1982, p. 115.
115. R. Klein and W. Hess, *Faraday Discuss. Chem. Soc. 76*: 137 (1983); *Progr. Colloid & Polymer Sci. 69*: 174 (1984).

3

Stabilization of Inverse Micelles by Nonionic Surfactants

STIG E. FRIBERG Chemistry Department, University of Missouri at Rolla, Rolla, Missouri

I. INTRODUCTION

In contrast to the case of ionic surfactants [1], whose phase behavior shows only limited and regular changes with temperature, systems with hydrocarbon, water, and nonionic surfactants of the type polyethylene glycol alkyl ethers display the most pronounced variation of the phase pattern with temperature. This fact was observed early by Shinoda in relation to emulsion preparation and stability in his introduction of the phase inversion temperature (PIT) or hydrophilic-lipophilic (HLB) temperature system for emulsifier selection [2]. The system was subsequently used to study the phase behavior of the isotropic liquid parts of such systems by Shinoda and co-workers [3–8] and others [9–29].

In the present contribution, an introductory description of the entire phase variations for a model system with an aliphatic hydrocarbon will be given, followed by a more detailed discussion of two structural problems in hydrocarbon solutions with water solubilized by nonionic surfactants.

II. TEMPERATURE DEPENDENCE OF ISOTROPIC SOLUTION REGIONS

The early contributions by Shinoda and others [2–10] emphasized the fact that the nonionic surfactant changed from being preferentially water-soluble to soluble in the hydrocarbon phase in the HLB temperature range. In the intermediate range, a three-phase system was observed with the addition of a "surfactant phase" to the water and oil phases. The phase behavior at constant surfactant concentrations could be described according to Fig. 1 for an aliphatic hydrocarbon.

At temperatures far below the HLB level, the interactions are limited to water and surfactant with moderate solubilization of

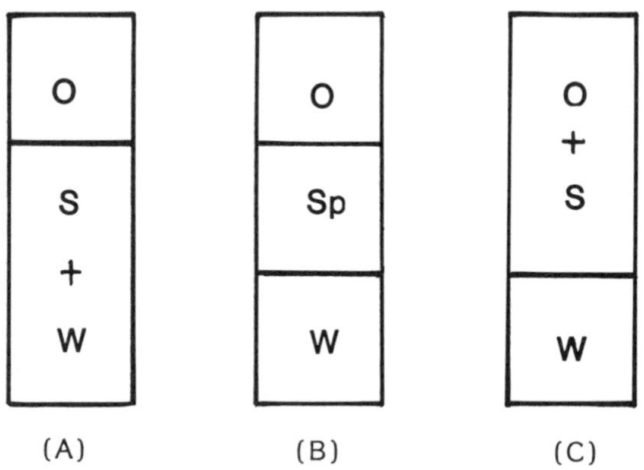

FIG. 1 The principal phase behavior with temperature in a water/hydrocarbon/polyethylene glycol alkyl ether system at low (≈ 4% by weight) and constant emulsifier concentration. S + W, aqueous phase with dissolved surfactant; O, hydrocarbon phase; W, pure water; Sp, surfactant phase; O + S, hydrocarbon phase with dissolved surfactant. (A) Low-temperature range; surfactant mainly dissolved in the water. (B) The HLB temperature range; a separate phase containing most of the surfactant is formed. (C) High-temperature range; surfactant dissolved in the hydrocarbon.

hydrocarbon. All the phases are in equilibrium with the hydrocarbon (Fig. 2A). An example of this behavior is found in the system water, hexadecane, and a commercial octaethylene glycol nonylphenol ether [30].

With increasing temperature, the first change in the general system from Fig. 2A is an increase of the mutual solubility of hydrocarbon and surfactant leading to the pattern in Fig. 2B. The next significant change is found at the cloud point of the surfactant. At temperatures in excess of that value, the infinite solubility of water in the micellar solution ceases to exist, but compositions in excess of a certain hydrocarbon/surfactant ratio give a continuous solubility region to 100% water. Hence, a small two-phase area is found close to the water corner and a corresponding dent is observed at high surfactant concentrations (Fig. 2C).

Both the dent and the two-phase region increase with temperature, leading to a complete separation into two areas (Fig. 2D). The hydrocarbon-rich phase may be called an oil-in-water (O/W) microemulsion, while the solubility area along the water/surfactant axis is appropriately described as a micellar solution. It should be observed that the O/W microemulsion and the micellar solution are not in equilibrium with each other. Instead, the lamellar liquid crystalline phase is involved in the equilibria. A case of such a separation is found for hexaethylene glycol dodecyl ether with water and decane at 40°C [21].

With further temperature increase, the O/W microemulsion region is separated from the aqueous corner and the phase equilibria become rather complicated, as shown by Fig. 2E. The two isotropic solutions are not in equilibrium and the structural changes brought forward by addition of hydrocarbon to the aqueous micellar solution are illustrative of the complex relationships found for high water content. A closer examination of the structural changes during addition of hydrocarbon to the aqueous micellar solution is justified.

Addition of hydrocarbon to the aqueous solution with composition A (Fig. 2E) initially results in a direct solubilization of the oil into normal micelles. When the composition reaches point B (Fig. 2F) maximum solubilization of oil in the aqueous micellar solution has been reached. Further addition of hydrocarbon in the range B–C (Fig. 1) leads to gradual replacement of the micellar solution with two phases: (1) pure water and (2) a lamellar liquid crystal. It is interesting to observe that all the initial surfactant and the hydrocarbon added to the micellar solution in range A–B and past point B are transferred to the liquid crystalline phase in the process. Further addition of hydrocarbon in the range C–D leads to a direct continuation of this process; all the added hydrocarbon is included in the lamellar liquid crystalline phase. Addition of oil in excess of point D reversed the process, characteristic of the range B–D. The

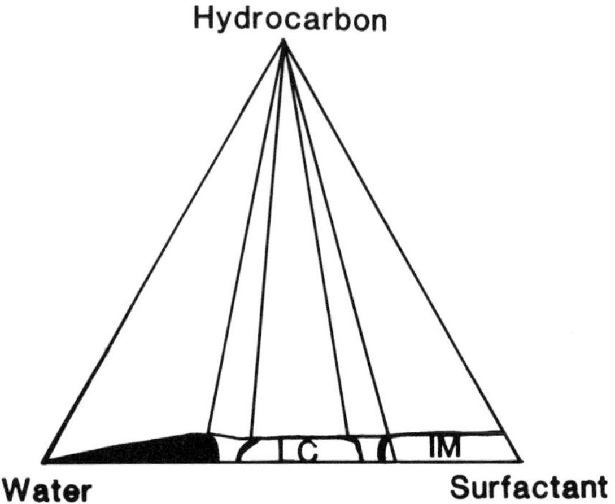

(A)

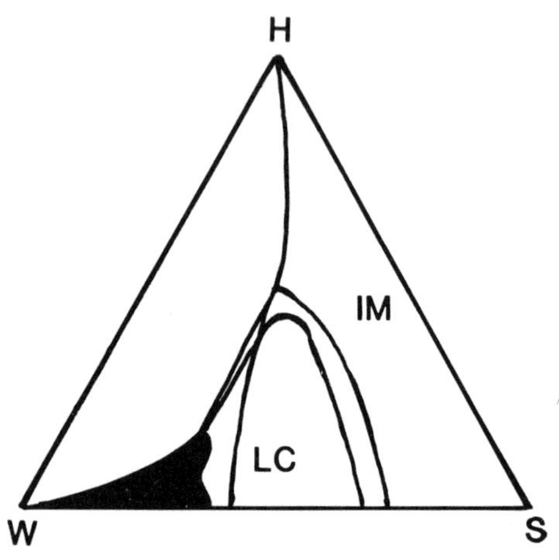

(B)

FIG. 2 Typical variations in phase behavior with temperature for the system water, polyethylene glycol alkyl ether, and an aliphatic hydrocarbon. The black phase region is the aqueous micellar region and its temperature-dependent variations. LC, lamellar liquid crystalline phase; IM, hydrocarbon/surfactant solution with solubilized water. (A) Temperature range far below the cloud point. (B) Temperature range immediately below the cloud point. (C) Temperature range immediately above the cloud point. (D) Temperature range for O/W microemulsions. (E) Temperature range between that of O/W

lamellar liquid crystal and the water phase now disappear; all the water, hydrocarbon, and surfactant are collected into the O/W microemulsion. The changes experienced during further addition of hydrocarbon are trivial.

Returning to the changes in the general phase equilibria caused by the temperature rise, the next significant point of change is found at the HLB temperature. At this temperature level (Fig. 2F) the O/W microemulsion region has been moved to an O/W ratio of approximately one, as well as parallel to and closer to the water/

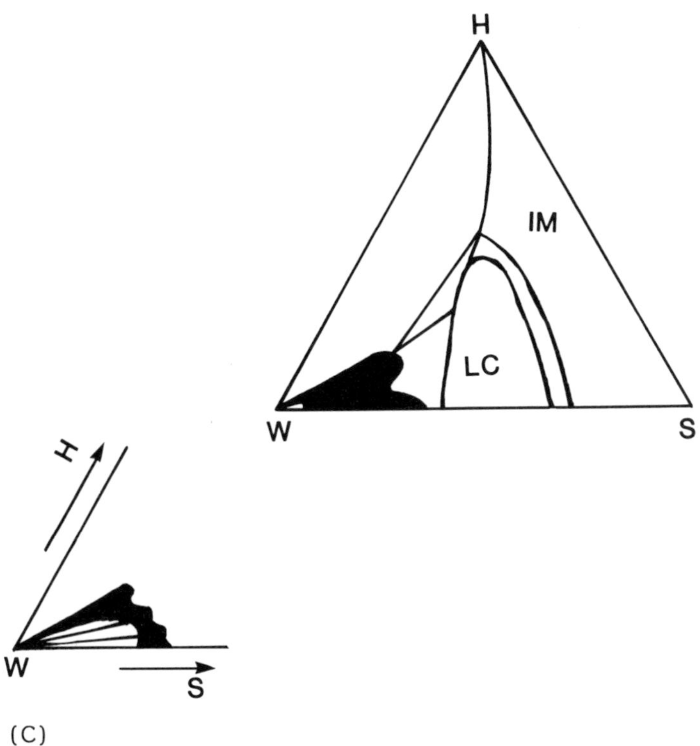

(C)

FIG. 2 (continued)

microemulsion and the HLB temperature. (F) Temperature range for the HLB temperature. (G) Temperature range for W/O microemulsions. (H,I) Temperature range for inverse micellar solution. (J) Highest temperature range.

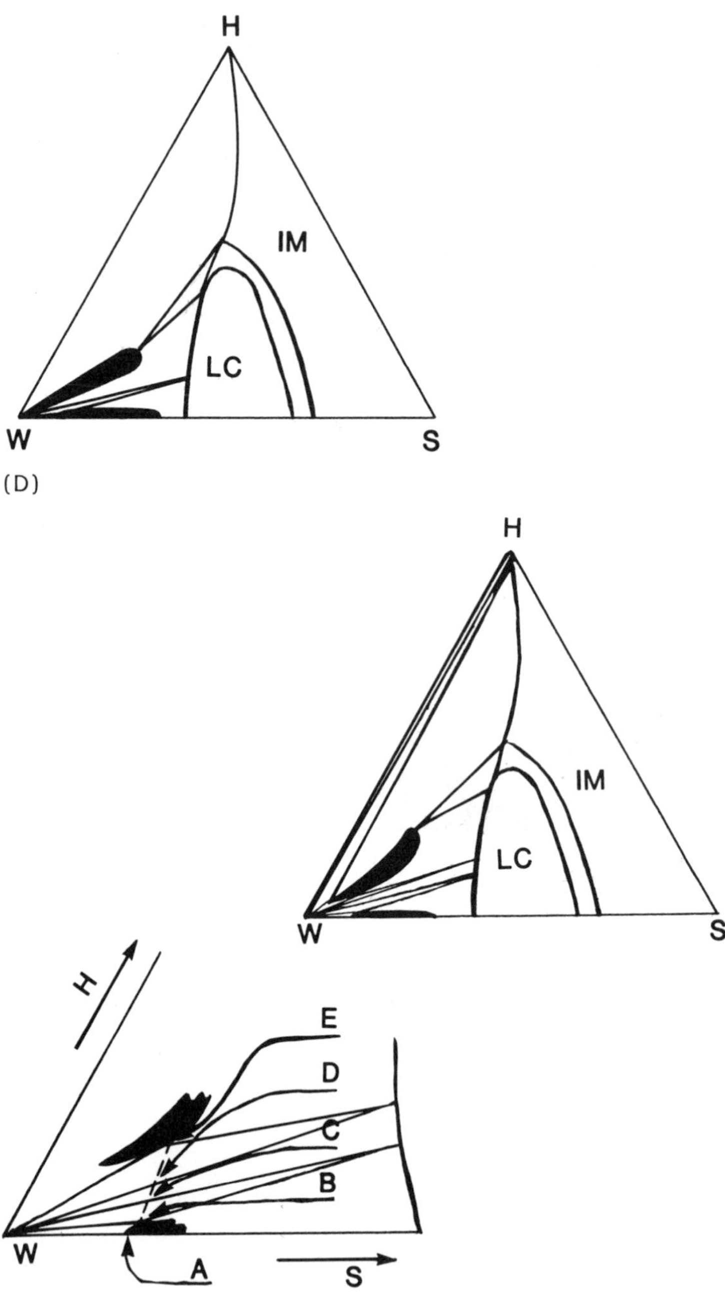

FIG. 2 (continued)

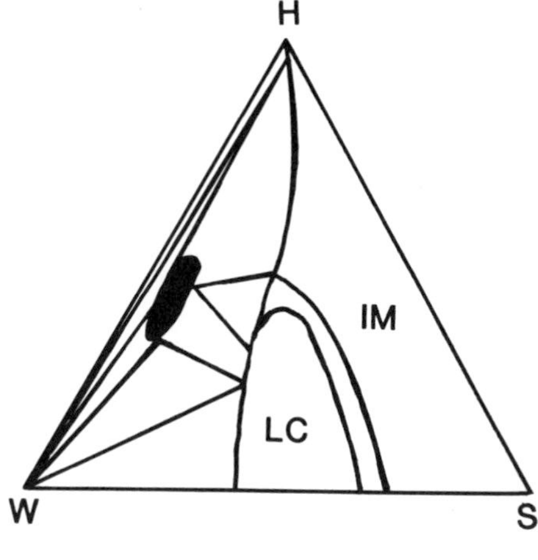

(F)

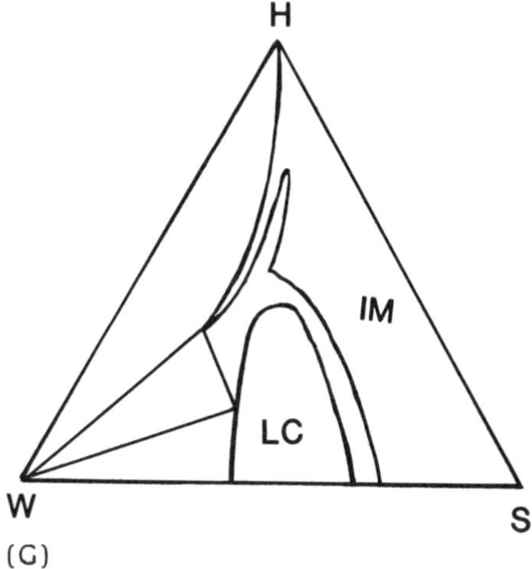

(G)

FIG. 2 (continued)

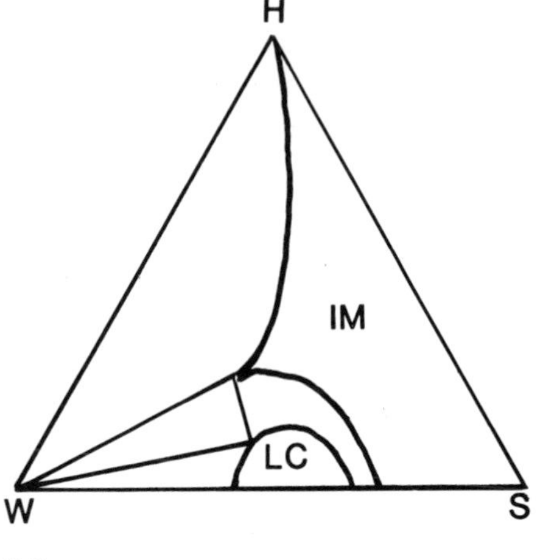

(H)

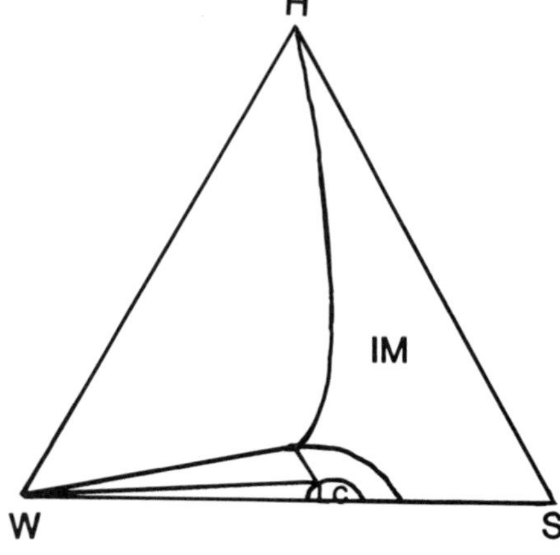

(I)

FIG. 2 (continued)

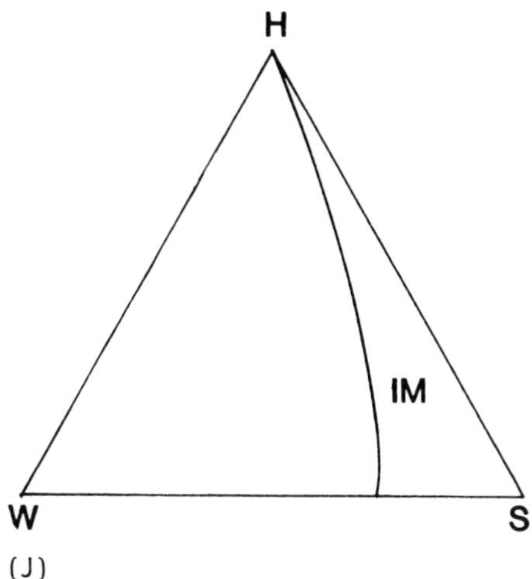

(J)

FIG. 2 (continued)

hydrocarbon axis. It is the temperature of maximum microemulsifying efficiency of the emulsifier. Numerous examples of this behavior are found in [16] and [18].

Higher temperatures cause this surfactant phase to move toward the hydrocarbon/surfactant solution region and to coalesce with it to form a water-in-oil (W/O) microemulsion region such as the one in Fig. 2G. This W/O microemulsion area is exemplified in the system water, tetraethylene glycol dodecyl ether, and decane at 20°C [16]. Further increase in the temperature leads to a gradual lowering of the hydrocarbon/surfactant ratio for the peak showing maximum water solubility in the W/O microemulsion region, with a commensurate reduction of the maximum hydrocarbon content of the liquid crystalline phase (Fig. 2H,I). In the final state (Fig. 2J), the liquid crystalline phase has disappeared and the surfactant/hydrocarbon solution with dissolved water remains.

As a final feature, it is instructive to compare the phase patterns in Fig. 2A and Fig. 2J. At the low temperature, water/surfactant interactions are the only significant forces to form association structures and the hydrocarbon is passively solubilized. In the high-temperature range, on the other hand, only hydrocarbon/surfactant interactions are important and now the water dissolved is attached to the polar groups of the surfactant. With these phase patterns, as with all other amphiphilic association structures, it is essential to realize that the variations in structure with the phase changes (such as the ones along the water/surfactant axis in Fig. 2A) are only a reflection of geometric factors emanating from the elongated shape of the surfactant with the polar part attached to a long hydrocarbon chain [26,27].

These variations with temperature are similar for all kinds of hydrocarbons and structure of surfactants, and the influence of different factors has been evaluated [2]. As an approximate rule of thumb, the following values may be useful. A system of water, pentaethylene glycol dodecyl ether, and decane has an HLB temperature of about 30°C. Changes introduced by modification of the hydrocarbon structure are as follows: (1) an increase in the hydrocarbon chain length by five methylene groups leads to an increase of the HLB temperature by 6–8°C (and a corresponding reduction of chain length leads to a similar reduction), and (2) a change from an aliphatic to an aromatic hydrocarbon leads to a reduction of the HLB temperature by approximately 50°C. Addition of one acyethylene group to the surfactant polar part raises the HLB temperature by 10–15°C.

A good illustration of the pronounced influence of hydrocarbon aromaticity on the phase behavior is in the system water, octaethyleneglycol nonylphenol ether, and hydrocarbons [16]. With hexadecane, the system is typical of one far below the HLB temperature (see Fig. 2A), and with *p*-xylene the behavior illustrates conditions at temperatures far in excess of the HLB temperature (Fig. 2H,I).

With these temperature-dependent phase relations described, a reasonable basis is obtained for a discussion of some phenomena in the W/O microemulsion or inverse micellar region.

III. INVERSE MICELLAR REGION

The inverse micellar region emanates from the hydrocarbon/surfactant solution reaching in a peak toward the water corner according to Fig. 3A. It is tempting and, as a matter of fact, common to treat this peak as a maximum water solubilization limit, and its shape lends itself readily to thermodynamic speculations.

It is important to realize that such efforts, however stimulating from an intellectual, or at least algebraic manipulative, point of view, have only insignificant relations to reality.

In reality, this peak is not a solubility maximum; it reveals the surfactant/hydrocarbon ratio at which the association structure is transformed from an isotropic solution to a lamellar liquid crystal, as shown in Fig. 3B. It is important to realize that the different phase regions in Fig. 3B may be considered with advantage as parts of one solubility area; the essential border is the one pointing toward the water/hydrocarbon axis. The broad line in Fig. 3C illustrates this concept; it reveals the similarity to the diagrams for simple systems such as water, ethanol, ethyl acetate.

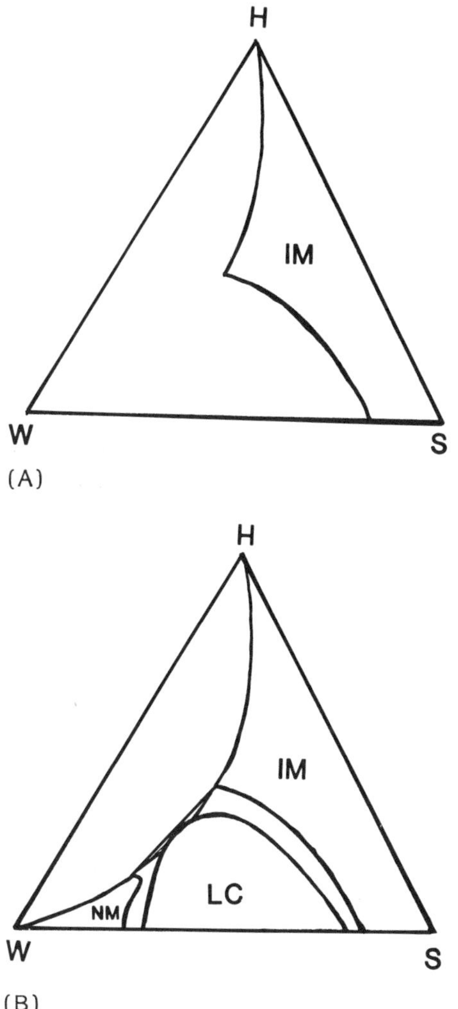

FIG. 3 The "solubility maximum" of water found in the inverse micellar (IM) region (A) is in reality the point at which a transfer to a liquid crystalline (LC) phase takes place (B). The three phases, the inverse micellar region, the liquid crystalline phase, and the normal micellar (NM) solution, may be considered as parts of a one-phase region (C). The phase changes observed are due to geometric constraints by the elongated shape of the surfactant.

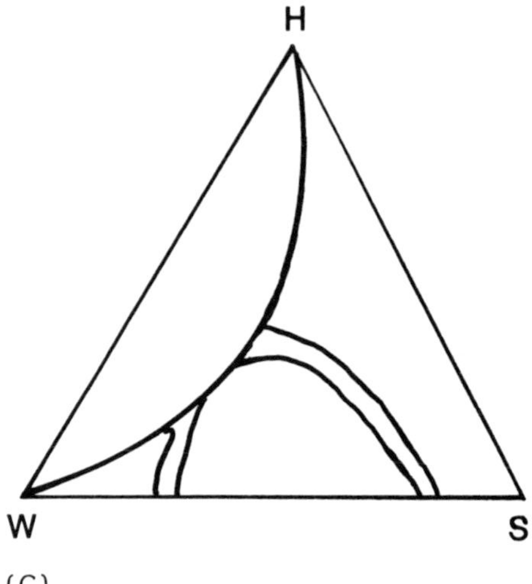

(C)

FIG. 3 (continued)

This relation between the liquid crystal and the isotropic solution makes meaningful an analysis of the structures in the W/O microemulsion region such as the one in Fig. 2G.

A. Water-in-Oil Microemulsions

The structural differences between the hydrocarbon/surfactant solution and the W/O microemulsion (Fig. 2G) have been extensively investigated by Ravey and collaborators [33,34].

The conditions in the hydrocarbon/surfactant solutions were clarified by a combination of light- and neutron-scattering methods [33]. The evaluation of neutron-scattering data was made by using the hard-sphere model with implicit small attraction effects [35]. It was noted that a small value of the hard-sphere radius may be explained by a cylindrical particle with the hydrophobic chains *parallel* to the axis [36] or by polydispersity.

The binary systems—the organic solvent/surfactant solutions without water—were divided into three categories depending on the structure of the solvent [33]. In the first category were polar and hydrogen-bonding liquids such as methanol and chloroform, as well as aromatic hydrocarbons such as benzene. For these liquids, no association in excess of dimers was found. The hydrogen bonds with methanol and chloroform and the charge transfer complexes with benzene prevent the formation of surfactant intermolecular associations. This interpretation is supported by the fact that little or no water solubilization was found in the surfactant concentration range (0–15% by weight) investigated.

In the second category, including heptane and cyclohexane, a gradual association with oligomers took place with increased concentration of surfactant, as shown in Table 1. The structure is considered "hanklike" with pronounced interpenetration of the polar chains and predominantly parallel packing of the hydrocarbon chains.

The third category was characterized by long-chain hydrocarbon solvents such as decane to hexadecane and short-chain hydrocarbons such as cyclohexane in combination with a surfactant with a long polar chain. Now the aggregation members are enhanced and a sudden increase is observed at low concentrations (Table 2). For this category the expression "a critical concentration" may be justified.

Nowhere in this region were inverse micelles with spherical form observed. This is a reasonable result, as shown by the following arguments [33]. With aggregation numbers in the range 10–25, spherical packing would imply an unrealistic area per polar head of the magnitude of 70 $Å^2$ per molecule. Such a value is reasonable or even large for a double-chain surfactant, but obviously too great by at least a factor of 2 for a single-chain surfactant. In addition, the large depolarization ratio found in these solutions [29,33] excludes isotropic particles. In this context, it should be mentioned that isotropic solutions of monoglycerides [39,40] show associations of nonspherical shape.

Addition of water to the binary system caused the aggregation number to increase in a pronounced manner. As shown in Fig. 4,

TABLE 1 Morphological Parameters of Aggregates in Solvents of the Second Class[a] [33]

Solvent	Surfactant	C_s	N
Heptane	$C_{12}(EO)_4$	6	
		10	5
		15	6
		20	8
Cyclohexane	$C_{12}(EO)_4$	2	1
		4	2
		6	2
		10	3
		15	4

[a] C_s is the surfactant concentration (weight percent), N the aggregation number.

TABLE 2 Morphological Parameters in Solvents of the Third Class[a] [33]

Solvent	Surfactant	T (°C)	C_s	N
Decane	$C_{12}(EO)_6$	10	3	22
			5	25
			7	28
			11	30
		20	6	23
			8	23
			10	23
			12	28
		45	4	10
			6	10
			8	10
			10	10
Decane	$C_{12}(EO)_4$	20	3	1–2
			7	9
			11	9
			15	10

[a]C_s is the surfactant concentration (weight percent), N the aggregation number.

the aggregation number reaches 1000 when the number of water molecules per surfactant reaches 9. These high aggregation numbers occur close to the phase limit against high water content. Calculations of the area per molecule show a value close to 42 Å^2, which remains constant for the entire region for compositions close to demixing.

This value is also the one found in the lamellar liquid crystalline phase and Ravey [38] has been able to rationalize the structures in the following manner. In the isotropic solution region the initial hank-shaped aggregates will separate the polar groups in a transverse direction with increasing water content, forming lamellae. These lamellae grow with increased water content, and at approximately nine water molecules per surfactant molecule, the layered inverse micelles cannot accommodate higher amounts of water and separation to a lamellar liquid crystalline phase occurs. In this structure, the water content of the amphiphilic layer remains constant and added water forms a liquid film between the layers. It is important to note that the dimensions of the lamellae remain identical when passing from the isotropic solution region to the lamellar liquid crystalline phase.

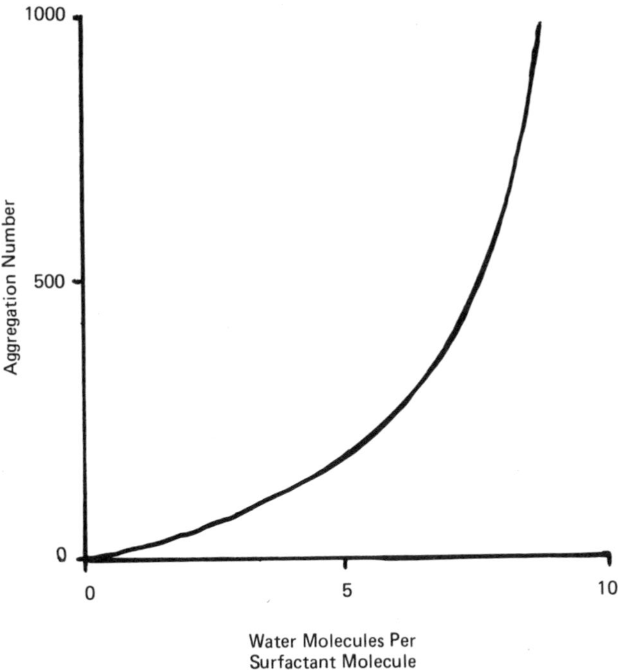

FIG. 4 Aggregation number of surfactant versus number of water molecules per surfactant molecule for pentaethylene glycol dodecyl ether in decane at a surfactant/hydrocarbon ratio of 0.25 [33].

The O/W microemulsion region contains aggregates of oblate shape with water in the center. These association structures are formed under three conditions: low surfactant concentrations, high water content, or high temperature. Their relation to the liquid crystalline structure is intuitively evident; they are formed under conditions at which the regular lamellar packing of the liquid crystal is perturbed sufficiently to make it unstable. Too low a surfactant concentration in hydrocarbon causes instability of the liquid crystalline phase due to excess hydrocarbon, and too high a water content has a similar influence on the structure. Enhanced temperature causes increased magnitude of the vibration pattern from the temperature-dependent oscillations, also causing instability. The reduction of the hydrocarbon content of the liquid crystalline phase with enhanced temperature (Fig. 2G–I) exemplifies this last phenomenon.

In general, these aggregates, their shape, and their equilibria are mainly determined by the volume ratios discussed by Ninham and collaborators [22,23] and individual bonds have less importance.

However, in addition to temperature and composition, there is another factor which has equal importance for the phase behavior of water, hydrocarbon, and surfactant systems. The difference between the phase diagram with an aliphatic (Fig. 5A) and an aromatic (Fig. 5B) hydrocarbon is striking [45]. The phase diagram with the former (Fig. 5A) is typical of conditions at temperatures far below the HLB temperature, while the one with latter (Fig. 5B) is characteristic of the pattern found at higher temperatures.

This phenomenon lent itself readily to spectroscopic investigations; the following section will delineate the extent to which the patterns could be directly related to molecular interactions.

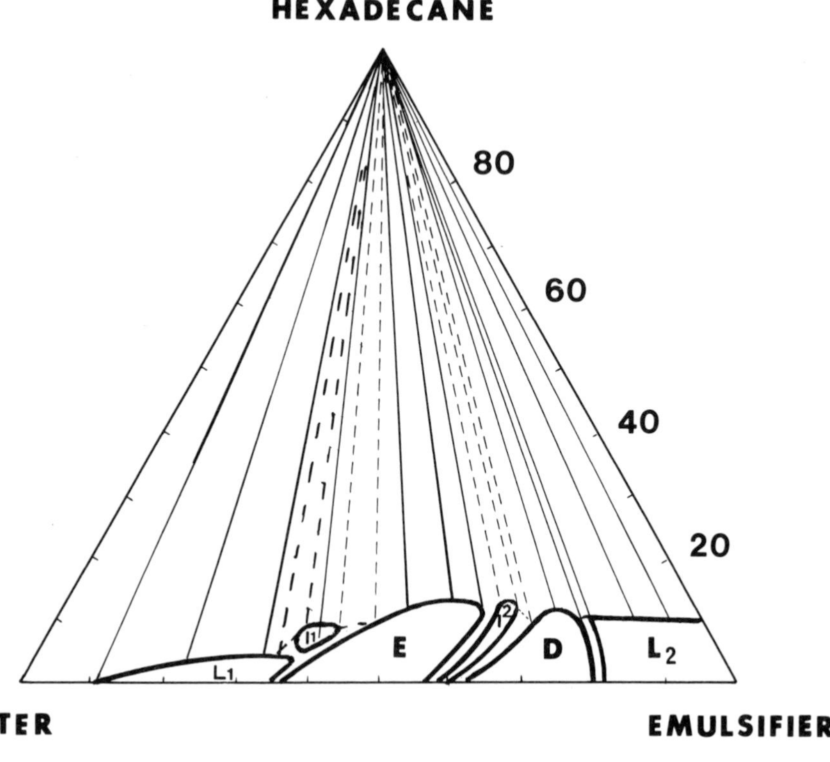

(A)

FIG. 5 The phase pattern is changed in a pronounced manner when an aliphatic hydrocarbon is replaced by an aromatic one [30].

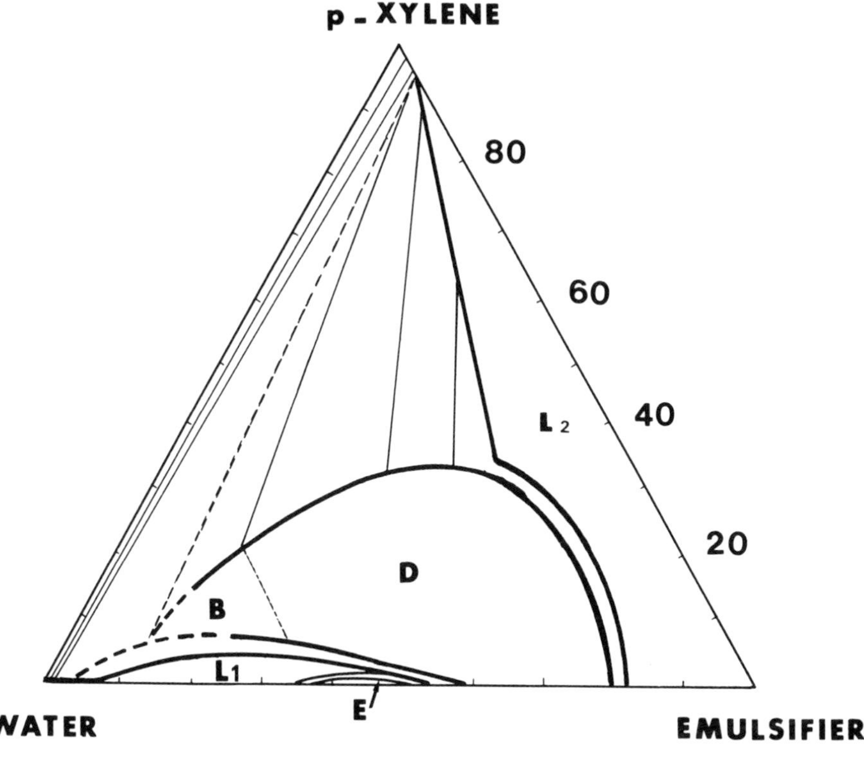

(B)

FIG. 5 (continued)

B. Aromaticity and Association Structures

The presence of aromatic hydrocarbons means a change of phase pattern to the one typical of high temperatures (Fig. 2B,J). A surfactant with a shorter ethylene oxide chain combined with an aliphatic hydrocarbon will show a pattern typical of temperatures in excess of the cloud point even at room temperature (Fig. 6, solid line), but an aromatic hydrocarbon will change the pattern to the one expected for even higher temperatures (Fig. 6, dashed line). The change is, in essence, due to a higher requirement for surfactant concentration for association to inverse micelles in the presence of an aromatic hydrocarbon than in the presence of an aliphatic one. The initial associations are amenable to spectroscopic investigations. Hence, a description of the characteristics of nuclear magnetic resonance (NMR) spectra of simple nonionics will be given before the specifics of the association are covered.

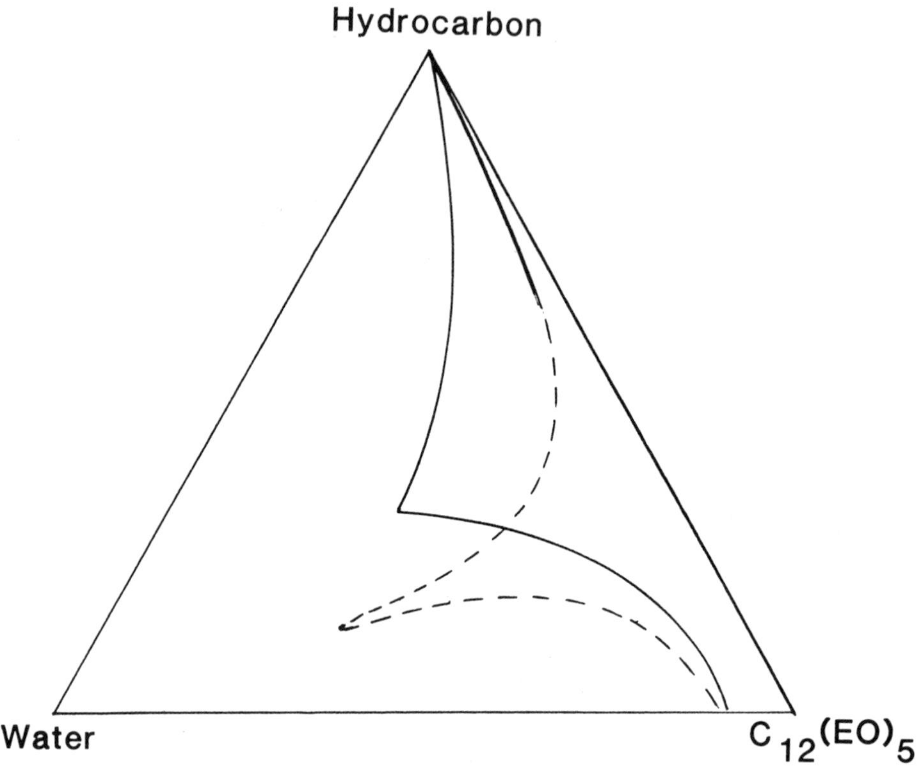

FIG. 6 The inverse micellar solution region in the system water, pentaethyleneglycol dodecyl ether $C_{12}(EO)_5$, and hydrocarbon at 30°C changed when cyclohexane (solid line) was replaced by benzene (dashed line).

C. NMR Spectra of Polyethylene Glycol Alkyl Ethers

The ethylene oxide groups of nonionic surfactants give separate resonance frequencies in the NMR range and the spectra have been analyzed in detail [41–45]. For the present discussion, the results from proton NMR form a sufficient basis.

The pentaethylene glycol dodecyl ether shows four separate resonance peaks (Fig. 7) in dilute solution in an aromatic hydrocarbon [44]. Their assignment is given in Fig. 8. The methylene group in the hydrocarbon chain adjacent to the first ether oxygen forms the resonance peak at the highest field, followed by the two peaks from the first two ethylene groups in the polyoxyethylene chain and the one large resonance maximum from the two subsequent

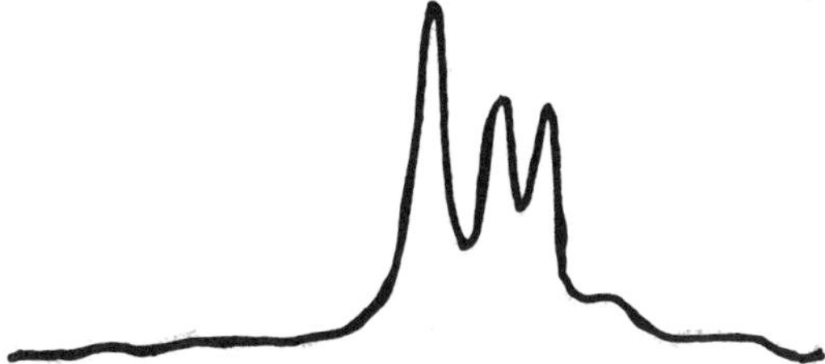

FIG. 7 The pentaethylene glycol dodecyl ether shows three main proton resonance peaks in dilute solution in benzene [44].

ethylene groups. The terminal ethylene group is masked by these pronounced resonance peaks. These assignments have been used to determine the association pattern in the presence of hydrocarbon and water.

D. NMR and the Association in Hydrocarbons

Increase of the surfactant content in benzene leads to coalescence of the three predominant resonance frequencies in Fig. 7. Following

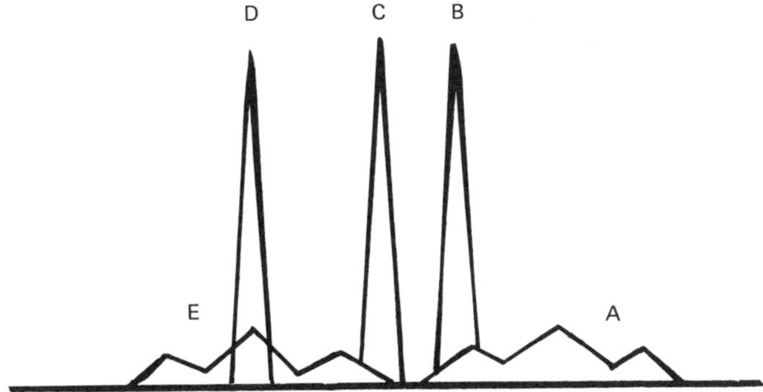

FIG. 8 Assignment of the resonance frequencies of pentaethylene glycol in benzene [44]: (A) methylene group in the hydrocarbon chain adjacent to the first ether oxygen in the oxyethylene chain, (B) oxyethylene group adjacent to the hydrocarbon chain, (C) second oxyethylene group counted from the hydrocarbon chain, (D) oxyethylene groups three and four counted from the hydrocarbon chain, and (E) terminal oxyethylene group.

the changes with increased surfactant concentration (Fig. 9), the frequencies can no longer be observed separately at surfactant concentrations in excess of 60% by weight.

Excluding orientational [46] and coupling [44] effects on the location of resonance frequencies, a reasonable interpretation would involve penetration of aromatic hydrocarbon between the polar chains, which are interassociated at high surfactant concentrations. The spectra of pentaethylene glycol in an *aliphatic* hydrocarbon [44] showed no changes from the spectrum of the surfactant in its native state even at high dilutions.

These results have been interpreted as a proof of a stronger interaction between benzene molecules and the polar part of the surfactant than is the case for an aliphatic hydrocarbon. This view is supported by the fact that infrared (IR) spectra [44] revealed a larger amount of non-hydrogen-bonded surfactant in the benzene solutions. Early investigations [47–50] of the benzene/polar group interactions showed a transient orientation of the benzene molecule by local solute dipoles to solvate the positive end of a dipole.

The difference in the hydrocarbon/surfactant interactions results in a distinct alteration of the association behavior when water is introduced.

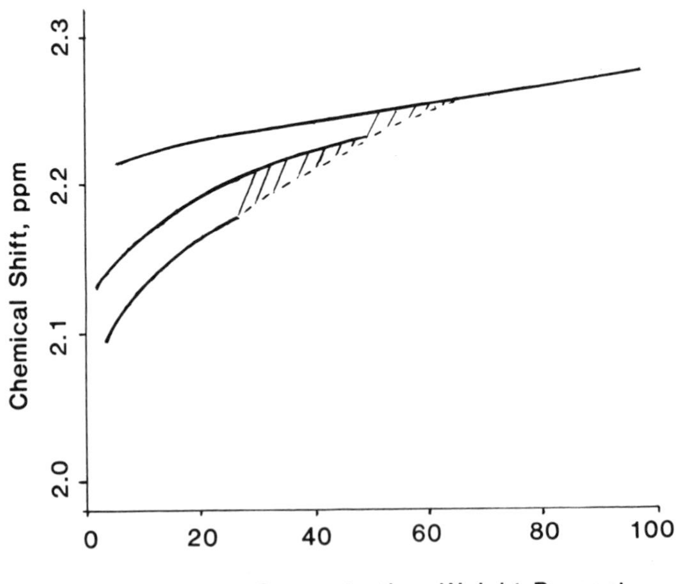

FIG. 9 The chemical shift versus TMS of the oxyethylene groups of pentaethylene glycol dodecyl ether in benzene [44].

E. Association in the Presence of Water

Addition of water to a dilute surfactant solution in benzene [44] leads to a general shift of the resonance frequencies in the downfield direction and to separation of the degenerate peak representing oxyethylene groups 3 and 4. This was interpreted as a sign of the water molecules being predominantly concentrated close to the terminal ethylene group of the polyoxyethylene chain. Such a behavior is characteristic for the conditions in the concentration range where no appreciable solubilization of water takes place. Solubilization occurs first when the signals in the water-free system (Fig. 9) have coalesced, indicating self-association of the surfactants. Addition of water to such samples leads to a new separation of signals [45]. This downfield separation was largest for the terminal groups and smaller for groups toward the hydrocarbon chain [45]. Addition of water to systems with aliphatic hydrocarbon [51] gave a downfield shift but no separation of the single resonance peak, which was observed at *all* hydrocarbon/surfactant ratios.

The interpretation of these results is straightforward and gives a plausible explanation for the divergent association behavior in aromatic and aliphatic hydrocarbons.

F. Association with Inverse Micelles in Aliphatic and Aromatic Hydrocarbons

In an aromatic hydrocarbon the surfactant molecules are separated from each other to a high degree due to interaction with the aromatic hydrocarbon. This aromatic hydrocarbon interaction prevents the water molecules from spreading along the polyoxyethylene chain; instead, they are localized close to the terminal hydroxide group. No inverse micelles are formed in the low surfactant concentration range and the water solubilization is extremely small—on the average less than approximately one water molecule per surfactant molecule (compare Figs. 6 and 9).

The inverse micelles are formed first after the surfactants have commenced self-association in the water-free hydrocarbon (Fig. 9). When this has occurred, the water molecules are able to interact with the entire polar chain and multiple associations with inverse micelles can take place to form the hanklike figures described by Ravey et al. [33].

In the presence of an aliphatic hydrocarbon, on the other hand, no barrier to water attachment to the entire polar chain exists and the presence of water will enable formation of micelles at lower surfactant concentrations. In this case, addition of water will induce the association of surfactant molecules.

Against this background, the difference between the behaviors in aromatic and aliphatic hydrocarbons appears reasonable and expected. Extrapolation to the general behavior is straightforward;

any difference in interaction will give a lower HLB temperature for the surfactant with the greater interaction. In short, the enhanced competition between water and a solvent with stronger interaction capacity for the polar groups of the surfactant leads to acceleration of the temperature-dependent dehydration of the surfactant polar groups.

IV. SUMMARY

The inverse micellar solutions stabilized by nonionic surfactants of the polyethylene glycol alkyl(acryl) ether type were related to the temperature-dependent behavior of the aqueous isotropic micellar solution of the system at low temperatures. The structures of the aggregates in the inverse micellar region and the W.O microemulsion were related to the lamellar structure of the liquid crystalline phase, and the dependence of the association behavior on the aromaticity of the hydrocarbon was explained in terms of interactions with the polar group of the surfactant.

ACKNOWLEDGMENT

Support from Department of Energy grant DE-AC02-83ER13083 is gratefully acknowledged.

REFERENCES

1. P. Ekwall, in *Advances in Liquid Crystals*, G. H. Brown, ed., Academic Press, New York, 1974, p. 1.
2. K. Shinoda and H. Arai, *J. Phys. Chem. 68*: 3485 (1964).
3. K. Shinoda, *J. Colloid Interface Sci. 24*: 4 (1967).
4. K. Shinoda, *J. Colloid Interface Sci. 34*: 278 (1970).
5. K. Shinoda and T. Ogawa, *J. Colloid Interface Sci. 24*: 56 (1967).
6. H. Saito and K. Shinoda, *J. Colloid Interface Sci. 24*: 10 (1967).
7. H. Saito and K. Shinoda, *J. Colloid Interface Sci. 32*: 647 (1970).
8. K. Shinoda and H. Kunieda, *J. Colloid Interface Sci. 42*: 381 (1973).
9. T. Nakagawa and K. Tori, *Kolloid-Z. 168*: 132 (1960).
10. T. Nakagawa, K. Kuriyama, and H. Inove, *J. Colloid Interface Sci. 15*: 168 (1960).
11. K. Kunyama, *Kolloid-Z. 180*: 55 (1962).

12. K. Kon-no and A. Kitahara, *J. Colloid Interface Sci. 37*: 469 (1971).
13. H. -F. Eicke, *J. Colloid Interface Sci. 68*: 440 (1979).
14. J. E. Vinatieri and P. D. Flemming, III, Society of Petroleum Engineers Symposium Proceedings, Tulsa, Oklahoma, Series 7057, 1978.
15. S. G. Frank and G. J. Zografi, *J. Colloid Interface Sci. 29*: 27 (1969).
16. S. Friberg and I. Lapczynska, *Prog. Colloid Polym. Sci. 56*: 16 (1975).
17. J. B. Brown, I. Lapczynska, and S. Friberg, *Proc. Int. Conf. Colloid Surface Sci.*, Vol. 1, E. Wolfram, ed., Budapest, 1975, p. 507.
18. S. Friberg, I. Lapczynska, and G. Gillberg, *J. Colloid Interface Sci. 56*: 19 (1976).
19. T. A. Bostock, M. P. McDonald, G. J. T. Tiddy, and L. Warring, in *Surface Active Agents*, Society of Chemical Industry, 1980.
20. E. J. Staples and G. J. T. Tiddy, *J. Chem. Soc. Faraday Trans. I 74*: 2530 (1978).
21. S. Friberg, I. Buraczewska, and J.-C. Ravey, *Proc. Int. Symp. Micellization, Solubilization, Microemulsions* (1977), p. 901.
22. G. J. T. Tiddy, K. Rendall, and P. Galsworthy, *Mol. Cryst. Liq. Cryst. 72*(Letters): 147 (1982).
23. C.-U. Hermann, U. Wurz, and M. Kahlweit, *Ber. Bunsenges. Phys. Chem. 82*:560 (1978).
24. M. Kahlweit, *J. Colloid Interface Sci. 90*: 197 (1982).
25. M. Kahlweit, E. Lessner, and R. Strey, *J. Phys. Chem. 87*: 5032 (1983).
26. P.-G. Nilsson and B. Lindman, *J. Phys. Chem. 86*: 271 (1982).
27. P.-G. Nilsson and B. Lindman, *J. Phys. Chem. 87*: 4756 (1983).
28. M. Kahlweit, E. Lessner, and R. Strey, *Colloid Polym. Sci. 261*: 954 (1983).
29. P.-G. Nilsson, H. Wennerström, and B. Lindman, *J. Phys. Chem. 87*: 1377 (1983).
30. S. Friberg, L. Mandell, and K. Fontell, *Acta Chem. Scand. 23*: 1055 (1969).
31. J. N. Israelachvili, D. J. Mitchell, and B. W. Ninham, *J. Chem. Soc. Faraday Trans. II 72*: 1525 (1976).
32. D. J. Mitchell and B. W. Ninham, *J. Chem. Soc. Faraday Trans. II 77*: 601 (1981).
33. J. C. Ravey, M. Buzier, and C. Picot, *J. Colloid Interface Sci. 97*: 9 (1984).
34. S. E. Friberg, I. Buraczewska, and J. -C. Ravey, in *Micellization, Solubilization and Microemulsions*, K. L. Mittal, ed., Plenum, New York, 1977, pp. 901–911.

35. J. K. Percus and G. J. Yevick, *Phys. Rev. 110*: 1 (1958).
36. D. Glatter, in *Small Angle X-ray Scattering*, O. Glatter and O. Kratky, eds., Academic Press, New York, 1982.
37. P. Van Beurten and A. Vrij, *J. Chem. Phys. 74*: 2744 (1981).
38. J.-C. Ravey, personal communication, to be published.
39. K. Larsson, *J. Colloid Interface Sci. 72*: 152 (1979).
40. K. Fontell, L. Hernqvist, K. Larsson, and J. Sjöblom, *J. Colloid Interface Sci. 93*: 453 (1983).
41. A. A. Ribeiro and E. A. Dennis, *Biochemistry 14*: 3746 (1975).
42. A. A. Ribeiro and E. A. Dennis, *J. Phys. Chem. 80*: 1746 (1976).
43. A. A. Ribeiro and E. A. Dennis, *J. Phys. Chem. 81*: 957 (1977).
44. H. Christenson and S. E. Friberg, *J. Colloid Interface Sci. 75*: 276 (1980).
45. H. Christenson, S. E. Friberg, and D. W. Larsen, *J. Phys. Chem. 84*: 3633 (1980).
46. J. A. Pople, W. G. Schneider, and H. J. Bernstein, *High Resolution Nuclear Magnetic Resonance*, McGraw-Hill, New York, 1959.
47. J. E. Anderson, *Tetrahedron Lett. 51*: 4713 (1965).
48. D. H. Williams, J. Ronayne, H. W. Moore, and H. R. Huldrey, *J. Org. Chem. 33*: 998 (1968).
49. K. J. Lin, *Macromolecules 1*: 213 (1968).
50. H. Tadokoro, *Macromol. Rev. 1*: 119 (1967).
51. S. E. Friberg, H. Christenson, and G. Bertrand, in *Reverse Micelles*, P. L. Luisi and B. E. Straub, eds., Plenum, New York, 1984, p. 105.

4

Enhanced Oil Recovery with Microemulsions

C. A. MILLER Chemical Engineering Department, Rice University, Houston, Texas

S. QUTUBUDDIN Chemical Engineering Department, Case Western Reserve University, Cleveland, Ohio

I. GENERAL BACKGROUND

The escalation of oil prices after 1973 and the widespread recognition that the world's conventional petroleum reserves cannot indefinitely support consumption at the present rate have led to increased emphasis on improving the efficiency of petroleum recovery from underground reservoirs. There is ample room for improvement. The petroleum already recovered in the United States plus that producible at present prices with conventional oil field technology amounts to only about one-third of the 450 billion barrels which have been discovered [1]. Enhanced recovery processes, including those employing surfactants, offer promise of enabling the world's petroleum resources to be used more fully.

The low recovery efficiency with present technology stems from the facts that petroleum (crude oil) is found in small pores of rocks buried well below the earth's surface, that oil must flow along tortuous pathways through interconnected pores to the sites of widely spaced wells, and that the permeability of the rocks to flow usually varies greatly with position. Some feeling for pore structure in petroleum reservoirs may be obtained by noting that typical oil-bearing sandstones have porosities in the range 0.15 to 0.30 and average pore sizes of the order of 10 μm.

Not only is one constrained to work with the highly inhomogeneous reservoirs provided by nature, but it is not feasible with present techniques to obtain a detailed mapping of permeability and other key reservoir properties as a function of position. The best that can be done is to combine the results of detailed laboratory tests on a few samples of the reservoir rock with those of various tests which can be made in the field to obtain local values of some properties near individual wells or average values for the entire region between adjacent wells.

Interfacial phenomena play an important role in oil recovery because reservoir pore space is almost always occupied by two or more immiscible fluids with the result that numerous fluid interfaces are present. Even in the initial undisturbed condition before any wells are drilled, the reservoir contains both oil and *connate water*, the latter normally a brine occupying some 10 to 35% of the pore space.

The locations of oil and water in the rock are determined by wetting properties. The wetting fluid occupies the smaller pores and the nonwetting fluid the larger pores. Few if any pores contain comparable amounts of both fluids, although thin films of the wetting fluid may exist at the surfaces of pores filled mainly with the nonwetting fluid. Some reservoirs are strongly water-wet, as might be expected for sandstones and limestones. Others are of intermediate wettability or even oil-wet owing to adsorption of long-chain compounds from the crude oil. Still others are described as having mixed wettability; i.e., part of their pore surface is water-wet and part oil-wet.

Substantial quantities of both oil and brine are invariably present during oil recovery processes. Indeed, formation of various phases containing oil, brine, and surfactant is a central feature of recovery processes utilizing surfactants. Therefore, this chapter, unlike most others in this volume, is not limited to oleic phases. Instead, all phases which significantly influence oil recovery performance are discussed as necessary.

Several previous reviews have discussed various aspects of surfactant flooding. Reed and Healy [2] summarized their pioneering work on phase behavior in microemulsion systems and its relationship to oil displacement. Gogarty [3,4] provided an extensive discussion of process design and field application of surfactant flooding as well as economic considerations controlling its implementation. Mattax et al. [5] discussed the advantages and disadvantages of using various types of surfactants and polymers. They also reviewed recent field experience. Shah and co-workers [6–9] and Lake [10] have reviewed certain fundamental aspects of surfactant recovery processes. While much of the present review is an updated account of phase behavior in oil-water-surfactant systems, an effort is also made to describe in simple terms other important factors influencing process performance and to indicate approaches that are being considered for improving process effectiveness.

II. MECHANISM OF OIL TRAPPING

During the *primary* recovery stage immediately following drilling, oil is driven to individual wells by natural forces in the reservoir. An important factor influencing primary production in many reservoirs is natural gas, which escapes from solution in the oil and helps maintain reservoir pressure. In this case it should be noted that three fluid phases are present in the pore space: gas, oil, and water.

Following primary recovery operations, the usual practice for crude oils of relatively low viscosity is to begin waterflooding. This *secondary* recovery process involves injecting water into some wells to drive oil through the reservoir to other wells. Even after completion of a successful waterflood, however, about half of the original oil typically remains in the swept area. Most of it is present as immobile, discrete globules or "ganglia" of residual oil, an example of which is shown in Fig. 1.

To understand the origin of these trapped ganglia, one must first recognize that capillary effects are much larger than viscous effects for waterflooding and indeed for most cases of two-phase flow in oil reservoirs. This result should not be surprising in view of

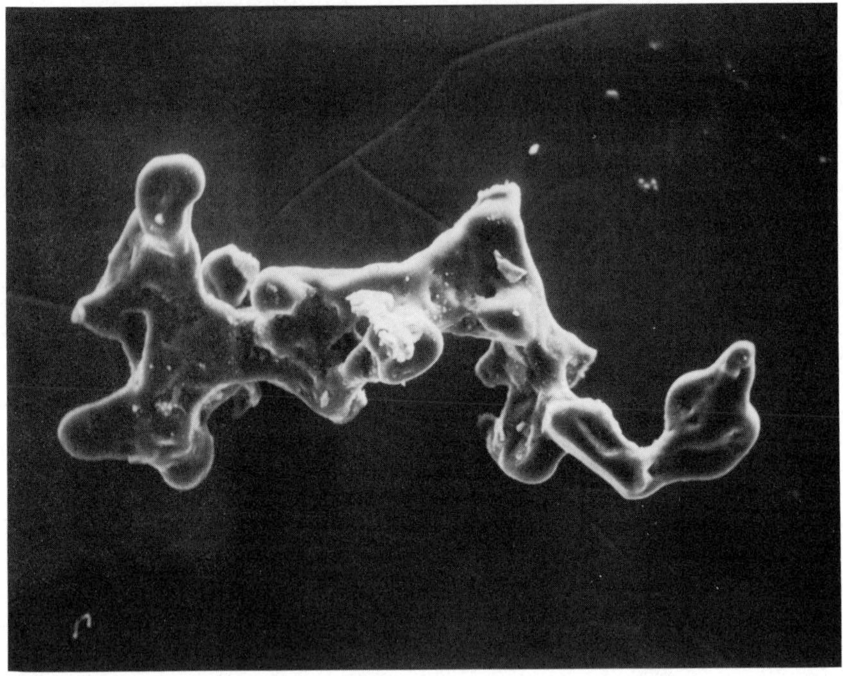

FIG. 1 Residual oil ganglion. (Photograph by courtesy of Norman Morrow.)

the relatively high curvatures of the interfaces in the small pores and hence the relatively large pressure drops across these interfaces. Moreover, viscous effects are small because the high resistance to flow provided by the several hundred feet of reservoir rock typically separating adjacent wells leads to extremely low fluid velocities, in the range of 1 foot per day (3.5×10^{-4} cm/sec). Under these conditions interfacial phenomena (wetting properties) determine the positions of the fluids within the pore space, and flow occurs along separate continuous but tortuous paths for oil and water.

Suppose that water is the wetting fluid and thus occupies the smallest pores. As the local volume fraction of water increases during waterflooding and that of oil decreases at some position in the reservoir, water invades larger and larger pores. Eventually a point is reached where the invading water severs the continuous oil filaments, so that the oil phase at that location has the form of individual ganglia of various shapes. In sandstones this discontinuous oil phase typically occupies between 20 and 40% of pore volume.

Once formed, the ganglia are immobile because the viscous forces exerted on them by the water flowing past are insufficient to overcome the capillary forces opposing flow. The sources of the latter are (1) the nonuniform size of the pathways through the porous medium and (2) contact angle hysteresis. As Fig. 2a shows, the radius of curvature at the rear of an oil drop entering a constriction is greater than that at the front of the drop. If the drop is immobile and its pressure uniform, application of the Young–Laplace equation at positions 1 and 2 shows that a finite pressure difference must exist in the water phase between these points to advance the drop or even to maintain it in the position shown. Similarly, a finite pressure difference between points 3 and 4 (Fig. 2b) is required to advance a drop along a channel of uniform diameter when there is contact angle hysteresis, i.e., when the advancing contact angle θ_A near point 3 exceeds the receding contact angle θ_R near point 4. Such pressure differences can arise only from viscous effects, in other words from the pressure drop accompanying flow of water along alternative pathways through the porous medium not shown in Fig. 2.

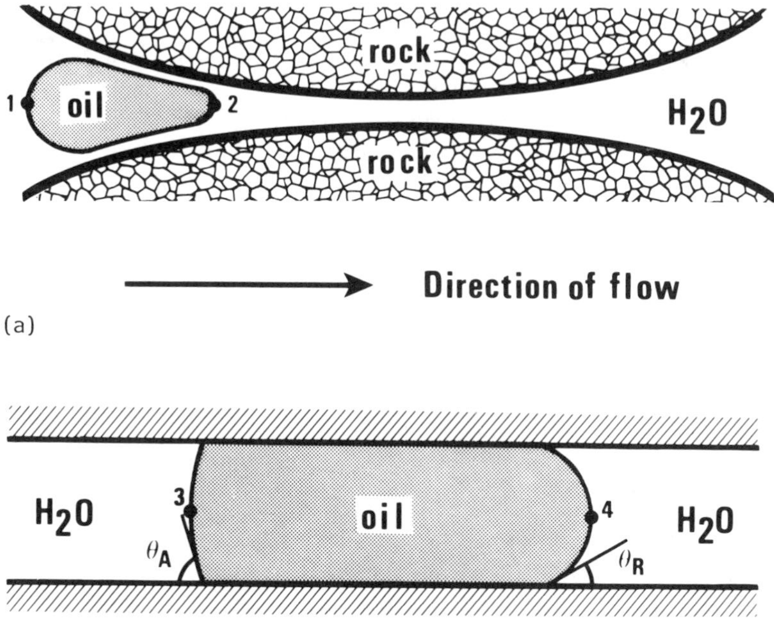

FIG. 2 Effects of (a) nonuniform pore size and (b) contact angle hysteresis on trapping of oil drops.

The ratio between viscous and capillary forces can be characterized by a dimensionless capillary number N_{ca}, which can be defined as follows:

$$N_{ca} = \frac{\mu_w u_w}{\gamma \phi} = \frac{k k_{rw}}{\gamma \phi} \frac{\Delta P_w}{L} \tag{1}$$

Here k and ϕ are the permeability and porosity of the porous medium, γ is the oil-water interfacial tension (IFT), μ_w is the water viscosity, u_w and $(\Delta P_w/L)$ are the superficial velocity and pressure gradient of the flowing water, and k_{rw} is the *relative permeability* of water for the given flow conditions. This last quantity is defined by the usual extension of Darcy's law to two-phase flow:

$$u_w = \frac{k k_{rw}}{\mu_w} \frac{\Delta P_w}{L} \tag{2}$$

Equation (2) was employed in Eq. (1) to replace u_w. For a given porous medium and fixed fluid compositions, k_{rw} is a function of the water *saturation*, i.e., the fraction of the pore space occupied by water. More detailed discussion of the capillary number and trapping of oil ganglia may be found in [11–13].

Figure 3, from the work of Foster [14], shows the results of waterflooding experiments carried out over a wide range of capillary numbers in sandstone cores. At low capillary numbers in the range 10^{-7} to 10^{-6}, residual oil saturation is high and independent of the value of N_{ca}. Under these conditions all oil ganglia are immobile. Such values of N_{ca} are typical of field conditions for waterflooding, as may easily be seen by substituting $\mu_w = 1$ cP, $u_w = 3.5 \times 10^{-4}$ cm/sec, $\gamma = 30$ dyne/cm, and $\phi = 0.25$ into Eq. (1).

As N_{ca} increases, some ganglia can move and the amount of residual oil decreases. When N_{ca} reaches a value of about 10^{-2}, viscous forces become large enough to render virtually all ganglia mobile, and residual oil saturation approaches zero. Key features of the curve in Fig. 3 can be explained by using concepts from percolation theory [15,16].

Figure 3 indicates that if N_{ca} could be increased by a factor of some 10^4 over the values which occur during waterflooding, most of the residual oil remaining after waterflooding could be recovered. Equation (1) suggests that such an increase could be achieved by increasing the pressure gradient $(\Delta P_w/L)$ in the reservoir, and hence the water velocity u_w. The problem with this approach is that it would require very high pressures at the injection wells, which would exceed the mechanical strength of the reservoir rock and induce fracturing. Because fractures would provide less resistance to

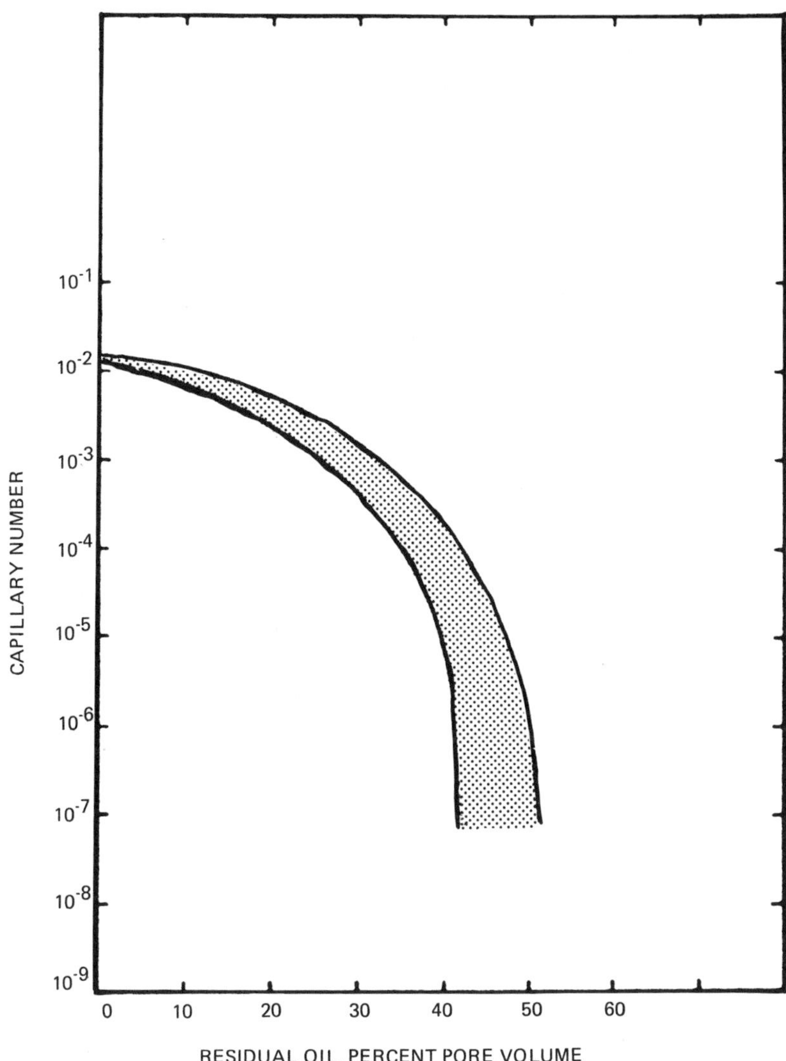

FIG. 3 Dependence of residual oil saturation on capillary number N_{ca}. (From [14] with permission.)

flow than the pore network of the rock, there would be extensive bypassing of oil, leading to poor recovery.

It is clear, therefore, that if conditions favoring low values of residual oil are to be achieved, a drastic reduction in capillary forces is required. That is, IFT must be reduced by several orders of magnitude from about 30 dyne/cm, a typical value between crude oil and brine, to the range 0.001 to 0.01 dyne/cm. Such values are, of course, extremely low by the usual standards of surface chemistry and are not easily achieved even with surfactants. Workers at some major oil companies discovered independently in the 1960s that IFT values in this range did exist in practical systems [14,17,18]. Much research carried out during the past decade or so has dealt with the origin of "ultralow" values (<0.01 dyne/cm) of IFT and systematic means for achieving them.

III. FRONTAL INSTABILITIES

As stated above, the trapping of residual oil ganglia is the chief limitation on waterflooding efficiency in the portion of the reservoir actually contacted by water. But owing to reservoir heterogeneities, well patterns, and, in some cases, frontal instabilities, water never reaches certain parts of the reservoir. For example, permeability often varies significantly and somewhat randomly with depth owing to changes over geologic time in depositional conditions. Indeed, reservoirs are frequently modeled as stacks of rock layers having different permeabilities. In this case water flows preferentially through the most permeable layers. Once these layers have been reduced to residual oil conditions, most of the injected water continues to flow through them with only a small amount of water actually contributing to waterflooding of the less permeable layers. Usually, continuation of waterflooding until all oil is recovered from the less permeable layers cannot be justified economically.

Even when permeability is uniform, however, displacement of oil by water can be highly nonuniform and much of the oil can be bypassed if frontal instabilities develop. As Fig. 4 indicates, an unstable front is highly irregular instead of smooth. It can be shown [19,20] that such instabilities can be expected when the *mobility ratio* M exceeds unity, with M defined by

$$M = \frac{\lambda_w}{\lambda_o} = \frac{k_{rw}/\mu_w}{k_{ro}/\mu_o} \tag{3}$$

Here λ_w and λ_o are the *mobilities* of the water behind and the oil ahead of the displacement front which develops during waterflooding, i.e., the narrow zone where oil and water saturations change rapidly

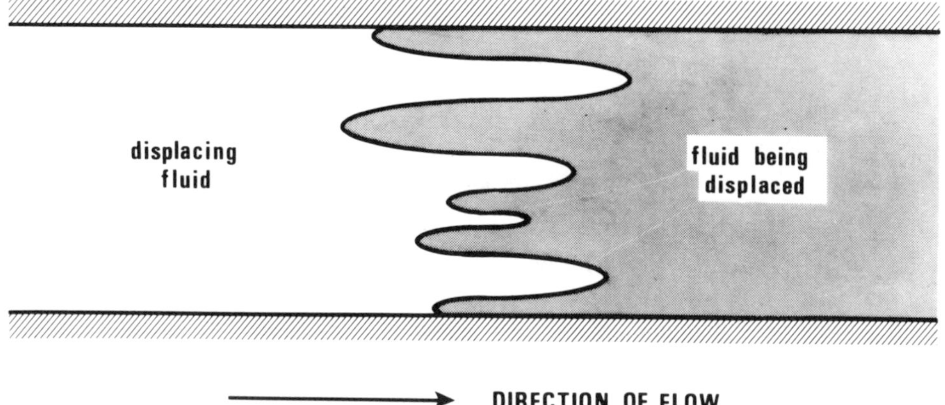

FIG. 4 Schematic diagram of an unstable displacement front.

[21]. Since water content typically varies somewhat behind the front, k_{rw} is an average relative permeability of water in this region. It is clear from Eq. (3) that frontal instabilities accompanied by reduced waterflooding performance can be expected when oil viscosity is sufficiently high.

IV. CHEMICAL FLOODING PROCESSES

A. Mechanisms of Oil Recovery

Several possible mechanisms exist by which surfactants can increase recovery of oil from porous media [13,22]. First, adsorption of surfactants can cause displacement of oily compounds from the rock surfaces with a resulting increase in wettability to water. Experiments in sandstone cores have shown that waterflood residual oil decreases as the rock becomes more water-wet [23,24]. Of course, the amount of residual oil is substantial even for water-wet rock, so only a limited amount of additional oil can be obtained by wettability alteration. It is possible that slight improvements in recovery achieved in a few waterfloods by adding small amounts of surfactant were produced by this mechanism [25].

Wettability alteration is also believed to be one mechanism of recovery in caustic flooding processes where the caustic is injected to convert organic acids present in the crude oil to in situ surfactants [26]. The increase in pH makes the rock more negatively charged while simultaneously transforming organic acids adsorbed on the rock to negatively charged soaps. As a result, extensive desorption occurs, and the reservoir becomes more water-wet.

Another recovery mechanism is solubilization of residual oil in a surfactant solution which flows past trapped ganglia. Owing to the relatively high quantity of surfactant required, solubilization is not attractive as a principal mechanism of recovery, but it nevertheless acts as a secondary mechanism in some cases.

When the surfactant is preferentially oil-soluble, however, water (or brine) transfers into the oil phase and becomes solubilized. The result is an increase in the volume of the oil phase, which can result in coalescence of ganglia into a continuous oil network along which flow can occur. Unless interfacial tensions are ultralow, a residual oleic phase is ultimately left behind in the reservoir in such a process. Because of the solubilized brine, however, the volume of oil in the trapped ganglia is less than that after an ordinary waterflood. Again, the requirement of relatively large amounts of surfactant renders this mechanism of limited value if used alone. But it does provide a means for initial reconnection of trapped ganglia, which may prove useful in some situations.

Emulsification of oil and subsequent blocking of some pores by emulsion drops may decrease the effective mobility of water and lead to less bypassing of viscous oils as a result of frontal instabilities [26]. This mechanism is considered important in some caustic floods.

Finally, surfactants can allow ultralow IFTs to be achieved so that oil ganglia become mobile, as discussed in Sec. II. This mechanism is the dominant one in surfactant recovery processes currently being investigated. In an idealized process the ganglia first mobilized near an injection well move ahead and coalesce with trapped ganglia, thereby converting oil from a trapped to a continuous phase. An *oil bank* is thus formed, i.e., a region in which both oil and water flow as continuous phases ahead of the advancing surfactant solution. It is necessary that ultralow IFTs be maintained only at the rear of the oil bank to prevent redeposition of ganglia when oil content drops below that required for flow as a continuous phase. Most of the oil recovered during such an ideal process is from the oil bank and thus is largely free of surfactant and its associated potential problems, such as formation of stable emulsions. The situation is more complicated, of course, for processes in actual reservoirs.

B. Fluids Injected During Chemical Flooding

First-generation surfactant processes employing petroleum sulfonate surfactants were developed independently by several oil companies [14,17,18,27–29]. Some variation among these processes exists, but a typical one might involve sequential injection of the following:

1. *A preflush of low-salinity brine*. One of its main purposes is to displace the reservoir brine, which frequently provides an adverse environment for petroleum sulfonates because of its high

salinity and high content of divalent cations. Another is to remove divalent cations from clay minerals by an ion exchange process. The preflush may contain small quantities of chemicals with various purposes such as wettability alteration or conditioning of the rock surface to reduce subsequent adsorption of surfactant. It may also contain polymer to reduce its mobility and thus increase the fraction of a heterogeneous reservoir contacted (see below for further discussion of polymers). A preflush may not be required in some situations, e.g., a surfactant slug formulated with reservoir brine.

2. *A surfactant slug.* It may take the form of a "soluble oil" containing the surfactant and an alcohol cosolvent but little or no brine, a microemulsion having substantial contents of both oil and brine, or an aqueous "solution" of the surfactant and alcohol containing little or no added oil. Soluble oils and oil-external microemulsions were initially developed as solvents for displacing oil miscibly; i.e., they were not deliberately designed for producing ultralow IFTs. Careful examination of the process reveals, however, that ultralow IFTs are necessary for successful process performance. Economic factors dictate use of as little surfactant as possible and hence a relatively small slug. If the oil-continuous slug is not to be trapped and thus rendered ineffective during displacement by the aqueous polymer solution which follows, an ultralow IFT is required between the slug and polymer solution.

   The use of water-continuous slugs with little oil has the advantage that it minimizes injection of expensive oil along with the expensive surfactant. As typical surfactant contents are relatively high (in the range 2 to 10% by weight) the surfactant-alcohol-brine mixtures comprising such slugs often contain liquid crystalline material, either dispersed or as a single phase [30,31]. The viscosity of the surfactant slug should be high enough to provide a favorable mobility ratio ($M<1$) when it displaces an oil bnak. Otherwise, frontal instabilities develop at the boundary between oil bank and slug. When dispersions of liquid crystals are present, a high effective slug viscosity can often be obtained by varying composition to alter the macroscopic structure of the dispersion [32,33]. In other cases polymer may be added to increase slug viscosity. The surfactant slug is usually in the range of 5 to 20% of the pore volume to be swept by the process. Larger slugs are normally used with lower surfactant concentrations, the total amount of surfactant required being largely determined by surfactant losses due to adsorption, precipitation, retention in trapped oil, etc.

3. *A dilute aqueous polymer solution.* A favorable mobility ratio ($M<1$) is also required at the rear of the surfactant slug. This condition can be met by injecting a polymer solution. It typically contains a high molecular weight, water-soluble polymer at concentrations ranging from a few hundred to about 2000 parts per

million. The two most commonly used polymers are partially hydrolyzed polyacrylamides and polysaccharides such as xanthan gum. The total polymer slug may be 50% of the pore volume to be contacted or even more. Usually polymer concentration is greater in the first parts of the slug injected than in the last [34,35] in order to provide a gradual increase in mobility between the surfactant slug and the injected brine.
4. *Brine.* Following the polymer slug, brine is injected until the rate of oil production is too low to justify continuing the process.

It is worth emphasizing that one of the challenges of chemical flooding with surfactants is to devise processes which will be effective under diverse conditions. Wide composition differences exist among crude oils and reservoir brines. The latter, for instance, range from relatively fresh water to brines with total concentrations of dissolved solids far exceeding that of seawater. Reservoir temperatures range from near ambient conditions in shallow reservoirs to over 100°C in deep reservoirs. At the high temperatures stability of surfactants and polymers over process lifetimes of several years must be achieved. Although different chemicals are thus required for different reservoirs, the basic sequence of fluids injected is that given above.

Physical properties, e.g., porosity and permeability, and chemical composition of reservoir rocks vary widely. At present the high content of divalent cations and high levels of adsorption are thought by many to preclude chemical flooding in limestone formations. However, a few field experiments involving injection of surfactants into limestone reservoirs have recently been carried out with some degree of success [36]. Even for sandstones, variations in the composition and content of clay minerals strongly influence the adsorption capacity of the rock for surfactant and the amount of ion exchange between the rock and the injected fluids.

In the following sections equilibrium phase behavior of oil-brine-surfactant systems is reviewed, and its implications for achieving ultralow IFTs and hence for developing successful surfactant flooding processes are considered. Then additional factors involving flow and transport phenomena which influence the behavior of chemical flooding processes employing surfactants are discussed.

V. PHASE BEHAVIOR OF SURFACTANT-OIL-BRINE SYSTEMS

A. Introduction

The term *microemulsion* is used in this chapter to denote a "stable translucent micellar solution of oil, water that may contain electrolytes, and one or more amphiphilic compounds," as defined by Healy and

Reed [37]. Microemulsions need not be transparent and are not required to contain electrolytes, cosurfactants, or cosolvents. Research on microemulsions has continued to be extensive in view of their interesting thermodynamics, chemistry, and applications [38–43].

Phase behavior studies of surfactant-oil-brine systems provide an important tool in understanding and predicting the formation and displacement of microemulsions and other phases in underground reservoirs as a function of composition, temperature, pressure, and other variables. Important physical properties which influence oil displacement, such as IFT and viscosity, are strongly related to phase behavior.

The surfactant formulations used for enhanced oil recovery can be in the form of microemulsions, aqueous micellar solutions, liquid crystalline dispersions, and soluble oils, as previously mentioned. Differences among these are detailed in [2]. Typically, the microemulsion systems are multicomponent, consisting of surfactant, cosurfactant or cosolvent (e.g., short-chain alcohol), water, electrolytes, and oil. Only the equilibrium aspects of microemulsion phase behavior as particularly relevant to enhanced oil recovery are discussed in this section. However, it should be recognized that the effectiveness of a chemical flood depends on compositions established in the mixing zones, particularly those between the oil bank and chemical slug and between the slug and the polymer buffer drive. Nonequilibrium phenomena are discussed briefly in a later section.

Physicochemical aspects of the phase behavior of microemulsion systems containing commercial petroleum sulfonates have been documented by Healy et al. [2,37,44,45] and others [46,47]. The microemulsion system for such anionic surfactants is termed "simple" by Reed and Healy [2] when it behaves as though composed of three pure components having ternary diagrams as illustrated in Fig. 5. Figure 5a at low salinities shows a two-phase region in which microemulsions along the binodal curve exist in equilibrium with oil containing molecularly dispersed surfactant. The surfactant is partitioned predominantly into the aqueous phase. This system is equivalent to Winsor's type I [48], and the microemulsion has been called a lower phase by Healy et al. [44]. Nelson and Pope [49] prefer to call this phase behavior "underoptimum" or "type II (−)." Figure 5c at high salinities indicates a two-phase region where microemulsions exist in equilibrium with excess water, i.e., Winsor's type II [48]. Such microemulsions have been called upper phase by Healy et al. [44] and "overoptimum" or "type II (+)" by Nelson and Pope [49]. The surfactant is partitioned mainly into the oil phase. Usually a type II (−) multiphase region is skewed to the right, and a type II (+) region is skewed to the left as shown in Fig. 5.

Figure 5b at an intermediate salinity illustrates Winsor's type III microemulsion in the lower triangle. Any composition of type III has

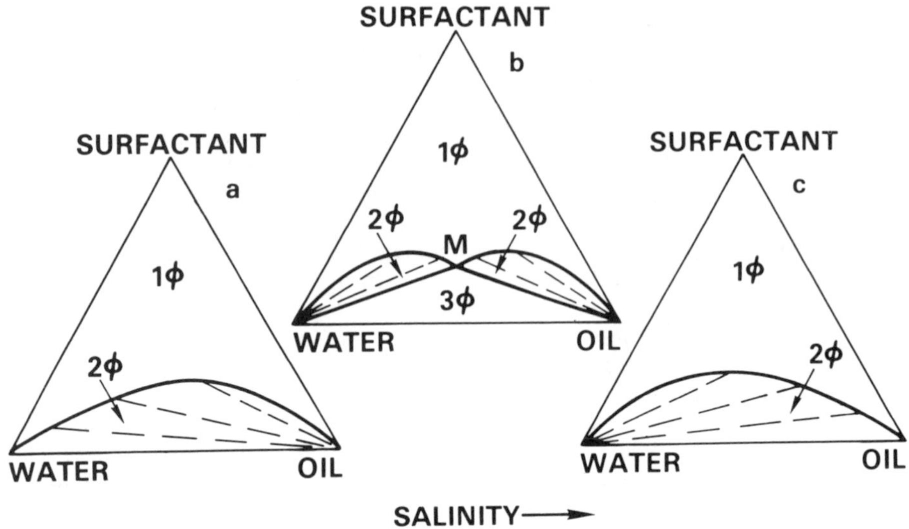

FIG. 5 Illustration of microemulsion phase behavior. (From [2] with permission.)

three phases in equilibrium: a microemulsion corresponding to composition M, excess brine, and excess oil. Type III microemulsions have been designated as middle phase by Healy et al. [44]. Shinoda and Friberg's term for them is the "surfactant" phase [38]. As salinity increases, there is a steady progression from lower phase [type II (−)] to middle phase (type III) to upper phase [type II (+)] microemulsions. Such transitions are also possible due to changes in other parameters such as alcohol concentration or temperature, as will be discussed later. Although not all microemulsion systems conform qualitatively to Fig. 5, "simple" behavior appears to be a good first approximation for numerous systems relevant to enhanced oil recovery. It is possible to have four coexisting phases, two of which may be microemulsions in equilibrium with one another [50], or a microemulsion in equilibrium with a liquid crystalline phase. A liquid crystalline phase may also exist in equilibrium with only brine and oil. It should be noted that the pseudoternary diagrams are slices at constant salinity of the tetrahedron necessary to display the phase boundaries of a four-component system.

Several factors which have been shown to affect the phase behavior of surfactant-oil-brine systems include salinity, surfactant type and concentration, cosolvent type and concentration, pH, oil composition, temperature, and pressure. The following sections will deal with the influence of these variables in systems containing commercial surfactants (often with broad molecular weight distributions and sometimes containing as much as 40% inactive ingredients) as well as in systems with pure surfactants. Model systems containing pure components have been investigated by several groups [51–56] in attempts to understand the fundamentals of microemulsion phase behavior. The behavior in pure surfactant systems is basically the same as for commercial mixtures [53]. Although the effects of different variables are described below separately, it is important to realize that in many cases the effects are interrelated. The changes in phase behavior can generally be understood in terms of relative strengths of hydrophilic and hydrophobic properties of the surfactant/cosurfactant films of the drops or other entities in the microemulsion phases.

B. Effect of Salinity

There are two bulk interfaces in the middle phase and one in the lower phase or upper phase microemulsion system. Thus, one or three values of IFT may be measured depending on system composition: (1) γ_{mo} between microemulsion and excess oil phases, (2) γ_{mw} between microemulsion and excess brine phases, and (3) γ_{ow} between excess oil and brine phases. Phase volumes and, consequently, the volumes of oil (V_o) and brine (V_w) solubilized in the microemulsion depend on various parameters which control the phase behavior. The solubilization parameters are defined as V_o/V_s and V_w/V_s, where V_s is the volume of the surfactant in the microemulsion phase. These can be determined from phase volume measurements if it is assumed that all the surfactant is in the microemulsion phase.

A typical example of the variation of IFT and solubilization parameters with salinity is shown in Fig. 6 for a microemulsion system containing a petroleum sulfonate (Witco's TRS 10-410) with a short-chain alcohol (isobutanol) as the cosolvent. It may be noted that ultralow IFT can be measured by using the sessile drop [37,55–57], spinning drop [58], or light scattering [59,60] technique. The magnitude of γ_{mo} decreases as V_o/V_s increases, i.e., as more oil is solubilized. Similarly, the magnitude of γ_{mw} decreases as V_w/V_s increases. The salinity at which the values of γ_{mo} and γ_{mw} are equal is called the optimal salinity based on IFT, C_γ [2]. Likewise, the intersection of V_o/V_s and V_w/V_s defines the optimal salinity based on phase behavior, C_ϕ. An optimal salinity based on miscibility, C_m, has been defined as the salinity that minimizes the height of the multiphase region in the ternary diagram at a water-oil ratio (WOR) of one. Generally, these three optimal values are nearly equal for a given system. On the basis of IFT and oil recovery data, Reed and Healy [2] concluded that C_γ very nearly corresponds to the best oil recovery. The displacement of oil by the microemulsion

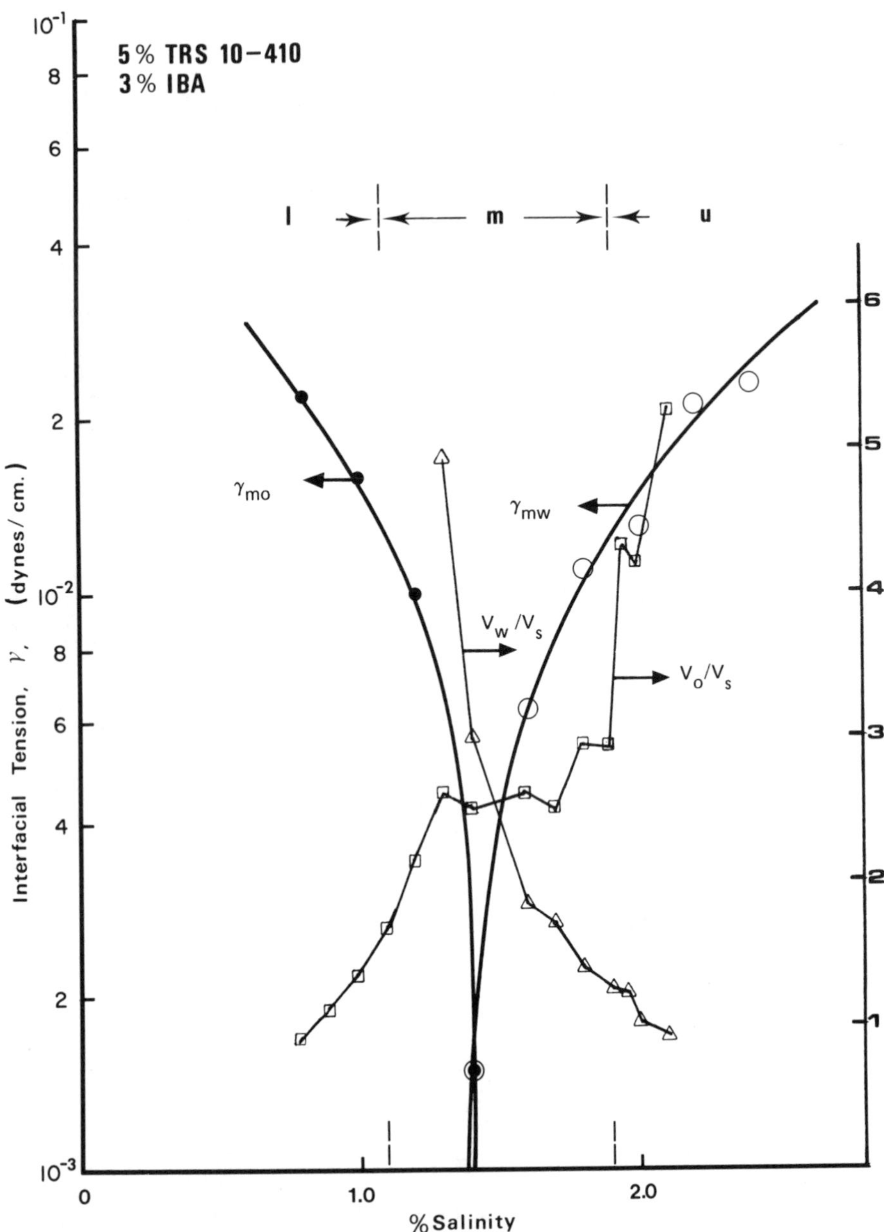

FIG. 6 Variation of interfacial tensions and solubilization parameters with salinity for a microemulsion system containing a petroleum sulfonate, TRS 10-410. An aqueous solution containing 5 wt. % TRS 10-410 (Witco), 3 wt. % isobutanol, and varying salinity was equilibrated with an equal volume of *n*-dodecane at 22°C to obtain the microemulsion.

phase is governed by γ_{mo} and the displacement of the microemulsion phase by the water phase is governed by γ_{mw}. The values of IFT at the optimal salinity should be ultralow ($<10^{-2}$ dyne/cm) for prevention of trapping and hence effective displacement of oil. The optimal salinity concept is a useful guide in screening surfactant formulations. Core tests are necessary, however, to optimize surfactant formulations for specific field conditions.

Figure 7 [56] shows the variation of IFT with salinity for a microemulsion containing pure sodium 4-(1'-heptylnonyl) benzenesulfonate (Texas-1) as the surfactant and *n*-propanol as the cosolvent. The variation of IFT and solubilization parameters with salinity in a pH-dependent microemulsion system containing oleic acid is illustrated in Fig. 8 [56]. It is evident from Figs. 6–8 that in both commercial and pure component systems, ultralow values of IFT necessary for efficient oil recovery are obtained only over a narrow range of salinity in the middle phase region. Obviously, it would be desirable to obtain middle phase microemulsions having high solubilization and exhibiting ultralow IFT over a wide range of salinity. Attempts to achieve that goal are discussed later.

Healy et al. [2,44] have shown that IFT between microemulsion and excess phases can be correlated with the solubilization parameters when salinity or overall composition is varied. Similar correlations have also been reported by Robbins [61] and Shah and co-workers [62,63]. However, such correlations are not always obeyed [64]. Theoretical models which provide a general relationship between solubilization and IFT between microemulsion and excess phases have been developed by Huh [65,66].

Controversy exists about the magnitude of the IFT between excess oil and brine phases (γ_{ow}) in the middle phase region. Reed and Healy [2] reported that γ_{ow} was on the order of 0.1 dyne/cm in petroleum sulfonate systems. Bellocq et al. [67] observed significantly lower values of γ_{ow} in a pure benzenesulfonate system. However, in both cases the thermodynamic inequality [68] $\gamma_{ow} \leqslant \gamma_{mo} + \gamma_{mw}$ was not satisfied. On the contrary, Pouchelon et al. [69] observed that γ_{ow} was close to the higher value of γ_{mw} or γ_{mo} in a pure surfactant system. Ultralow values of γ_{ow} have also been reported in commercial systems (e.g., see [70]). It seems that the apparent violation of the above thermodynamic inequality in some cases could be due to deviations from thermodynamic equilibrium or to variations in experimental technique. The very low surfactant concentrations in excess oil and brine phases impose demanding requirements on the experiments.

Bae and Petrick [70] observed that at the low-salinity end of the middle phase region of a petroleum sulfonate blend, not only values of γ_{mw} (between microemulsion and brine) but also values of γ_{ow} (between oil and brine) were lower than those at the optimal salinity. With such low tensions, both middle phase and oil should be effectively displaced by brine. Indeed, they found for this system

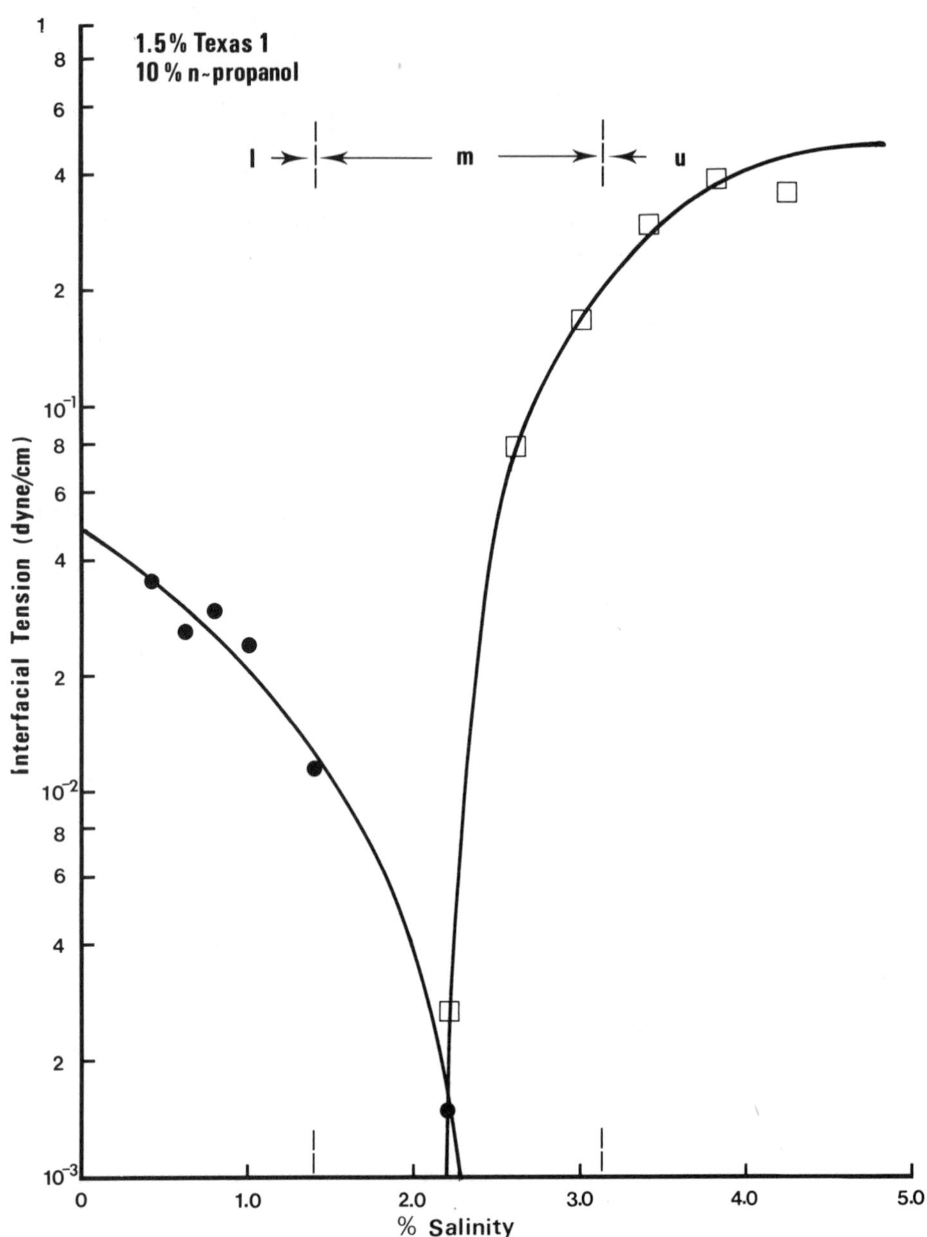

FIG. 7 Variation of interfacial tension with salinity for a microemulsion system containing pure Texas-1. An aqueous solution containing 1.5 wt. % Texas-1, 10 wt. % *n*-propanol, and varying salinity was equilibrated with an equal volume of *n*-decane at 22°C to obtain the microemulsion.

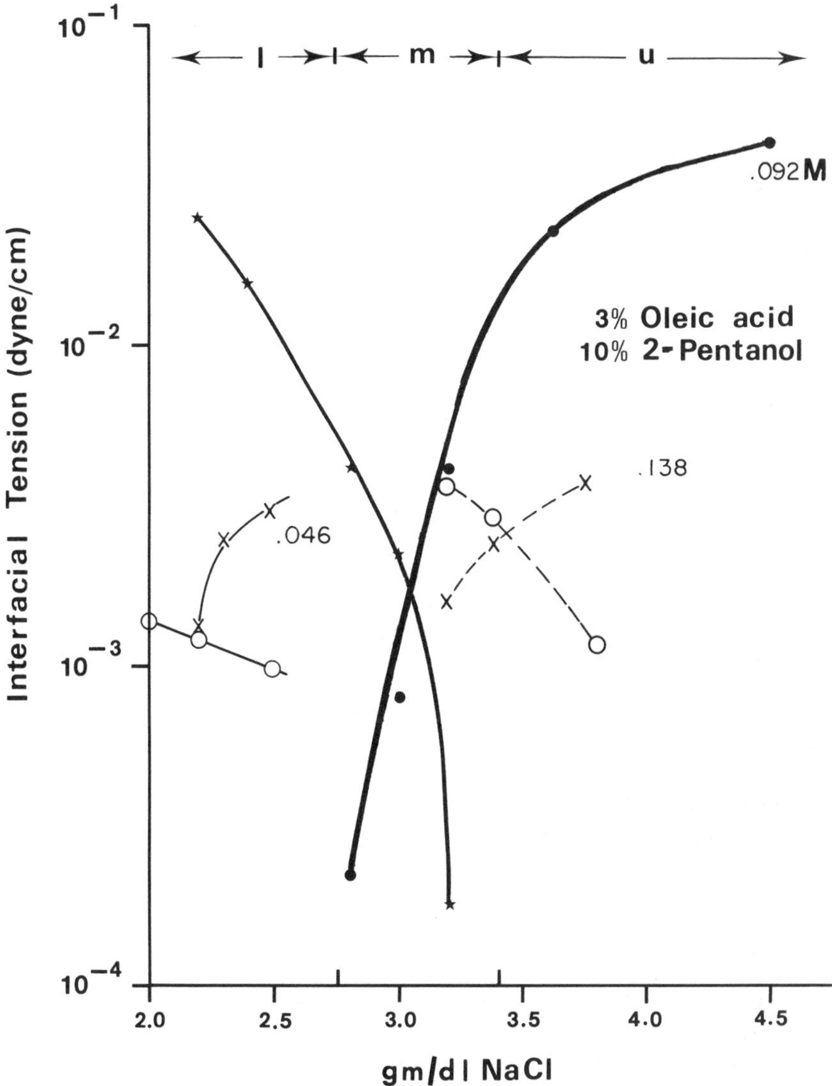

FIG. 8 Variation of interfacial tension with salinity for a pH-dependent microemulsion system containing oleic acid. An aqueous solution containing 3% (v/v) oleic acid, 10% (v/v) 2-pentanol, a constant NaOH concentration (0.046, 0.092, and 0.138 M), and varying salinity was equilibrated with an equal volume of *n*-decane at 22°C to obtain the microemulsion.

that the maximum oil recovery in core flooding tests was obtained at a salinity below the optimal salinity. It should be recognized, however, that core flooding results can be influenced by factors other than phase behavior and IFT, e.g., by fluid mobilities and surfactant retention.

1. Phase Continuity, Drop Size, and Phase Separation

Lower phase microemulsions are water-continuous while upper phase microemulsions are oil-continuous. The phase continuity of microemulsions was confirmed by ultracentrifuge experiments [71,72]. Several light-scattering investigations followed (e.g., see [54,73]). Figure 9 shows the variation in drop size of the dispersed phase for a petroleum sulfonate (TRS 10-410) microemulsion system. At low salinities the drop size increases with increasing salinity, which is consistent with the observed increase in solubilization shown in Fig. 6. As salinity increases further, the middle phase appears and is initially water-continuous. The microemulsion undergoes inversion near the optimal salinity. At higher salinities the middle phase becomes oil-continuous. After the transition from three to two phases, the microemulsion remains oil-continuous with drop size decreasing with increasing salinity. Drop size measurements in microemulsions with pure components (e.g., [53,54] confirm the trends shown in Fig. 9. Values of drop size near the optimal salinity in Fig. 9 should be viewed with some skepticism since middle phase microemulsions near the inversion point are of unknown structure (see below) and moreover are quite sensitive to the application of centrifugal fields.

Miller et al. [74] suggested a theory for phase separation in microemulsions which accounts for the salient features observed experimentally. The oil-in-water microemulsion drops are attracted to each other by London-van der Waals forces and, if ionic, repelled by electrical double-layer forces. The repulsion term dominates at low salinity. The drop size increases and the double-layer thickness decreases with increasing salinity. Accordingly, the drops can approach each other more closely and attraction forces become more important. Eventually, attraction dominates and the microemulsion separates into two water-continuous phases, a middle phase rich in drops and an excess brine phase lean in drops. Miller et al. [74] also calculated IFT values between these phases using the approach of Cahn and Hilliard [75]. The theory predicts that IFT values between two water-continuous or two oil-continuous phases can be ultralow even when the system is not close to the consolute point. This prediction was later confirmed experimentally [32].

Near-critical conditions are observed at the salinities corresponding to transitions from lower to middle phase and middle to upper phase. The correlation length diverges at the critical point, causing critical opalescence, which is observed experimentally. The IFTs

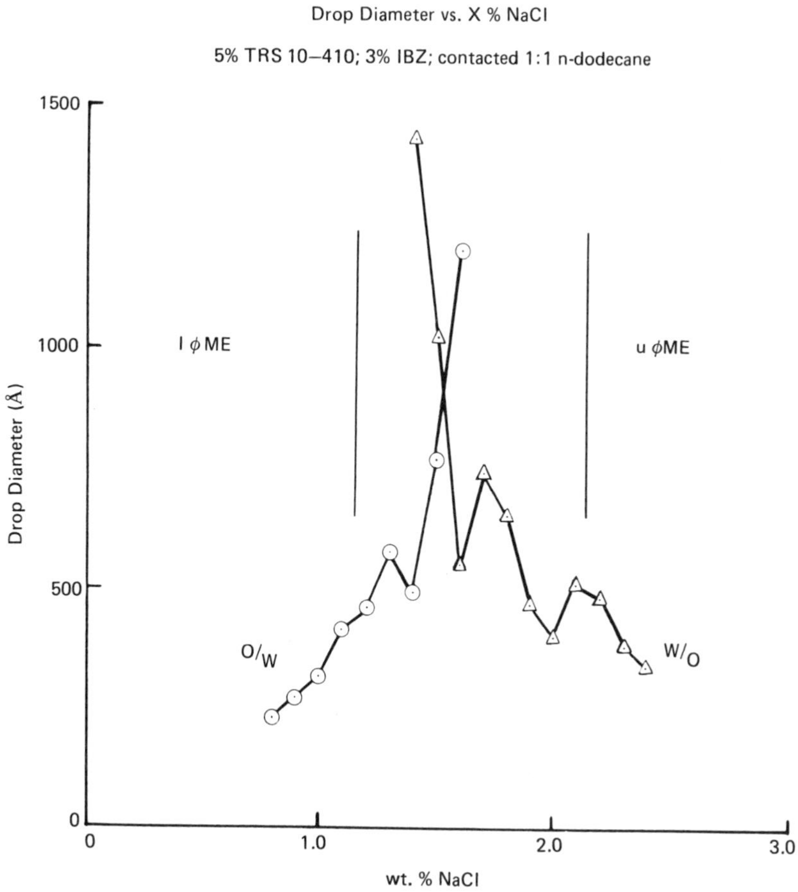

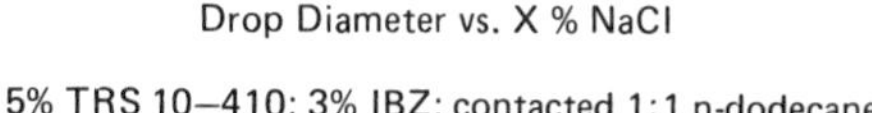

FIG. 9 Variation of drop size with salinity as determined by ultracentrifugation for a microemulsion containing a petroleum sulfonate, TRS 10-410. An aqueous solution containing 5 wt. % TRS 10-410 (Witco), 3 wt. % isobutanol, and varying salinity was equilibrated with an equal volume of *n*-dodecane at 22°C to obtain the microemulsion.

between the middle phase and the new excess phase are extremely low, as evident in Figs. 6–8 discussed earlier. In fact, the IFT is zero at a true critical end point, where the volume of the phases remains constant and the interface disappears. Near the optimal salinity both γ_{mo} and γ_{mw} are influenced by respective critical end points. In addition to salinity, other variables, e.g., temperature, which produce phase transitions in the neighborhood of a consolute point will also cause critical behavior [76].

Critical phenomena are implicit in the thoery of Miller et al. [74]. The importance of critical phenomena was made clear and quantitative treatments of phase behavior and IFT based on the scaling laws were presented by Fleming and Vinatieri [77,78]. Using light-scattering techniques, several investigators [60,76,79–82] provided further insight into the role of critical phenomena in microemulsion phase behavior. More work is needed on the significance of measured critical exponents. Light-scattering evidence [80] of thick interfaces (>100 Å) in the regions of low IFT is consistent with the theoretical models of Miller et al. [74] and Talmon and Prager [83].

The structure of the middle phase microemulsion in the inversion zone has not yet been conclusively determined. It could be bicontinuous on a microscopic scale, as suggested by Scriven [84] and analyzed by Talmon and Prager [83] and others [41,85,86]. Alternatively, it could be bicontinuous on a macroscopic scale as suggested by Miller et al. [74,54], or even a "molecular dispersion" with most of the oil and water present in a dissolved state [54]. Contrary to this, Bansal and Shah [10] proposed that the middle phase is a water-external microemulsion with monodispersed oil drops suspended in a continuous aqueous phase. Shinoda and Friberg [38] and Huh [65] considered the middle phase to consist of alternating thin layers of oil and brine containing surfactant molecules oriented at each planar interface according to their amphiphilic character. Although such order is considered to prevail on a local scale, thermal fluctuations continually produce changes in layer orientation at various locations so that no long-range order exists.

It may be noted that the existence of microstructure in microemulsions is not universally accepted [87]. Analogy is usually made to oil-brine-alcohol systems, which show phase behavior similar to that of microemulsions. Thus, transitions from two-phase to three-phase to two-phase systems are possible as alcohol solubility shifts from aqueous phase to oil with increasing salinity [88]. However, the alcohol concentration is high compared to the surfactant concentration in microemulsions. Thus, while the alcohol-rich phases are molecular dispersions, it is difficult to envision the absence of microstructure in a microemulsion consisting mainly of oil and water with but a few percent surfactant. Of course, the low surfactant content of microemulsions makes them much more attractive than alcohols for application to enhanced oil recovery. Interfacial tensions are lower as well.

2. Multivalent Ions

Reservoir brines contain multivalent ions, the actual concentrations varying from field to field. Reed and Healy [2] reported that addition of Ca^{2+} to a surfactant formulation in small amounts can significantly reduce the optimal salinity. The reduction is greater than would be expected from the usual ionic strength relationships. Thus, there is no optimal ionic strength which applies for all ratios of divalent to monovalent ions. The effect on IFT and the extent of the multiphase region were found to be negligible in the systems investigated [2]. Glover et al. [89] showed that the change in optimal salinity of a system containing divalent cations can be modeled by considering equilibrium among the various species in the brine including sulfonate-cation complexes. They found optimal salinity to be a linear decreasing function of the mole fraction of sulfonate ions associated with divalent cations. It should be noted that ion exchange processes in the reservoir may significantly alter the concentration of different ions during a surfactant flooding process [90–92].

C. Effect of Surfactant

1. Surfactant Structure

Figure 10 [2] shows IFT versus salinity for three alkyl chain lengths of *o*-xylene sulfonates. The surfactant becomes more hydrophobic with an increase in the alkyl chain length (N). Both the optimal salinity (C_γ) and IFT at C_γ decreases with increasing N. However, the range of salinity where middle phase microemulsions (and ultralow IFT) exist is also decreased. Puerto and Reed [93] confirmed this effect of surfactant lipophile structure on microemulsion phase behavior. Similar results are expected with surfactants containing head groups other than sulfonate. It should be noted that above a certain lipophile length, liquid crystalline structures may form and even predominate, particularly in the absence of a cosolvent and at low temperatures [93,94]. The theoretical model of Huh [66] correctly predicts a narrow salinity range for ultralow IFT when the optimal salinity is low, and an increase in the width of the middle phase region with increasing optimal salinity as observed with *o*-xylene sulfonates [2] and other anionic surfactants [94,95].

The effect of branching of the lipophile on the solubilization parameter and optimal salinity has been discussed in the literature [93,94]. In general, the more nearly linear the lipophile, the more effective will be the surfactant in terms of minimal IFT, provided liquid crystals or gels do not form. If additional lipophile branching is necessary to avoid gel formation, Puerto and Reed [93] suggest branching of the alkyl chain rather than adding methyl groups to the benzene ring.

A large number of pure and commercial surfactants have been tested for their potential in enhanced oil recovery in the laboratory

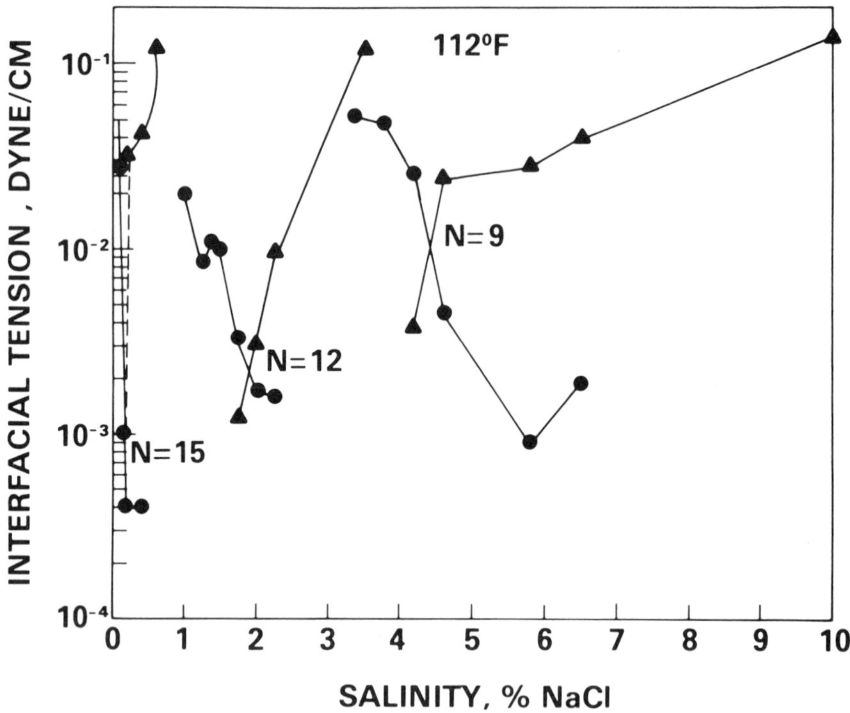

FIG. 10 Dependence of interfacial tension on salinity for a microemulsion system containing *o*-xylene sulfonates as the surfactant; N denotes the alkyl chain length of the surfactant. (From [2] with permission.)

of Wade and Schechter [94,96–98]. These studies led to correlations between the width of the middle phase region and the solubilization parameter as well as the IFT at optimal conditions for specific surfactant structures [95]. The removal of benzene rings from sulfonates improves the solubilization and increases the optimal salinity. Work of Barakat et al. [94] indicates that secondary alkane and α-olefin sulfonates can be used to produce microemulsions with high salt tolerance, i.e., that exhibit low values of IFT at high salinities.

2. Surfactant Concentration and Phase Behavior

The optimal salinity is independent of surfactant concentration and water-to-oil ratio if the microemulsion contains an isomerically pure surfactant with no cosolvent [99]. But if an alcohol is added, its partitioning affects the optimal salinity, as does any partitioning of components in surfactant mixtures. The phase behavior will depend on the hydrophobic/hydrophilic nature of both surfactant and alcohol.

Reed and Healy [2] recognized that overall surfactant concentration affects the optimal salinity in a microemulsion with an alcohol cosolvent. In the case of *o*-xylene sulfonates with only monovalent ions present, they observed a reduction of optimal salinity with increasing surfactant concentration at fixed WOR and surfactant-to-alcohol ratio. Similar results have been reported by Glover et al. [89], Baviere et al. [100], and Hirasaki [101]. Such behavior can be explained in terms of the partitioning of alcohol between the bulk phases and the interfacial film. In the *o*-xylene sulfonate-tertiary amyl alcohol system [2], an increase in surfactant-alcohol concentration leads to a higher alcohol concentration in the interfacial film, making it more hydrophobic and thus lowering the optimal salinity. However, with most conventional petroleum sulfonates, the optimal salinity increases with surfactant concentration at fixed surfactant-to-alcohol ratio. This behavior is due to increased partitioning of hydrophilic disulfonate or low molecular weight alcohol molecules into the interfacial film at high surfactant concentrations.

Figure 11 [56] illustrates the effect of surfactant concentration and salinity on the phase behavior of a pure component microemulsion containing Texas-1, which is a rather hydrophobic surfactant. At a fixed alcohol concentration, the width of the middle phase region decreases with increasing surfactant concentration and so does the optimal salinity. A qualitative explanation is that at a higher surfactant-to-alcohol ratio the system becomes relatively more hydrophobic, and less salt is needed to drive the microemulsion into an upper phase.

The optimal salinity increases with increasing surfactant concentration when divalent cations are present [89,92,101,102]. This effect could be due to a decrease with increasing surfactant concentration in the mole fraction of sulfonate complexed with divalent cations (see previous discussion). The change in optimal salinity with divalent cations present may, of course, also be affected by the partitioning of alcohol and surfactant [100,101].

The effect of surfactant concentration on phase behavior when both electrolyte and alcohol concentrations are fixed is shown in Fig. 12 [56] for the Texas-1 microemulsion system. Middle phase microemulsions are observed at concentrations as low as 0.005 wt. % Texas-1 with an appropriate amount of alcohol. At very low surfactant concentrations the middle phase was seen to exist as a lens at the brine-oil interface; i.e., it was not a perfectly wetting phase.

The amount of oil solubilized in the microemulsion increases with the surfactant concentration since again the increase in the surfactant-to-alcohol ratio renders the system more hydrophobic. For a given WOR, all the oil may be solubilized at a high surfactant concentration as seen in the last sample in Fig. 12.

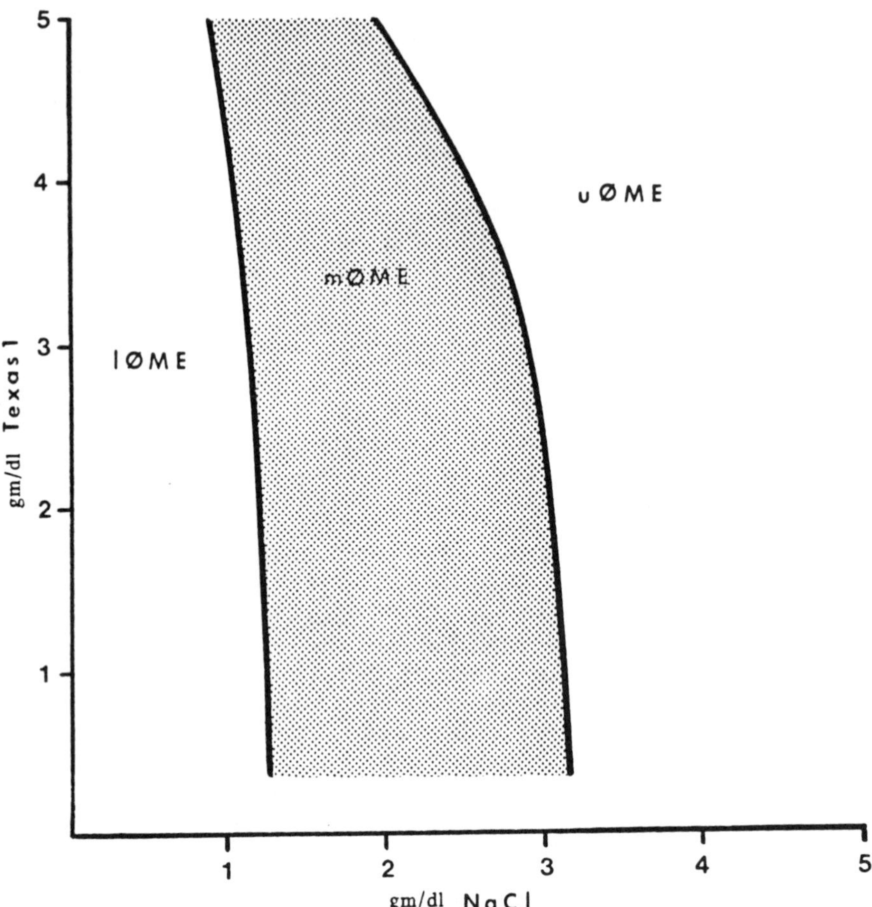

FIG. 11 Effect of surfactant and salt concentrations on microemulsion phase behavior of Texas-1 system with 10 g/dl *n*-propanol. An aqueous solution containing the surfactant, alcohol, and salt was equilibrated with an equal volume of *n*-decane at 32°C.

3. Water-to-Oil Ratio (WOR)

The effect of overall WOR on microemulsion phase behavior, particularly optimal salinity, can be understood in terms of the partitioning of the cosolvent or cosurfactant [2,100]. In addition, changing WOR to very large or very small values may lead to the formation of liquid

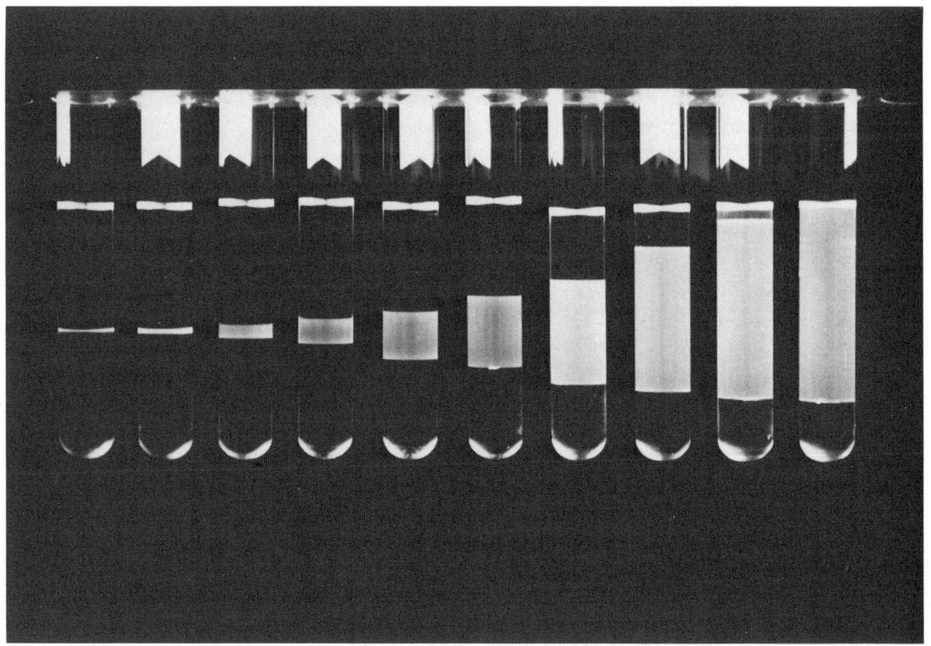

FIG. 12 Photograph of a surfactant scan with 7.5 g/dl *n*-propanol and 1.0 g/dl NaCl for the Texas-1 system at 32°C.

crystalline or other phases [103,104]. Addition of oil to surfactant-alcohol-brine mixtures causes transition from lamellar liquid crystal to microemulsion phases. These transitions have been observed to depend on salinity. Like the general aspects of the phase behavior, the dependence on other variables like alcohol concentration or temperature should be predictable. At low salinities, the transition is through a two-phase region with coexisting liquid crystal and oil-in-water microemulsion phases. Slightly below the optimal salinity, two three-phase regions are observed: two microemulsion phases exist in equilibrium with a liquid crystalline phase in one and with brine in the other. At high salinities, the transition to water-in-oil microemulsion with decreasing WOR involves a streaming birefringent isotropic phase. The effect of WOR, particularly the existence of multiphase regions, is relevant to enhanced oil recovery because of the change in WOR as oil is displaced by the surfactant slug in porous media.

4. Surfactant Concentration and Ultralow IFT

When the surfactant concentration is greater than about 1 wt. % the recovery technique is generally termed micellar or microemulsion flooding, in contrast to surfactant water flooding when the surfactant concentration is lower than about 1 wt. %. The mechanism of ultralow IFT in the two regions of surfactant concentration has been the subject of some controversy [105]. It was mentioned earlier that ultralow IFT values are obtained when middle phase microemulsions exist with large solubilization parameters. The fact that middle phase microemulsions can be formed at very low surfactant concentrations (Figs. 11 and 12) indicates that the same mechanism for ultralow IFT may be valid independent of the surfactant concentration. Differences in behavior attributed to surfactant concentration effects are probably due to differences in alcohol content and WOR in the systems investigated. At low surfactant concentrations problems with adsorption and low microemulsion phase saturation may become more important than IFT. Of course, whether or not microemulsions form at a very low surfactant concentration depends on a number of factors such as the surfactant type and presence of cosolvent. Liquid crystalline phases may sometimes prevail in the absence of a cosolvent, as has been suggested by Puig et al. [105] based on nonequilibrium studies in pure component systems.

Foster [14] investigated the IFT behavior of a petroleum sulfonate-sodium chloride-water system against crude oil. A typical IFT contour map is shown in Fig. 13 [14]. Cayias et al. [106] and Chan and Shah [107,108] also reported that at low surfactant concentrations the IFT first decreases and then, after reaching a minimum value, increases with surfactant concentration. Chan and Shah [107] attempted to relate the surfactant concentration for minimum IFT to the critical micelle concentration (CMC). However, the behavior shown in Fig. 13 can also be explained by the effect of surfactant concentration on optimal salinity as discussed earlier.

Puig et al. [105] argued that neither conventional adsorption of surfactant at the oil-brine interface nor formation of micelles is responsible for ultraflow IFT. Rather, they attributed the origin of ultralow IFT in the nonequilibrium systems investigated to the formationof a surfactant-rich third phase between the oil and water. They suggested that the third phase may be a liquid crystalline phase or a microemulsion depending on conditions such as temperature and concentration of alcohol.

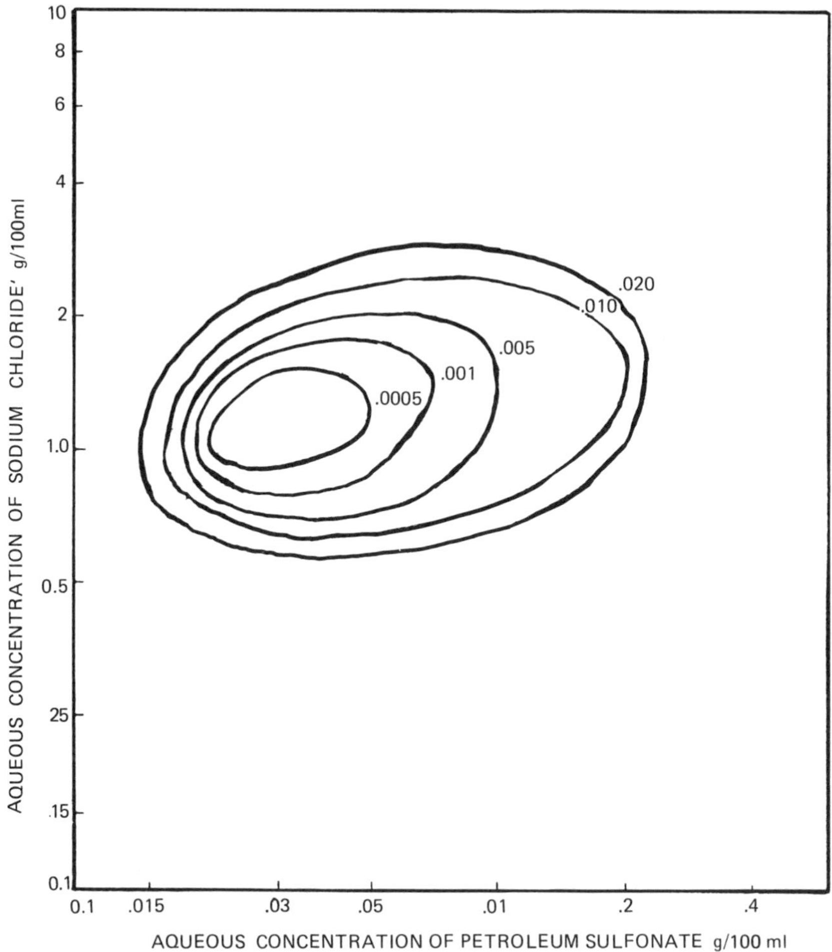

FIG. 13 Interfacial tension contour map of a petroleum sulfonate, NaCl, water system against an intermediate-paraffinic crude. (From [14] with permission.)

5. Surfactant Types and Blends

Investigations related to surfactant flooding initially employed conventional petroleum sulfonates. Later work also included several studies of "synthetic sulfonates," which are derived from the small oligomers, e.g., trimers or tetramers, of olefins such as propylene or butene. Some of the disadvantages of conventional petroleum sulfonates are: (1) problems in quality control due to changes in composition of the petroleum feedstocks, (2) chromatographic separation in the reservoir due to broad equivalent weight distributions, and (3) partitioning into excess phases resulting in increased surfactant loss. Synthetic sulfonates have relatively narrow molecular weight distributions, cost about the same as petroleum sulfonates in some cases, but have fewer disadvantages.

Blends of petroleum as well as crude oil sulfonates have been studied in attempts to optimize phase behavior and improve the efficiency of oil recovery [47,106–110]. The optimal salinity of surfactant blends depends on the composition. Puerto and Gale [109] used linear mole fraction averaging, whereas Salager et al. [47] suggested logarithmic mole fraction averaging. It may be noted that it is also possible to minimize alcohol requirements by proper choice of surfactant mixtures [110].

Synthetic alkylaryl sulfonates can be tailored to obtain the desired optimal salinity while maintianing good solubilization and yet avoiding unfavorable phases like gels, precipitates, and persistent emulsions. Optimal salinity and solubilization parameters generally decrease as branching increases, as indicated above, while the more linear lipophiles promote condensed phases. Alkylaryl sulfonates can be effective over a wide range of temperatures but only at low to moderate salinities.

Problems with both conventional and synthetic petroleum sulfonates in reservoirs with high salinities and high divalent ion contents have prompted a search for other surfactants. Nonionic surfactants such as alcohols with an ethylene oxide chain are relatively insensitive to salinity. However, the nonionics suffer from high adsorption loss and temperature sensitivity. Shinoda and Kuneida [111–113] extensively studied the phase behavior of nonionic surfactant systems. Microemulsions containing ethoxylated nonylphenols were investigated in the laboratory of Wade and Schechter [114,115]. Middle phase microemulsions and ultralow IFT were reported in such systems. Ethylene oxide groups increase the hydrophilicity of the surfactant, and hence the optimal salinity goes up.

Ethoxylated sulfates and sulfonates have received considerable attention because of the advantage that the range of optimal salinity can be widened by varying the ethylene oxide content. Thus, high optimal salinities corresponding to reservoir brine compositions can be achieved. However, ethoxylated sulfates are limited in application to temperatures below about 50°C because of hydrolysis problems

[5]. Ethoxylated sulfonates can tolerate high temperature as well as high salinity, but their high cost has hindered widespread commercial application to date. Other surfactants considered for tertiary oil recovery include petroleum-based organic acids [116], tall oil derivatives [117,118], ethoxylated carboxymethylates [119,120], and pH-dependent surfactants such as carboxylates [121].

Blends of different types of surfactants have been explored in attempts to broaden the region of ultralow IFT and/or shift the optimal salinity [62,108,122-128]. Hayes et al. [124] defined synergistic IFT behavior as occurring when the IFT minimum of the mixture is significantly lower and/or broader than that obtained with individual surfactants alone. Such systems are difficult to obtain [124,126]. Shifting the IFT minimum to higher optimal salinities without necessarily broadening the range of ultralow IFT is more common. This has been achieved by using mixtures of petroleum sulfonates with ethoxylated sulfonates [122,123], ethoxylated nonylphenols [125], and other nonionics [126,127]. Qutubuddin et al. [128] obtained middle phase microemulsions over a wide salinity range using a mixture of synthetic sulfonate and carboxylic acid with pH adjustments. However, the existence of ultralow IFT throughout the salinity range was not confirmed.

Two potential problems exist with blends of ionic and nonionic surfactants. These are (1) chromatographic separation of the sulfonates and nonionics and (2) fractionation among the nonionic species with different partitioning behavior between oil and brine phases [129,130]. Although significant advances have been made in developing surfactants for different field conditions, there is still a need for cost-effective surfactants for high-temperature, high-salinity environments.

6. pH and Interfacial Charge

The possible role of pH in surfactant flooding seems to have been largely overlooked. Investigations of the effect of pH on microemulsion phase behavior are rare. Gillberg and Eriksson [131] reported the influence of pH on the solubilization capacity of mixtures of nonionic and ampholytic surfactants as a function of temperature. Qutubuddin et al. [56,128,132] developed a model microemulsion system which exhibits pH-dependent behavior, using oleic acid as the surfactant.

Figure 14 [132] illustrates the phase behavior of a microemulsion containing oleic acid and 2-pentanol as a function of NaOH and NaCl concentrations at fixed surfactant and alcohol content. The dependence of phase behavior on salinity at constant pH is similar to that of of sulfonate systems. The variation of solubilization parameters and IFT with salinity in the oleic acid system was discussed in Sec. V,B. Depending on the salinity, it is possible to observe upper to middle to lower phase as well as upper to middle to upper phase transitions by changing NaOH concentration at fixed salinity. The sequence of

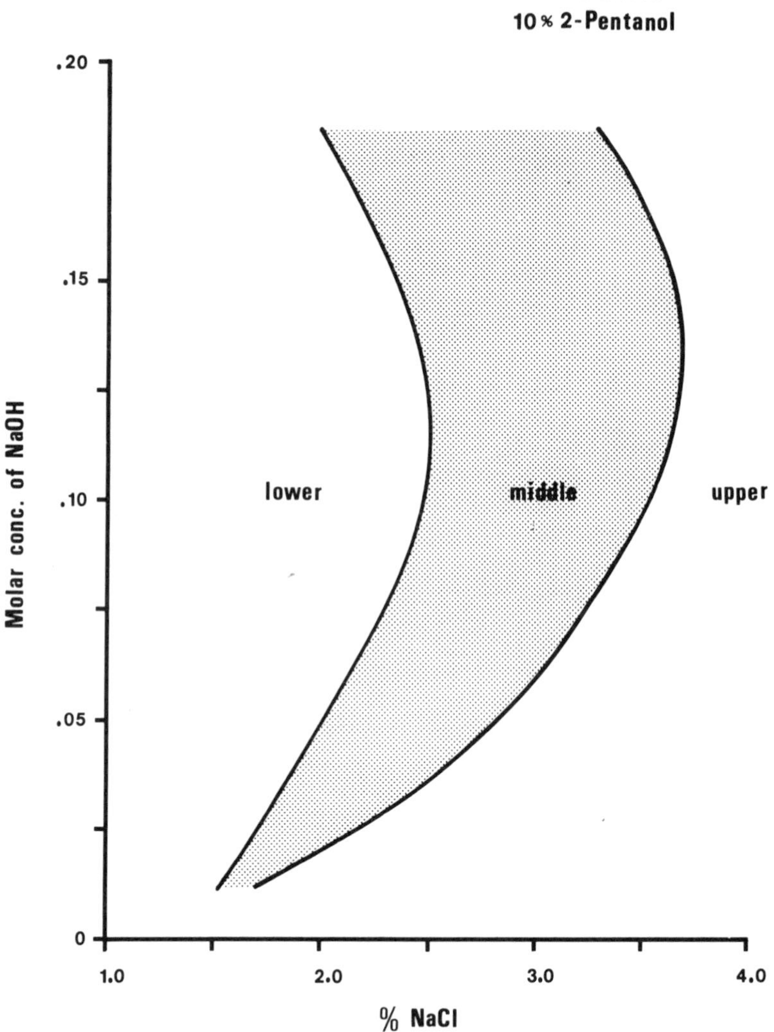

FIG. 14 Effect of NaCl and NaOH concentrations on oleic acid microemulsion system. An aqueous solution containing 3% (v/v) oleic acid, 10% (v/v) 2-pentanol, and varying salinity and NaOH concentrations was equilibrated with an equal volume of *n*-decane at 22°C to obtain the microemulsion.

phase transitions at low salinity (e.g., 2.2 wt. % NaCl) is opposite to that observed on increasing salinity. The transitions from upper to middle to lower phase microemulsions can be explained by the increase in the degree of ionization of the surfactant with increasing NaOH concentration and hence pH. Thus, the surfactant becomes more hydrophilic with an increase in pH, as opposed to becoming hydrophobic with an increase in salinity. The higher the salinity the larger is the amount of NaOH necessary for middle phase formation.

Above a specific NaOH concentration or pH, however, all the surfactant molecules are ionized. Then the counterbalancing effect of pH against salinity does not prevail. Thus, at concentrations above about 0.1 M NaOH (Fig. 14), the added NaOH acts primarily as an electrolyte causing transitions from lower to middle to upper phase microemulsions. Thus, middle phase microemulsions can be observed at a lower NaCl concentrations with an increase in NaOH concentration. This explains the curved shape of the middle phase region in Fig. 14.

Analogous to an optimal salinity, an optimal pH or NaOH concentration can be defined as the intersection point of the solubilization parameter or IFT curve. For high salinities, an NaOH scan corresponding to upper to middle to upper phase transitions will yield two intersection points, thereby defining an optimum range of pH or NaOH concentration where ultralow IFT values can be obtained.

As mentioned previously, the salinity range where middle phase microemulsions exist can be extended significantly by using pH-dependent microemulsions containing mixed surfactants instead of sulfonates alone [56,128]. At a constant NaCl concentration, one may observe transitions from upper to middle to lower to middle phases with increasing NaOH concentration. The possible use of such pH-dependent systems in surfactant flooding is discussed in Sec. VI.

Nelson et al. [133] suggested the use of a cosurfactant in alkaline flooding, which can be viewed as a type of chemical flooding in which the surfactant is formed in situ as the alkali converts petroleum acids in the crude oil to soaps. It should be recognized that several mechanisms are involved in enhanced oil recovery by alkaline flooding [26,134]. A discussion of the mechanisms is outside the scope of this chapter. The problem in alkaline flooding of having to choose between low IFT and alkaline concentration high enough to propagate the alkaline bank at a reasonable rate can be avoided by using a cosurfactant in the alkaline slug. The net system is a mixed one containing the injected cosurfactant and the "primary" surfactant formed in situ, which is sensitive to pH in addition to salinity. Such systems have the potential to enhance oil recovery by both low- and high-pH alkalies [133].

D. Effect of Cosolvent

Cosolvents or cosurfactants are used in microemulsion formulations for one of two reasons: (1) to eliminate the formation of or reduce the viscosity of liquid crystalline phases in surfactant solutions, both with and without oil, or (2) to shift the optimal salinity by making the films at the drops or other entities in the microemulsions more hydrophilic or hydrophobic. As mentioned earlier, cosolvents are not absolutely necessary for the formation of microemulsions. Short-chain alcohols have been the primary cosolvents used in microemulsion formulations for tertiary oil recovery, particularly at low salinities. Ethoxylated alcohols have found use as cosolvents to increase the optimal salinity and thereby allow petroleum sulfonates to be used at higher salinities and divalent cation contents.

The partitioning of alcohol between the drop surfaces and bulk phases contributes to the inadequacy of representing microemulsion phase behavior on pseudoternary diagrams. Alcohol partitioning cannot be neglected in modeling the displacement of oil by microemulsions in porous media [101].

Read and Healy [2] and Jones and Dreher [135] reported a decrease in microemulsion viscosity with alcohol addition. Also, formation of liquid crystalline phases can be avoided if sufficient alcohol is added.

The effects of alcohol on phase behavior of liquid crystalline phases and dispersions in aqueous solutions with or without hydrocarbon have been investigated by Benton, Miller, and co-workers [30–32,53]. The optimal salinity, the solubility and partitioning of the surfactant-cosolvent mixture, and the extent of surfactant adsorption all depend on the type and concentration of the cosolvent [92,93]. Jones and Dreher [135] demonstrated that various alcohols influence phase behavior differently. Salter [136] identified the relative solubility or partitioning of an alcohol between oil and brine as the key parameter affecting optimal salinity. Thus, hydrophilic alcohols increase the optimal salinity for a given surfactant, while hydrophobic alcohols decrease the optimal salinity.

Shah [137] presented a correlation of optimal salinity with the solubility of alcohol in brine for different sulfonates. Baviere et al. [100] and Blevins et al. [138] investigated the relationship between alcohol concentration in microemulsions containing petroleum sulfonate and the surfactant/alcohol and surfactant/hydrocarbon ratios. They determined the partitioning of the alcohol among different phases. Bourrel and Chambu [139] discussed the effect of alcohol on solubilization in terms of lipophilic and hydrophilic interaction energies of the surfactant and alcohol.

The effect of alcohol on microemulsion phase behavior has also been investigated using pure rather than commercial surfactants [51–53,56,116,140]. Figure 15 [53] illustrates the effect of carbon

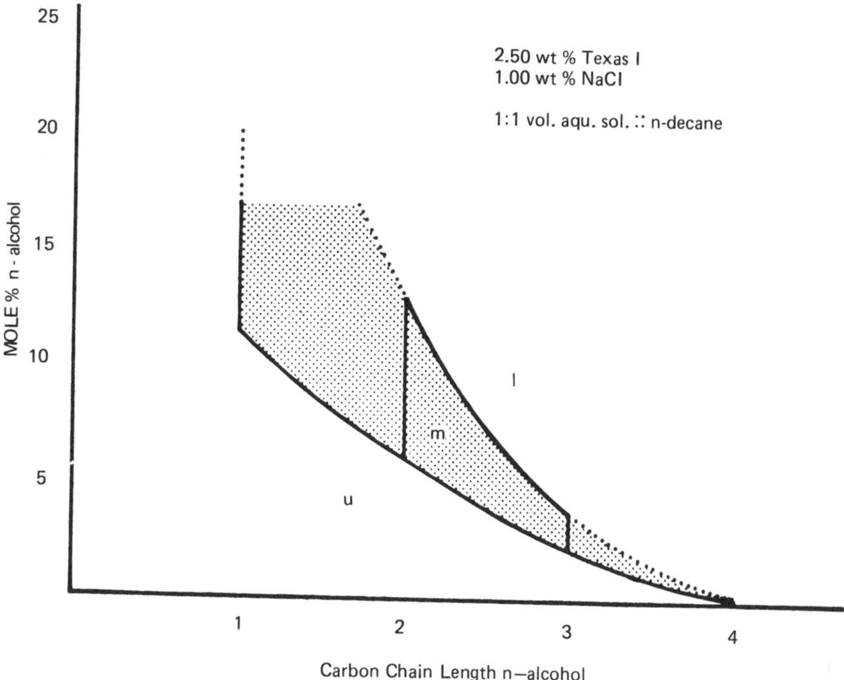

FIG. 15 Effect of cosolvent chain length on the phase behavior of Texas-1 microemulsion system.

chain length of a homologous series of short-chain *n*-alcohols on middle phase formation when Texas-1 surfactant and NaCl concentrations were fixed at 2.5 and 1.0 wt. %, respectively, and the aqueous phase was equilibrated with an equal volume of *n*-decane. In general, the microemulsion type changed from upper phase (oil-continuous) to middle phase to lower phase (water-continuous) with increasing alcohol content. Addition of short-chain *n*-alcohols can produce inversion to a lower phase as shown for the rather hydrophobic Texas-1 surfactant. However, addition of relatively long-chain alcohols, e.g., *n*-pentanol, cannot produce inversion at this salinity. The long-chain *n*-alcohols are too hydrophobic to form a water-soluble mixture of the surfactant and alcohol.

Both the molar concentration of alcohol required to produce a middle phase and the range of alcohol concentrations where the middle phase exists decrease with increasing alcohol chain length. These results, as shown in Fig. 15, can be attributed to the increase in surface activity of alcohols with a longer chain.

The effect of alcohol concentration on the phase behavior at different surfactant concentrations is illustrated in Fig. 16 [53]. For each surfactant concentration at a fixed salinity, a specific cosurfactant concentration is necessary to form an optimal middle phase ($V_O/V_W = 1$). The locus of optimal conditions is approximately indicated by the straight line "1" in Fig. 16. The intercept of this line with the horizontal axis at about 8% propanol can be interpreted as the amount of alcohol in the oil and brine, in contrast to the amount in the surfactant films, under optimal conditions. The slope of this line indicates that the molar ratio of surfactant to alcohol in the interfacial film is 1:2 for this system under optimal conditions [53].

Increasing the ratio of cosolvent to surfactant concentration generally increases the IFT at the optimal salinity while broadening the size of the middle phase region [136]. The increase in IFT and the chromatographic separation during flow through porous media constitute problems associated with the use of alcohols. As a result, most workers advocate that alcohols be used only to prevent formation of viscous gels and not to adjust optimal salinity. Even for avoiding gels, the use of branched-chain surfactants is ordinarily preferable to adding short-chain alcohols.

E. Effect of Oil

The aromaticity, chain length, and structure of oil influence the optimization of phase behavior to produce ultralow IFT necessary for tertiary oil recovery. Foster [14] reported that for a given petroleum sulfonate, the IFT minimum would shift to a lower salinity and a higher surfactant concentration for a naphthenic crude oil as compared to a paraffinic crude oil. Gale and Sandvik [18] observed lower IFT with oil of a higher aromatic hydrogen content. Increasing the aromaticity at constant salinity produces transitions from lower to middle to upper phase microemulsions [44].

There has long been an interest in replacing crude oil with a pure oil or mixture of pure oils so as to facilitate the design of surfactant systems for actual reservoir conditions. Investigators in the laboratory of Wade and Schechter [106,141] reported that a pure hydrocarbon can be modeled by replacing it with a particular *n*-alkane based on IFT behavior with a dilute surfactant solution. The variation of IFT with alkyl group carbon number of the given hydrocarbon series is compared with the *n*-alkanes. The alkyl group carbon numbers corresponding to the IFT minimum are considered equivalent. The number of carbon atoms in the alkane corresponding to the IFT minimum is the *equivalent alkane carbon number* (EACN) for the hydrocarbon.

The EACN concept has been extended to mixtures of pure hydrocarbons and to crude oils, and also to high surfactant concentrations

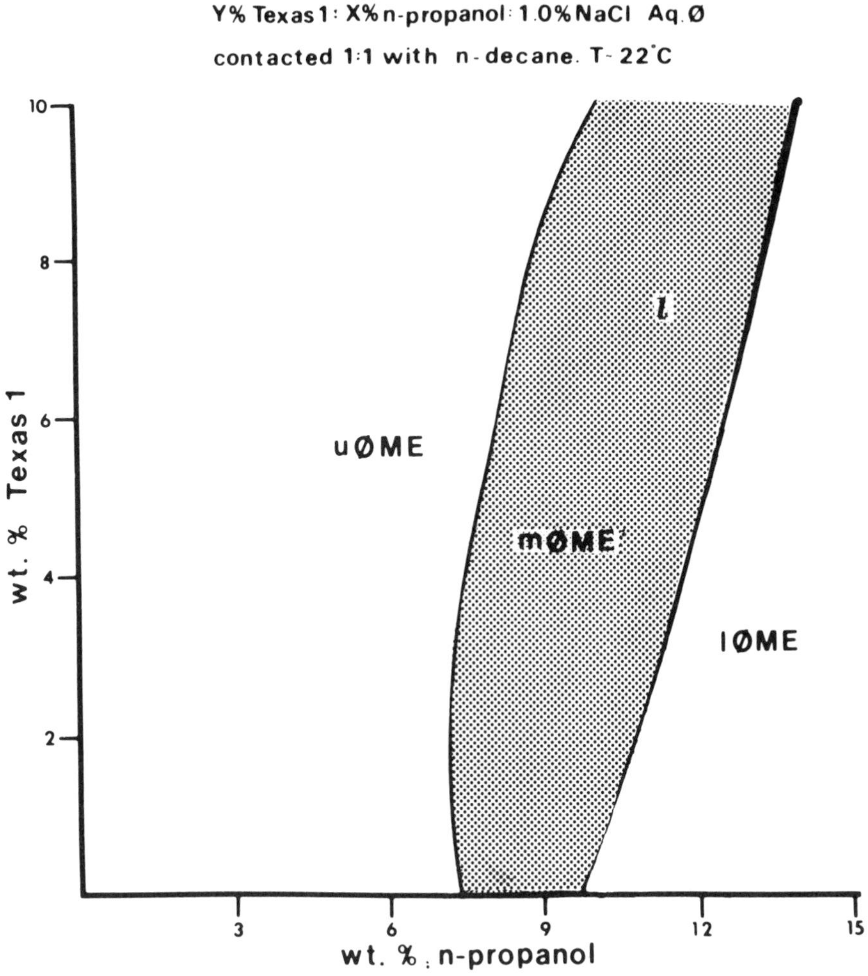

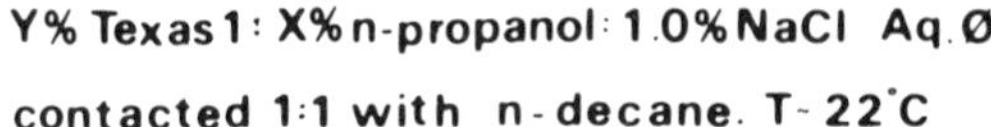

FIG. 16 Microemulsion phase behavior at constant salinity for the Texas-1/*n*-propanol system.

[47]. For high surfactant concentration systems, Salager et al. [47] denoted the midpoint of the salinity range over which a middle phase appears as the optimal salinity and also used a WOR of four. In subsequent applications of the EACN concept, the optimal salinity has been based on IFT data [142] or equal solubilization of brine and oil [143]. Irrespective of the definition, the optimal salinity is the only parameter used to evaluate the EACN. The EACN of the oil is the alkane number which matches the optimal salinity of the crude oil with a given surfactant-alcohol formulation at a specific temperature. However, two hydrocarbons having the same EACN do not necessarily exhibit the same phase behavior, IFT behavior, or core flooding performance.

Puerto and Reed [93] suggested the use of another parameter in addition to the optimal salinity to better characterize the interaction of oil with surfactant-alcohol formulations. In their "three-parameter" representation, optimal salinity is plotted against oil molar volume with curves of constant solubilization parameter under optimal conditions. Puerto and Reed [93] have shown that for many systems of moderate oil molar volume, choice of any two of the three parameters determines the third for a given surfactant-alcohol formulation and constant temperature. If two oils have equal molar volumes, optimal salinities, and solubilization parameters, they are called equivalent oils (Eqo's). Puerto and Reed [93] proposed the idea of Eqo's as a replacement for EACN. Eqo's having the same viscosities and similar phase behavior as a function of surfactant concentration, referred to as "equivalent oils for flooding," were shown to satisfactorily reproduce flooding results.

Experiments show that, for a given surfactant formulation and concentration, the optimal salinity decreases with decreasing chain length when the oil is a straight-chain hydrocarbon (e.g., see [47, 63]). Mukherjee and Miller [144] were able to predict this effect. Their theory indicates that a short-chain oil produces a greater entropy increase than a long-chain oil when it penetrates the surfactant film of a microemulsion drop. As a result, the film will incorporate more of a short-chain oil, a factor favoring upper phase microemulsions and hence lower optimal salinities. Huh's theory [66] makes similar predictions, though it differs in the details of the entropy calculation. Nelson [145], on the other hand, considered that interaction energy effects were also important and invoked concepts involving regular solution theory and solubility parameters.

F. Effects of Temperature and Pressure

Phase transitions from upper to middle to lower (or vice versa) have been discussed in previous sections with respect to a number of compositional variables such as salinity, pH, and alcohol concentration. All of these parameters influence the partitioning of the surfactant and cosolvent between the oil and water and thereby the composition of the interfacial film. Microemulsion phase behavior is influenced both by this partitioning and by the change in the surfactant hydrophilicity.

The effect of temperature on phase behavior can also be understood based on the above premise. An increase in temperature tends to increase the relative affinity of anionic surfactants for water, and the converse is true for nonionics. Thus, Healy et al. [44] reported that increasing temperature causes the phase behavior to change in the direction of upper to middle to lower for anionic systems, and the IFT between microemulsion and oil and solubilization parameter V_W/V_S increase. The phase inversion temperature (PIT) is defined as the temperature for a given composition where equal amounts of oil and brine are solubilized. The PIT increases with increasing salinity in ionic surfactant systems, both commercial and pure [44,47, 56,96]. Increasing temperature makes ionic surfactant systems more hydrophilic owing to an increase in electrical double layer thickness and hence a decrease in screening of the repulsion between adjacent surfactant ions. Thus, a higher temperature is needed to keep the system at optimum when the salinity is increased in anionic microemulsions. The effect of temperature on nonionic surfactants is in the opposite direction; i.e., the hydrophobicity increases with increasing temperature. Nonionic surfactant systems are also much more temperature sensitive than anionics [111–113].

Although the possible effect of pressure on phase behavior of microemulsion systems is well recognized, there have been only a few investigations [145–148]. Optimal salinity has been found to increase with increasing pressure [146]. Thus, high pressures make the system less hydrophobic, possibly because oil molecules have more difficulty penetrating surfactant films under these conditions. Many in-place crude oils contain a significant amount of dissolved gas that is absent in the stock-tank oil. It is possible that the surfactant will behave differently with live crude under reservoir temperature and pressure than with stock-tank oil [93,145]. Nelson [145] observed no changes in the phase behavior of a specific surfactant/brine/stock-tank oil system when pressurized alone or with methane. However, methane pressurization caused a change in phase behavior when a synthetic oil of lower molar volume was substituted for the stock-tank oil. The concept of equivalent oils mentioned above can be applied to replace microemulsion floods of live crude at high pressure with floods of appropriately diluted dead crude at low pressure. Further work is necessary to verify this concept and understand the effect of actual reservoir conditions on microemulsion phase behavior and oil recovery.

G. Polymers and Surfactant-Polymer Interactions

The role of the polymer slug, as indicated previously, is to provide mobility control. The two most commonly used polymers are the synthetically prepared partially hydrolyzed polyacrylamides and the biopolymers, primarily xanthan gum. The major advantages of the biopolymers over polyacrylamides are their good shear stability, good thickening power without causing loss in permeability, and low sensitivity to ionic environment. The major disadvantages of the biopolymers are high cost, difficulty of preparing solutions free of solid material that can plug the pore structure, and visocsity loss from biochemical or chemical reactions. Some of the problems associated with solution preparation and viscosity stabilization can be overcome [149], e.g., by using aqueous concentrates or broths.

The polyacrylamides are cheaper and work by both increasing viscosity and decreasing brine permeability. However, because they have a less rigid structure than biopolymers, they collapse rapidly and lose their thickening power upon addition of salt or divalent ions. The polyacrylamides are also subject to shear degradation. There is a need for the development of novel mobility control agents. So far, there have been only a few advances in this direction (e.g., see [150–152]).

The efficiency and economics of oil recovery can be adversely affected by interactions between surfactant and polymer. Such interactions occur because of mixing at the boundary between surfactant and polymer buffer solutions, and because residual surfactant adsorbed on the rock surface may later desorb into polymer solution. Mixing of polymer and surfactant may also occur thorughout the surfactant bank because of the "polymer inaccessible pore volume" effect [153]. Large polymer molecules are excluded from the smaller pores in the reservoir rock and travel faster than the surfactant. Thus, polymer molecules enter into the surfactant slug. Furthermore, in some operations the polymer may be added directly to the surfactant slug to make it more viscous.

The earliest investigations of surfactant-polymer interactions in enhanced oil recovery were carried out by Trushenski and co-workers [154,155]. They reported that low mobility and phase separation can occur due to such interactions. Szabo [156] studied several surfactant-polymer systems and found that aqueous mixtures of sulfonates and polymer solutions separated into two or three phases. Other studies have also dealt with the effect of polymers on the phase behavior of micellar fluids [128,157–162].

Qutubuddin et al. [128] proposed that the mechanism of phase separation is the "volume restriction" effect arising from the large size of the polymers currently used. The spacing in the bilayers in a lamellar liquid crystalline phase or between liquid crystalline particles is much smaller than the polymer molecules for most compositions of interest in surfactant flooding. Polymer molecules would lose configurational entropy when forced into the narrow spacings to conform to the structure of the surfactant solution. This is thermodynamically unfavorable, and hence phase separation occurs. Polymer was found to cause phase separation only when the surfactant solutions were liquid crystalline. In the absence of liquid crystals, such as at low salinities, xanthan gum and polyacrylamides were found to be compatible with petroleum sulfonate [128].

The effect of polymers on microemulsion phase behavior was first documented by Hesselink and Faber [157]. They described the surfactant-polymer phase separation in terms of the incompatibility of two different polymers in a single solvent, considering the microemulsion as a pseudopolymer system. Polymer may cause phase separation in an otherwise single-phase oil-in-water microemulsion. The basic mechanism of phase separation is similar to that in the absence of oil; i.e., it avoids loss of configurational freedom of the polymer molecules. Brine is transferred from the microemulsion to the polymer phase as salinity is increased. Qutubuddin et al. [128] presented a thermodynamic treatment which predicts the partitioning of water between the microemulsion and polymer phases. Huh [163] made similar calculations and showed that phase separation is enhanced by the repulsion between the microemulsion drops and polymer molecules.

VI. MAINTAINING ULTRALOW INTERFACIAL TENSIONS DURING SURFACTANT FLOODING

The need for ultralow interfacial tension to prevent formation of residual oil ganglia or to mobilize such ganglia once formed was discussed in previous sections. Conditions for which ultralow IFTs can be achieved between equilibrium phases have been described. The actual behavior accompanying displacement of a surfactant slug through a reservoir is rather complicated, however. In this section the effect on IFT of various phenomena associated with slug movement is discussed.

Ultralow IFT can typically be achieved only over rather narrow composition ranges, as indicated above. But as the trailing edge of the oil bank propagates through the reservoir, composition can be altered in several ways:

1. Overall surfactant composition can change as a result of preferential adsorption of some components in a surfactant mixture relative to others. As petroleum sulfonates and indeed most commercial surfactants are mixtures, this problem is a serious one.

2. If the salinity and divalent cation content of the brine used in the surfactant slug differ from those of the reservoir brine, the surfactant will experience changes in both variables as the slug advances. One reason is that, owing to reservoir heterogeneities, some mixing between the slug and reservoir brine is inevitable even if a preflush is employed. For example, cross-flow between layers of different permeability can provide such mixing. Another reason is that ion exchange occurs between the slug and the clays present in

the formation. For the case of a slug having a low content of divalent cations, ion exchange transfers divalent cations from the clays to the slug. For petroleum sulfonates some precipitation can result. In any case a change occurs in ionic composition, an important parameter affecting the achievement of proper phase behavior and ultralow IFTs for all ionic surfactants. In some reservoirs divalent cations can also enter the slug as a result of dissolution of certain minerals in the rock, e.g., $CaSO_4$.

3. Short-chain or ethoxylated alcohols, which are part of many surfactant formulations, have some surface activity and thus affect phase behavior and optimal salinity, as seen previously. But many alcohols have relatively high solubilities in bulk oil and/or brine phases. Consequently, the amount of alcohol present at the interfaces of the drops or other entities within microemulsion phases, and hence the optimal salinity as well, varies with the amounts of oil and brine mixed with a given quantity of a surfactant-alcohol mixture. Clearly, this effect can produce changes in optimal salinity as an alcohol-containing slug progresses through the reservoir, flowing at a different rate from and undergoing mixing with brine and oil which have different mobilities. Similar behavior occurs when a surfactant mixture contains components which are relatively soluble in brine and/or oil, e.g., the brine-soluble disulfonates often present in commercial petroleum sulfonates.

Because the phase behavior of surfactant systems is complex and the composition range exhibiting ultralow IFT typically small, because different phases present at the same place in the reservoir have different mobilities and hence advance at different rates, and because the effects of reservoir heterogeneity are always present, the design of a surfactant flooding process to maintain ultralow IFTs at the rear of an oil bank which must travel several hundred feet would appear to be a monumental task. Indeed, experience suggests that ultralow IFTs, even if achieved initially, are ultimately lost in the field. Three possible approaches which can be taken to eliminate, or at least to delay, this loss are discussed in the following sections.

A. Surfactant Slugs Made with Reservoir Brines

One method of maintaining ultralow IFT would be to use a surfactant with a narrow molecular weight distribution which requires no alcohol cosolvent and which can displace oil when formulated with reservoir brine. This last requirement entails that for many reservoirs the process must be effective under conditions of high salinity and high divalent cation content. Since petroleum sulfonates alone cannot meet this requirement, other surfactants have been sought which can achieve ultralow IFTs under these adverse conditions.

Leading candidates are anionic and nonionic surfactants containing ethylene oxide chains or, in some cases, both ethylene oxide and propylene oxide chains. As discussed previously, ethylene oxide groups increase the hydrophilic character of a surfactant and hence increase optimal salinity. Of course, such surfactants are generally more expensive than petroleum sulfonates. Their blends with petroleum sulfonates, which can be effective in slugs formulated with reservoir brine, thus have economic advantages. On the other hand, such blends are susceptible to chromatographic separation by preferential adsorption. Hence, maximizing the ability to maintain ultralow IFTs during the process argues for using ethoxylated surfactants alone.

Although some surfactants have proved effective in certain reservoir brines, development of improved surfactants for such processes remains an area of active research. If effective surfactants having lower costs or lower adsorption losses can be found, processes utilizing reservoir brines would become extremely attractive.

B. Systems with Broad Domains of Ultralow Interfacial Tensions

A second approach for maintaining ultralow IFTs would be to develop a surfactant system which would exhibit such IFTs over a relatively wide range of compositions. If successful, relatively inexpensive surfactants such as conventional petroleum sulfonates having broad equivalent weight distributions or synthetic sulfonates could be used. These surfactants are most effective in displacing oil at salinities and divalent ion contents well below those of many reservoir brines. Besides, adsorption of a given surfactant is generally less at low salinities, an obvious advantage in terms of process economics.

Since the ultralow values of IFT for a given system are usually found in the middle phase region, one might at first suppose that systems having broad middle phase regions should be sought. Unfortunately, as mentioned in Sec. V, experiments show that IFTs generally rise throughout the middle phase region as it becomes broader [96,162]. Thus, a trade-off seems to exist between having very low values of IFT over a narrow composition range or somewhat higher values of IFT over a broader range.

Factors favoring the former are surfactants with long, straight hydrocarbon chains, low salinities, short-chain oils, and low alcohol concentrations. All these factors contribute to reduced flexibility or greater rigidity of the surfactant films at the surfaces of the microemulsion drops or other entities. From another point of view they contribute to strong interactions either between the surfactant tails and oil or between the polar groups and brine [139]. Because films of low flexibility are not easily deformed to configurations of high curvature, they yield microemulsions with internal interfaces of low curvature, a situation which promotes greater solubilization. High solubilization is known both experimentally and theoretically to be associated with low IFTs [2,65,74,136].

Of course, if the films are too rigid, a lamellar liquid crystalline phase forms instead of a middle phase microemulsion [96,164]. Provided their viscosities are not too high, such liquid crystals may be useful for oil recovery [165]. This possibility has so far received relatively little attention, however, and is a highly attractive topic for further research.

It remains to be explained why microemulsions with relatively flexible films should be associated with broad three-phase regions. As indicated above, flexible films promote a decrease in drop size, which by itself should decrease the attraction between nearby drops due to London-van der Waals forces. Since attraction promotes aggregation of drops to form the middle phase [74], as discussed in a previous section, one might expect to find only a narrow range of existence of the middle phase for the small drops associated with flexible films.

On the other hand, flexibility permits close approach of drops and even interpenetration of their surfactant films in some cases. Light-scattering experiments with water-in-oil microemulsions indicate that an additional attractive force arises when interpenetration takes place [166,167]. Presumably an increase in the density of CH_2 groups within the region of overlap is the source of this effect. A few data even suggest that the broadening of the middle phase region caused by adding alcohol may be primarily at the expense of the upper phase region [136], a result which would support the importance of such interpenetration and the associated attractive forces.

Film flexibility is also favorable for drop coalescence, which presumably is a necessary factor in achieving the dynamic, bicontinuous morphology envisioned by Scriven [84,168] in a middle phase microemulsion. As the middle phase region is, in fact, broader for flexible films, it appears that the tendencies for interpenetration and coalescence promoted by flexibility outweigh any reduction in London-van der Waals forces.

In the light of this information one can consider possible ways of broadening the salinity range over which ultralow IFTs can be achieved. As the surfactant slug mixes with reservoir brine during its progress through the reservoir, its salinity increases and the surfactant becomes more oil-soluble or hydrophobic. If the system can be designed so that, at the same time, the concentration in the surfactant films of a rather hydrophilic compound increases or that of a rather hydrophobic compound decreases, the effect of the salinity increase can be at least partially offset. One might expect that short-chain alcohols could provide such behavior, though their adsorption and partitioning behavior is likely to be relatively insensitive to salinity because of their charge-free nature. So could surfactant mixtures provided that some components have appreciable solubility in one or the other bulk phase. Ionic surfactants would be most attractive for this purpose since their adsorption and partitioning behavior is strongly influenced by salinity. Indeed, the possibility of such behavior is one factor favorable to the use of crude oil sulfonates or other surfactants having broad molecular weight distributions.

Yet a third approach is to use a carboxylic acid or other pH-sensitive surfactant and design the system so that salinity changes are at least partially offset by pH changes which have compensating effects. For example, with a carboxylic acid surfactant, a salinity increase should be accompanied by a pH increase, which owuld increase the degree of ionization of the acid and thus provide a more water-soluble surfactant [132]. Practical implementation of this concept would, however, require that means be available for controlling reservoir pH.

Naturally, changing the composition of the surfactant films within microemulsion phases by any of these schemes can also change the film rigidity and hence modify IFT, as discussed previously. For lowest IFT the changes in film composition should act to increase rigidity with increasing salinity. Such behavior would be expected when an oil-soluble alcohol transfers from the films to the oil phase or when the degree of ionization of a carboxylic acid increased due to a rise in pH. In contrast, transfer of a water-soluble alcohol or surfactants of low molecular weight to the films should decrease the rigidity. It may be possible, however, to choose surfactant mixtures for which the surfactants of low molecular weight have considerable rigidity. It is known, for instance, that alkyl benzenesulfonates yield less rigid films than alkane sulfonates of comparable chain length [94]. Thus, transfer of water-soluble alkane sulfonates into a film consisting primarily of alkyl benzenesulfonates could provide a means of offsetting salinity increases without an accompanying loss in film rigidity and increase in IFT.

The above discussion, which is partly speculative, suggests that it should be possible to design a surfactant system having a broad range of ultralow IFTs at salinities well below those of reservoir brines. A few reports seem to support this possibility, although the systems investigated are rather complicated and the reasons for the observed behavior are not clear [169,170]. The search for such systems with broad regions of ultralow IFT should be continued utilizing the principles outlined above.

C. Processes Employing a Salinity Gradient

A third scheme for maintaining ultralow IFT during chemical flooding is to deliberately vary the concentration of one species during injection of the process fluids into the reservoir. This variaiton, in combination with the effects of hydrodynamic dispersion, produces a gradient in concentration of that species in the direction of flow. In particular, employment of a salinity gradient has been proposed [46, 92,171,172]. In such a process the preflush, if any, has a higher salinity than the surfactant slug. The latter has a higher salinity than the polymer solution. The basic idea is that even though composition changes occurring during the process alter the optimal salinity, the salinity gradient ensures that the actual salinity is equal to the optimal somewhere near the rear of the oil bank thorughout the process. Hence ultralow values of IFT are maintained and trapping of oil is reduced.

A key feature of the salinity graident scheme is a self-regulating mechanism which acts to keep much of the surfactant near the optimal salinity. If surfactant moves ahead into regions where salinity exceeds the optimal, it partitions into the oil and forms an upper phase microemulsion. As tensions are not ultralow under these conditions, the microemulsion is trapped, and the surfactant remains immobile until salinity drops to near the optimal value. If the surfactant falls behind into regions where the salinity is below optimal, it forms a lower phase microemulsion, which has a relatively high mobility since there is little trapped oil to impede its motion. Consequently, the surfactant tends to advance toward the region where the salinity is near-optimal.

Other arguments for the use of a salinity gradient are as follows:

1. The high salinity present at initial contact between surfactant and oil promotes swelling of the oil and reconnection of isolated ganglia to form an oil bank, a phenomenon discussed previously.
2. In most cases changes which occur in the reservoir reduce the optimal salinity below that corresponding to the initial slug composition. One such change is the increase in divalent ion content as a result of ion exchange processes. Another is the decrease in overall surfactant concentration as a result of adsorption and other retention processes. For many petroleum sulfonate formulations a decrease in overall concentration is accompanied by a decrease in the fraction of water-soluble species which adsorb in the surfactant films, and hence a decrease in the optimal salinity. Thus, a process in which the salinity of the injected fluids decreases with time should allow actual salinity to match the optimal somewhere in the reservoir at all times.

Oil displacement experiments in laboratory cores confirm that the salinity gradient approach can yield higher recovery efficiencies than simple injection of all fluids at some constant salinity different from that of the reservoir brine [49,92,171]. The basic problem with many constant-salinity processes is that once conditions are reached where the optimal salinity of the surfactant drops below the fixed salinity, not only does the IFT increase but the surfactant is stranded in the trapped ganglia of upper phase microemulsion which form. When combined with adsorption losses, the result is rapid depletion of the injected surfactant.

Success of the salinity gradient scheme depends on there always being some region within the reservoir where local salinity and local optimal salinity are nearly equal. This criterion can be met if the process design provides for maintenance throughout the process of a salinity higher than the local optimal near the leading edge of the surfactant slug and a salinity less than the local optimal near the trailing edge. If necessary conditions on fluid mobilities can also be satisfied, as discussed below, the salinity gradient method has great promise as a means of enabling surfactant floods to achieve high recovery efficiencies.

VII. MOBILITY CONSIDERATIONS

All three of the above schemes seek to maintain ultralow IFTs during chemical flooding processes employing surfactants. Achievement of this goal is a necessary condition for high oil recoveries but not a sufficient one. The mobilities of the various phases formed must also have suitable values. For example, when a salinity gradient is present, the oil in the zone where ultralow IFTs are achieved must advance at least as rapidly as the low-IFT zone itself to avoid being trapped as residual oil. In all processes surfactant-containing phases of very low mobility should be avoided because they virtually prevent advance of the surfactant through the reservoir.

More generally, the mobilities of the various injected fluids should be chosen to prevent the development of frontal instabilities which would severely reduce the fraction of the reservoir actually contacted by surfactant. In practice, the composition of the surfactant slug is adjusted to make its mobility less than the effective mobility of the oil bank which develops during the process.

Slugs of surfactant-alcohol-brine mixtures frequently contain liquid crystalline materials, as indicated previously. In this case composition changes can be used to modify the liquid crystalline structure and hence adjust the apparent viscosity. Experiments show that, whatever compositional variable is changed, apparent viscosity always passes through a maximum during transformation from a dispersion of liquid crystalline particles in a continuous aqueous medium to a single lamellar liquid crystalline phase [33]. Rheological behavior of both the dispersions and the liquid crystals is complex and rather sensitive to composition changes near the viscosity maximum. The apparent viscosity is a function of both shear rate and the duration of shearing.

In some cases polymer is added to the surfactant slug to make it more viscous and hence less mobile. The slug itself is followed by a polymer solution with yet a lower mobility. As indicated previously, the polymer used for mobility control purposes influences both phase behavior and the mobility of the phases formed. One advantage claimed for the salinity gradient process is that the salinity of the polymer slug is low enough to prevent undesirable phase separation when it mixes with surfactant.

Such phase separation may be unavoidable when the alternative approach is taken of injecting a surfactant which is effective in reservoir brine. Reed and Carpenter [173] suggested that with proper design the phase separation need not be a disadvantage. They propose injection of an emulsion consisting of small drops (about 0.1 μm in diameter) of aqueous polymer solution in a continuous microemulsion phase. Both phases are formulated with reservoir brine, which is at the optimal salinity for the surfactant used. The IFT between the phases is very low so that the emulsion travels intact through the reservoir without either phase being trapped. The mobility of the emulsion is low enough to prevent frontal instabilities from developing.

VIII. NONEQUILIBRIUM PHENOMENA

Because equilibrium phase behavior of oil-brine-surfactant systems plays a central role in design of chemical flooding processes, it has been discussed extensively above. In most analyses of chemical flooding it is assumed that fluid of a given overall composition present at a particular location and time instantaneously separates into its equilibrium phases, which then advance at rates depending on their respective mobilities. One justification for this approach is that the time required for convection of liquids through a typical pore generally exceeds the diffusion time. For instance, with a pore size a of 10 μm, a diffusivity D of 10^{-5} cm^2/sec, and a velocity v in the pores of 4 ft/day, corresponding to a superficial velocity $u = v\phi$ of 1 ft/day for a porosity ϕ of 0.25, convection time (a/v) is about 0.7 sec while diffusion time (a^2/D) is about 0.1 sec. Pore size is taken as the characteristic length because the truncated filaments making up oil ganglia have approximately this diameter. Equilibration of such ganglia with the injected surfactant solution may be considered as a typical dynamic process of interest.

Diffusion coefficients of surfactants in liquid crystals and microemulsions may be considerably less than the above value, however. Values in the range of 10^{-6} cm^2/sec have been measured in microemulsion phases by NMR techniques [174]. Moreover, rate processes other than diffusion may be important, e.g., the kinetics of phase formation and dissolution, or of coalescence of small droplets. Such effects are especially likely to be slow, and hence important, when liquid crystalline phases are present. Finally, some "dead-end" pores where there is no flow and transport occurs entirely by diffusion are thought to exist in most reservoirs. All these factors act to slow equilibration and hence to alter the above conclusion that local equilibrium is maintained during chemical flooding processes.

Of course, it is also possible that interfacial phenomena such as the Marangoni effect could speed up equilibration by inducing local flows which enhance mass transfer rates. Adsorbed surfactants normally strongly oppose development of Marangoni flow, but when adsorption and desorption are rapid, it may be that such flows can sometimes develop. Lam et al. [175] showed that in the absence of surfactants Marangoni flow can reduce the value of the capillary number N_{ca} required for mobilization of oil ganglia.

Hirasaki [176] developed a more detailed view of the equilibration process, taking into account the effect of hydrodynamic dispersion on concentration distributions within the reservoir. Suppose that a step change in injected concentration is carried out, e.g., initiation of surfactant injection. Hirasaki's calculations show that nonequilibrium effects can be of significance near the injection well, where concentration gradients are steep and hence concentration at a given location changes rapidly with time. But far from the injection well, where dispersion has caused substantial reduction in the concentration gradients, arrival of the surfactant is accompanied by little deviation from equilibrium.

Wasan and co-workers [177] investigated the process of oil bank formation. They showed that oil recovery decreases when oil bank formation is hindered by slow coalescence of oil drops.

Arnold [178] and Hirasaki [176] discussed the behavior expected when oil and an aqueous surfactant solution are brought into contact in the absence of flow. Although they obtained useful insights on the formation of intermediate phases and the occurrence of spontaneous emulsification, their results are incomplete because they did not consider full implications of the presence of liquid crystalline phases. These commonly occur at low oil contents in surfactant systems of interest for chemical flooding.

Benton et al. [179] and Raney et al. [180] considered the effect of liquid crystals and, in addition, observed contacting behavior with optical microscopy. A novel vertical cell arrangement and video techniques were used in the latter work. This scheme allowed the intermediate phases formed to be clearly seen and their growth rates measured. It also provided an excellent view of spontaneous emulsification and of certain instabilities which arose in cases where the intermediate phases were denser than the initial phases underlying them. It was found that the number and type of intermediate phases formed at various salinities and the occurrence or absence of spontaneous emulsification could be understood in terms of system phase behavior and of the "diffusion path" approach, which has proved useful in predicting conditions for spontaneous emulsification in simpler systems [181].

The implications for oil recovery of the rather complicated equilibration procedures observed by Raney et al. [180] are not presently clear. Such dynamic phenomena as mobilization, trapping, and breakup of oil ganglia may well be influenced by intermediate phase formation or by local flow near interfaces. These matters merit further investigation.

It should be emphasized that the nonequilibrium effects discussed in this section have not yet been proved to influence oil recovery in surfactant flooding processes. That is, assuming instantaneous establishment of phase equilibrium has not yet been shown to be inadequate for describing the results of laboratory core floods. On the other hand, incomplete understanding of such factors as surfactant retention and multiphase flow at high capillary numbers makes it difficult to determine whether differences between model predictions and experimental results stem from these shortcomings or from nonequilibrium effects.

IX. ADSORPTION AND PRECIPITATION

Surfactant loss in the reservoir due to adsorption, precipitation, and formation of surfactant-rich immobile phases is the primary determinant of the total amount of surfactant required in a chemical flooding process. Thus, these phenomena have a major influence on process viability. Nevertheless, they have begun to be understood only in recent years. Because phase behavior has been discussed previously, only adsorption and precipitation are considered in this section.

Figure 17 [182] shows adsorption isotherms of some isomerically pure sulfonate surfactants on kaolinite. Similar results were found for alumina. These two solids were chosen because of their availability in pure form and because they were expected to provide some insight regarding adsorption on clay minerals. Previous studies had indicated that most adsorption in a reservoir does occur on clays [18]. Hence, surfactant requirements are larger for sandstones with high clay contents.

Several features of Fig. 17 are of interest. One is that the amount of adsorption is constant for surfactant concentrations above the critical micelle concentration, an indication that micelles themselves do not adsorb. Another is that surfactants with longer chains adsorb more readily. The reason is apparently the favorable free energy change of transferring hydrocarbon chains from water to the surface, as shown in the top diagram of Fig. 18 [182]. This molecular configuration has been proposed for the dilute (linear) portion of the adsorption isotherms of Fig. 17 [182].

At somewhat higher concentrations isotherm slope increases. The middle diagram of Fig. 18 shows that in this region the hydrocarbon

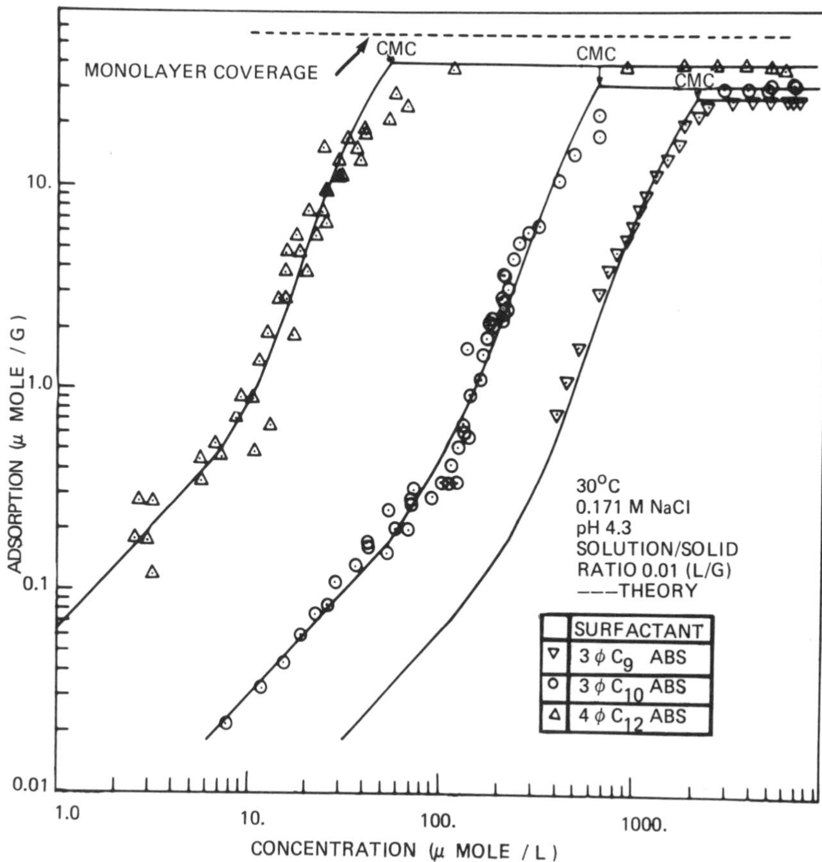

FIG. 17 Effect of alkyl chain length on adsorption of pure sulfonate surfactants on kaolinite. The nϕ in the surfactant designation indicates that the benzene ring is attached to the nth segment of the chain. (From [152] with permission.)

REGION I

REGION II

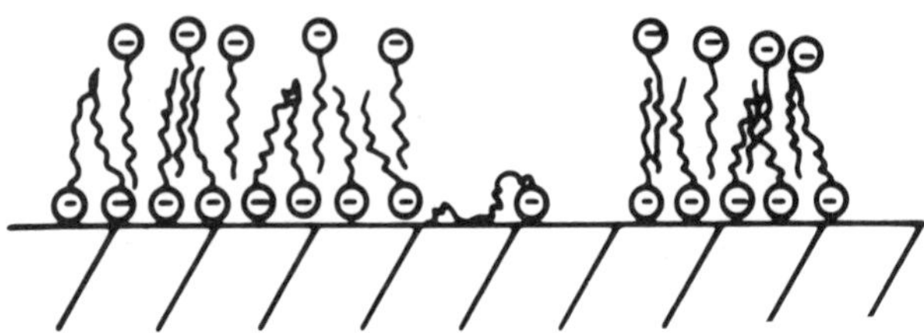

REGION III AND PLATEAU ADSORPTION REGION

FIG. 18 Configuration of adsorbed anionic surfactant molecules. (From [182] with permission.)

chains of many adsorbed molecules are believed to leave the surface and aggregate to form "hemimicelles." Finally, the isotherms bend over near the CMC, and the adsorbed film becomes a bilayer such as that in the bottom diagram of Fig. 18. A bilayer is a reasonable final state for the adsorbed film since polar groups are in contact with the bulk aqueous phase, whereas hydrocarbon chains would be in contact with the aqueous phase for an adsorbed monolayer.

The isotherms of Fig. 17 are for a constant pH of 4.3 and a constant salinity of 1% NaCl in the aqueous solution. Experiments show that adsorption of anionic surfactants increases substantially with both decreasing pH and increasing salinity [183]. The former effect is due to the change in surface charge of the clays to more positive values with decreasing pH. Clearly, surfactant floods at low pH should be avoided. Instead, the pH should be adjusted to neutral or preferably even basic values.

Recently, a theory was proposed to describe adsorption of isomerically pure anionic surfactants [184]. It employs a pseudophase separation model similar to that sometimes used to model micelle formation. That is, it considers the transition from individual adsorbed molecules to a bilayer to occur at a critical surfactant concentration at each point on the surface. The model includes hydrophobic and electrostatic effects on pseudophase formation and takes into account surface heterogeneity.

Petroleum sulfonates are, of course, complex mixtures of surfactants. Their adsorption isotherms often exhibit maxima (see Fig. 19) [29,185]. Trogus et al. [186] showed that with a mixture of surfactants a maximum in adsorption is theoretically possible because the concentrations of the various individual surfactant molecules in solution continue to change beyond the CMC. Somasundaran and Hanna [187] reported that whether or not a maximum exists depends on the composition of the inorganic electrolyte present. They

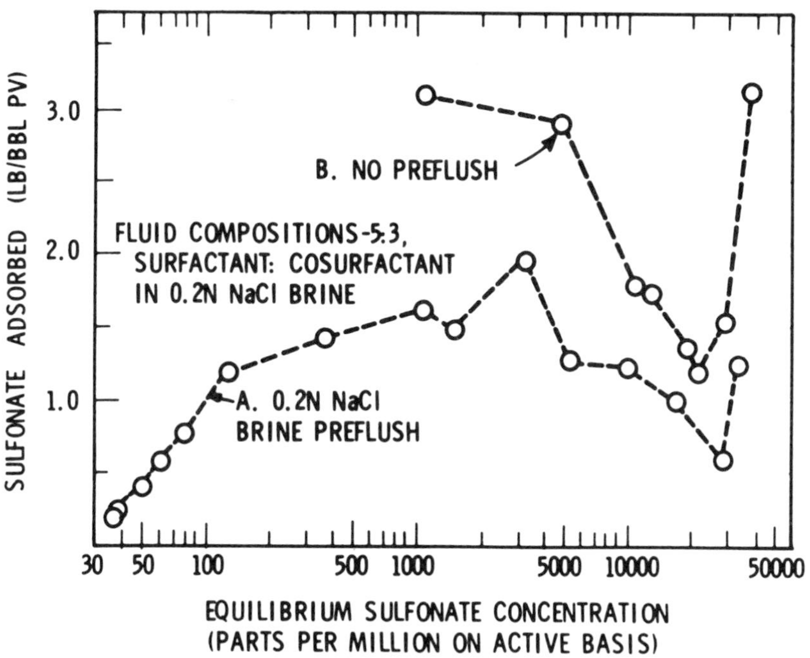

FIG. 19 Adsorption of a petroleum sulfonate on Berea sandstone at 110°F. (From [29] with permission.)

suggested that effects of water structure and of electrostatics may be involved.

Bae and Petrick [188] found that surfactant losses observed during flow of petroleum sulfonate solutions through sandstone cores were in some cases dependent on flow rate. They also noted rapid changes in certain physical properties of the solutions near the concentration of the adsorption maximum. Apparently these changes were associated with the presence of liquid crystalline material.

The results described so far are for oil-free surfactant solutions. Of course, oil is present as well during actual recovery processes. A study of adsorption of sodium dodecylbenzenesulfonate on alumina in the presence of *n*-dodecane indicated that surfactant adsorption sometimes exceeds oil-free values at low surfactant concentrations, perhaps because adsorbed oil favors hemimicelle formation [189]. The plateau value of adsroption above the CMC did not change appreciably, however. Some oil recovery tests in sandstone cores also showed little effect on total surfactant losses of the relative amounts of oil and brine present [190].

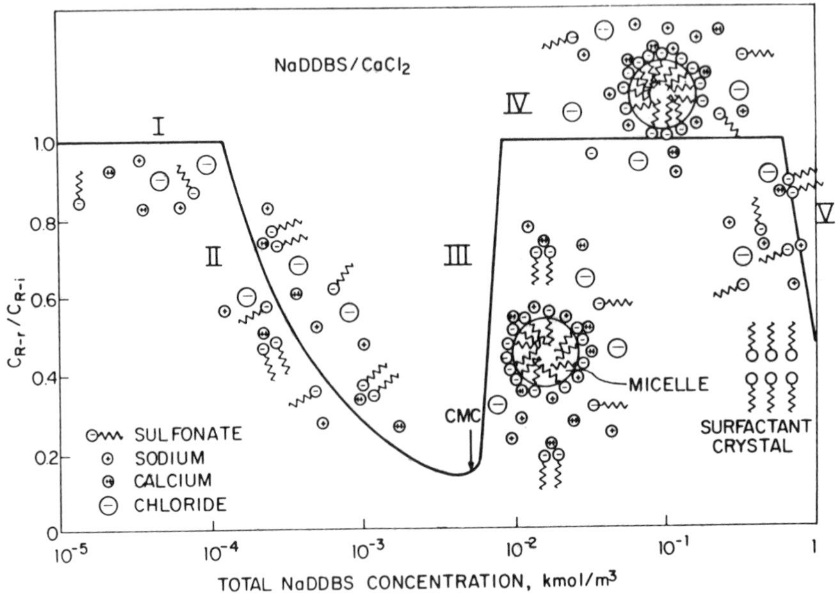

FIG. 20 Schematic diagram of precipitation-redissolution phenomena for anionic surfactants. (From [191] with permission.)

Surfactant precipitation by divalent cations can also be a source of surfactant losses. Figure 20 [191] illustrates the behavior that occurs when sodium dodecylbenzenesulfonate is added to an aqueous solution containing $CaCl_2$. A value of the ordinate equal to unity indicates that no precipitate is present. For values below unity a precipitate of calcium sulfonate is present, with smaller values of the ordinate corresponding to larger quantities of precipitate.

In region I all species are in solution as individual ions. When the solubility product for calcium sulfonate is reached, precipitation begins. It continues through region II until the CMC of the surfactant is reached. When micelles are present, the calcium ions prefer to adsorb at the micelle surfaces instead of remaining in the precipitate. As a result, the precipitate redissolves in region III. All the calcium sulfonate has redissolved, and a micellar solution exists in region IV. Finally, the micellar solution may become so concentrated that a crystalline or liquid crystalline phase forms as indicated in region V of Fig. 20.

As discussed previously, surfactant-brine mixtures frequently contain lamellar liquid crystalline material in the composition range proposed for injection. Presumably, divalent cations adsorb on the surfaces of the surfactant bilayers in the liquid crystalline phases just as they do on micelles. In at least some cases, however, it is possible to have separate precipitate and liquid crystalline particles dispersed in an aqueous phase [33]. When no separate aqueous phase is present, precipitate particles diffuse and collect along defects in the liquid crystalline structure [33]. When sufficient surfactant is added, the precipitate redissolves as described above for micellar solutions [192].

Two other points should be noted regarding calcium sulfonates. First, they are rather soluble in oil and hence can be transported with the oil through the reservoir or trapped in residual oil ganglia [193]. The second point is that surfactant concentrations corresponding to the adsorption and precipitation maxima are sometimes nearly the same, an indication that part of the "adsorption" losses reported in some studies with divalent cations present may actually be the result of precipitation [194].

One way of reducing surfactant adsorption and precipitation is to add relatively cheap "sacrificial agents" to the surfactant slug and/or to the preflush which normally precedes it [14,195]. These materials may occupy some sites on the surface and block surfactant adsorption, or they may form complexes with divalent ions and reduce precipitation. Examples of sacrificial agents are sodium carbonate, EDTA (ethylenediaminetetraacetic acid), sodium tripolyphosphate, and lignosulfonates. A recent report indicates that use of lignosulfonates in the preflush can reduce adsorption losses of petroleum sulfonates by about 50% [196].

X. PROCESS MODELING

A detailed discussion of numerical schemes for simulation of surfactant flooding processes is beyond the scope of this chapter. Nevertheless, a few remarks should be made, because development of satisfactory simulators is vital if surfactant flooding is to be widely applied.

The sequence of steps typically followed in designing a surfactant flooding process will be described in general terms. Early in the design process, data on phase behavior and IFT as a function of salinity are collected for various promising formulations. For those which seem most attractive based on these data, oil displacement experiments are carried out in linear or radial cores of Berea sandstone. Formulations which exhibit the best oil recovery are further tested in cores of the reservoir rock. Further data may also be collected, e.g., on viscosities of slug and microemulsion phases and on surfactant retention effects. Once the best formulation is chosen, it is used in a pilot test in the field involving a few wells. Finally, if the pilot is successful and if fieldwide application of the process is economically attractive, the process is expanded to commercial scale, incorporating any changes in process design shown by the pilot to be desirable.

Simulation is used during the design process in conjunction with data from the various experiments mentioned above. It is generally impractical to run core tests investigating the entire range of interest of all pertinent process variables. If a simulator which adequately describes the core tests actually performed is available, it can be used to predict performance for other conditions. Moreover, a simulator is required to apply linear core test results to the design of the pilot test, where several factors such as well patterns and reservoir heterogeneities must be considered. Finally, results of the pilot test are used to adjust the reservoir description parameters of the simulator before using it to establish final process characteristics for the commercial-scale project.

Generally speaking, two approaches to mathematical description of chemical flooding processes have been taken. The simpler approach involves use of the method of characteristics to solve the governing equations [197,198]. With this scheme diffusional effects such as hydrodynamic dispersion cannot be included. Incorporation of certain other effects such as three-phase flow is possible in principle. However, it has not yet been implemented owing to the substantial increase in the effort required for the solution. Hence, the chief value of this approach is to illuminate the main physical mechanisms which influence process performance, and it has proved to be most useful in this respect. Beneficial in the initial stages of process design, it is not adequate as a final design tool.

The second approach consists of direct numerical solution of the governing equations using finite-difference techniques [199–203].

Given unlimited computer time, complete understanding of all physical phenomena involved, sufficient information on physical properties of the phases which occur and on flow properties such as relative permeabilities, and detailed knowledge of the reservoir, this method is capable of accurately predicting oil recovery. Of course, none of these conditions is satisfied in practice. A very general formulation of the pertinent equations has been presented [204]. While some specific phenomena such as gravity override have been investigated by using two-dimensional simulators [205], most published numerical results to date have been limited to one-dimensional problems with inclusion of up to seven components and three flowing phases. Local equilibrium between phases is assumed, capillary pressure effects are neglected, but surfactant adsorption and ion exchange effects are included. In some simulators hydrodynamic dispersion is modeled explicitly, while in others its effect is presumed to be represented adequately by numerical dispersion inherent in the method of solution.

It is important to recognize the limitations on the reliability of simulation results imposed by the lack of adequate input information. Two-phase flow and three-phase flow at high capillary numbers, where at least some discrete ganglia are mobile, are not well understood. Nor is hydrodynamic dispersion in multiphase flow. Furthermore, many parameters are required to specify phase behavior, interfacial tensions, and fluid viscosities. Even Pope and Nelson's initial simulator [200] with only six components and no dispersion coefficients required specification of 64 parameters.

With all their limitations, however, existing simulators have been used to describe many key features of surfactant flooding. The importance of operating in the middle phase region where ultralow IFTs exist has been confirmed. In addition, simulators have shown that the actual slopes of the tie lines are more important when operating in the upper than in the lower phase region [201]. The value of the salinity gradient concept has also been confirmed by simulation [172]. Furthermore, it was demonstrated that the salinity of the surfactant slug itself is less important than those of the preflush and polymer solution [201]. A sensitivity study has indicated that the total quantity of surfactant injected has a stronger influence on oil recovery than does surfactant concentration in the slug [201]. The transport of calcium sulfonates in the oil phase as discussed in Sec. IX has been successfully modeled [206]. And simulation has been used to predict recovery achieved by two quite different processes in a given reservoir and thus enable a choice to be made between available options [203]. Such information is obviously of great value and illustrates the potential of simulation to identify aspects of process operation which might not be evident and to provide answers to questions more quickly and cheaply than could an experimental program.

XI. APPLICATION OF CHEMICAL FLOODING

Although it is beyond the scope of this chapter to discuss field experience with chemical flooding and process economics in detail, a few remarks are useful to provide perspective. Initial field experiments employing ultralow IFTs and involving only small areas and few wells were generally unsuccessful [4,207,208]. This is not a surprising result in view of the process complexity discussed above. Marathon conducted a series of tests in one field in Illinois, employing continuing process improvements in successive tests. At present the only commercial-scale surfactant flood yet undertaken is in progress in this field, jointly financed by Marathon and the U.S. government. Based on the results to date, it seems likely to be a technical success [209].

Numerous other field tests were initiated during the past decade as the rapid increase in oil prices between 1974 and 1981 provided hope of economic viability. Among other things, these tests have demonstrated that an accurate reservoir description is essential for proper design of a chemical flood. Otherwise oil may be displaced from one location to another in the reservoir but never recovered, owing to flow patterns different from those anticipated. The effects of reservoir heterogeneity and of rock mineral content on process performance have also been demonstrated. For instance, the extreme difficulty of achieving an adequate low-salinity preflush, in spite of heterogeneity and cation exchange between process fluids and clay minerals in the formation, is now more widely appreciated than in the past [207].

From the economic point of view, surfactant flooding has the inherent disadvantage that a large investment in chemicals is required early in the process. Moreover, no significant oil production can be expected until at least a year or two after chemical injection because of the low reservoir velocities mentioned earlier. Several more years are required to produce all of the additional oil.

Gogarty [3] presented the results of an economic study of surfactant flooding. He found that the rate of return is particularly sensitive to oil price, the amount of residual oil actually present in the formation (a quantity that is very difficult to determine accurately), and the number of wells per unit area. In most cases it is advantageous to drill additional wells beyond those used during water-flooding in order to decrease well spacing and hence reduce the time required for recovery of the displaced oil. Even with an optimized process, however, the absence of privately financed commercial projects indicates that surfactant flooding is not yet economically attractive, especially with oil prices below $20 per barrel as at the time this article was completed.

Nor is surfactant flooding yet a reliable process from a technical point of view. As indicated previously, actual behavior in the reservoir is very complicated, with multiphase flow, mass transfer, adsorption, and ion exchange all taking place simultaneously in a heterogeneous porous medium. Progress has been made in devising schemes for maintaining ultralow IFTs throughout the process, but the problem has not yet been solved. In addition, further research is needed on the application of surfactants and polymers under the adverse conditions present in many reservoirs, i.e., high temperatures, high salinities, and high divalent cation contents.

Nevertheless, there are encouraging factors as well. Results of some recent field tests have been favorable [210,211]. And as evident from this chapter, fundamental understanding of oil-brine-surfactant systems has vastly improved during the past decade. Because chemical flooding represents a major advance in technology over conventional recovery methods, time is needed to develop improved techniques for reservoir description, improved chemicals, new fundamental knowledge pertinent to process behavior, and more sophisticated simulation techniques required to design successful processes. With continued effort, further progress leading to better process performance and reliability should follow.

ACKNOWLEDGMENT

Although opinions expressed herein are those of the authors, many useful suggestions and comments were provided by J. H. Bae, S. P. Gupta, G. J. Hirasaki, and R. L. Reed. The authors also thank Ms. Valerie Del Torto for her assistance in typing the manuscript.

REFERENCES

1. C. W. Perry, *J. Pet. Technol. 33*: 2033 (1981).
2. R. L. Reed and R. N. Healy, in *Improved Oil Recovery by Surfactant and Polymer Flooding*, D. O. Shah and R. S. Schechter, eds., Academic Press, New York, 1977, p. 383.
3. W. B. Gogarty, *J. Pet. Technol. 30*: 1089 (1978).
4. W. B. Gogarty, *J. Pet. Technol. 35*: 1168, 1581 (1983).
5. C. C. Mattax, R. J. Blackwell, and J. F. Tomich, Recent Advances in Surfactant Flooding, preprint for the 11th World Petroleum Congress, paper, RTD2(3), 1984.
6. V. K. Bansal and D. O. Shah, in *Microemulsions*, L. M. Prince, ed., Academic Press, New York, 1977, p. 149.
7. V. K. Bansal and D. O. Shah, in *Micellization, Solubilization, and Microemulsions*, K. L. Mittal, ed., Plenum, New York, 1977, vol. 2, p. 87.
8. V. K. Bansal and D. O. Shah, in *Recent Developments in Separation Science*, N. N. Li, ed., CRC Press, Cleveland, 1981, vol. 7, p. 151.
9. A. N. Sunder Ram and D. O. Shah, in *Emulsions and Emulsion Technology*, K. J. Lissant, ed., Dekker, New York, 1984, vol. 3, p. 139.

10. L. W. Lake, in *Enhanced Oil Recovery for the Independent Producer*, Southern Methodist University, Dallas, 1983, p. 102.
11. G. Stegemeier, in *Improved Oil Recovery by Surfactant and Polymer Flooding*, D. O. Shah and R. S. Schechter, eds., Academic Press, New York, 1977.
12. J. C. Melrose and C. F. Brandner, *Can. J. Pet. Technol. 13*: 54 (1974).
13. J. J. Taber, in *Surface Phenomena in Enhanced Oil Recovery*, D. O. Shah, ed., Plenum, New York, 1981, p. 13.
14. W. R. Foster, *J. Pet. Technol. 25*: 205 (1973).
15. R. G. Larson, L. E. Scriven, and H. T. Davis, *Nature 268*: 409 (1977).
16. R. G. Larson, L. E. Scriven, and H. T. Davis, *Chem. Eng. Sci. 36*: 57, 75 (1981).
17. H. J. Hill, J. Reisberg, and G. L. Stegemeier, *J. Pet. Technol. 25*: 186 (1973).
18. W. W. Gale and E. I. Sandvik, *Soc. Pet. Eng. J. 13*: 191 (1973).
19. R. L. Chuoke, P. Van Meurs, and C. Van der Poel, *Trans. AIME 216*: 188 (1959).
20. P. G. Saffman and G. I. Taylor, *Proc. R. Soc. London Ser. A 245*: 312 (1958).
21. F. F. Craig, The Reservoir Engineering Aspects of Waterflooding, presented to the Society of Petroleum Engineering, AIME, Dallas, 1971.
22. R. G. Larson, H. T. Davis, and L. E. Scriven, *J. Pet. Technol. 34*: 243 (1982).
23. W. W. Owens and D. L. Archer, *J. Pet Technol. 23*: 873 (1971).
24. R. Ehrlich, H. H. Hasiba, and P. Raimondi, *J. Pet. Technol. 26*: 1335 (1974).
25. C. G. Inks and R. I. Lahring, *J. Pet Technol. 20*: 1320 (1980).
26. C. E. Johnson, Jr., *J. Pet. Technol. 28*: 85 (1976).
27. W. B. Gogarty and W. C. Tosch, *J. Pet. Technol. 20*: 1407 (1968).
28. L. W. Holm, *J. Pet. Technol. 25*: 1475 (1971).
29. S. P. Trushenski, D. L. Dauben, and D. R. Parrish, *Soc. Pet. Eng. J. 14*: 633 (1974).
30. W. J. Benton, C. A. Miller, and T. Fort, Jr., Structure of Aqueous Solutions of Petroleum Sulfonates, SPE Preprint 7579, presented at the Annual Fall Meeting, Houston, 1978.
31. W. J. Benton and C. A. Miller, *J. Phys. Chem. 87*: 4981 (1983).
32. C. A. Miller, S. Mukherjee, W. J. Benton, J. Natoli, S. Qutubuddin, and T. Fort, Jr., in *Interfacial Phenomena in Enhanced Oil Recovery*, D. Wasan and A. Payatakes, eds., AIChE Symp. Ser. 212, 1982, vol. 78.

33. W. J. Benton, S. K. Baijal, O. Ghosh, S. Qutubuddin, and C. A. Miller, Viscosity and Phase Behavior of Petroleum Sulfonate Solutions in the Liquid Crystalline Region With and Without Small Amounts of Added Hydrocarbons, SPE/DOE Preprint 12700, presented at the Symposium on Enhanced Oil Recovery, Tulsa, 1984.
34. A. C. Uzoigive, F. C. Scanlon, and R. L. Jewett, *J. Pet. Technol. 26*: 33 (1974).
35. E. L. Claridge, *Soc. Pet. Eng. J. 18*: 315 (1978).
36. W. T. Adams and V. H. Schievelbein, Surfactant Flooding Carbonate Reservoirs, SPE/DOE Preprint 12686, presented at the Fourth Symposium on Enhanced Oil Recovery, Tulsa, 1984.
37. R. N. Healy and R. L. Reed, *Soc. Pet. Eng. J. 14*: 491 (1974).
38. K. Shinoda and S. Friberg, *Adv. Colloid Interface Sci. 4*: 281 (1975).
39. S. L. Holt, *J. Dispersion Sci. Technol. 1*(4): 432 (1980).
40. R. A. Mackay, *Adv. Colloid Interface Sci. 15*: 131 (1981).
41. P. G. De Gennes and C. Taupin, *J. Phys. Chem. 86*: 2294 (1982).
42. I. D. Robb, ed., *Microemulsions*, Plenum, New York, 1982.
43. D. Wasan and A. Payatakes, eds., *Interfacial Phenomena in Enhanced Oil Recovery*, AIChE Symp. Ser. No. 212, 1982, vol. 78.
44. R. N. Healy, R. L. Reed, and D. G. Stenmark, *Soc. Pet. Eng. J. 16*: 147 (1976); *Trans. AIME 261*: 147 (1976).
45. R. N. Healy and R. L. Reed, *Soc. Pet. Eng. J. 17*: 129 (1977).
46. S. P. Gupta and S. P. Trushenski, *Soc. Pet. Eng. J. 19*: 116 (1979).
47. J. L. Salager, E. Vasquez, J. C. Morgan, R. S. Schechter, and W. H. Wade, *Soc. Pet. Eng. J. 19*: 107 (1979).
48. P. A. Winsor, *Solvent Properties of Amphiphilic Compounds*, Butterworths, London, 1954.
49. R. C. Nelson and G. A. Pope, *Soc. Pet. Eng. J. 18*: 325 (1978).
50. B. M. Knickerbocker, C. V. Pesheck, H. T. Davis, and L. E. Scriven, *J. Phys. Chem. 86*: 393 (1982).
51. A. M. Bellocq, J. Biais, B. Clin, A. Gelot, P. Lalanne, and B. Lamanceau, *J. Colloid Interface Sci. 74*: 311 (1980).
52. K. E. Bennett, C. H. Phelps, H. T. Davis, and L. E. Scriven, *Soc. Pet. Eng. J. 21*: 747 (1981).
53. W. J. Benton, J. Natoli, S. Qutubuddin, S. Mukherjee, C. A. Miller, and T. Fort, Jr., *Soc. Pet. Eng. J. 22*: 53 (1982).
54. J. Van Neiuwkoop and G. Snoei, Phase Behavior and Structure of a Pure-Component Microemulsion System, *Proc. Second European Symp. Enhanced Oil Recovery*, Paris, (1982).

55. J. Natoli, Ph.D. dissertation, Carnegie-Mellon Univ., Pittsburgh, Pa., (1980).
56. S. Qutubuddin, Ph.D. dissertation, Carnegie-Mellon Univ., Pittsburgh, Pa. (1983).
57. P. F. Wilson and C. F. Brandner, *J. Colloid Interface Sci.* *60*:473 (1977).
58. J. L. Cayias, R. S. Schechter, and W. H. Wade, *ACS Symp. Ser.* *8*: 234 (1975).
59. A. Pouchelon, J. Meunier, D. Langevin, and A. M. Cazabat, *J. Phys. Lett.* *41*: 239 (1980).
60. M. W. Kim, J. S. Huang, and J. Bock, *Soc. Pet. Eng. J.* *24*: 203 (1984).
61. M. L. Robbins, in *Micellization, Solubilization and Microemulsions* (K. L. Mittal, ed.), Plenum, New York, 1977, vol. 2.
62. V. K. Bansal and D. O. Shah, *Soc. Pet. Eng. J.* *18*: 167 (1978).
63. W. C. Hseih and D. O. Shah, The Effect of Chain Length of Oil and Alcohol as Well as Surfactant to Alcohol Ratio on the Solubilization Phase Behavior and Interfacial Tension of Oil/Brine/Surfactant/Alcohol Systems, SPE Preprint 6594, presented at the International Symposium on Oilfield and Geothermal Chemistry, La Jolla, Calif., June 1977.
64. P. K. Shankar, J. H. Bae, R. M. Erick, Salinity Tolerance and Solution Property Correlations of Petroleum Sulfonate-Cosurfactant Blends, SPE Preprint 10600, presented at the International Symposium on Oilfield and Geothermal Chamistry, Dallas, 1982.
65. C. Huh, *J. Colloid Interface Sci.* *71*: 408 (1979).
66. C. Huh, *Soc. Pet. Eng. J.* *23*: 829 (1983).
67. A. M. Bellocq, J. Biais, P. Bothorel, D. Bourbon, B. Clin, P. Lalanne, and B. Lamanceau, in *Microemulsions*, I. D. Robb, ed., Plenum, New York, 1982, p. 131.
68. B. Widom, *J. Chem. Phys.* *62*: 1332 (1975).
69. A. Pouchelon, J. Meunier, D. Langevin, D. Chatenay, and A. M. Cazabat, *Chem. Phys. Lett.* *76*: 277 (1980).
70. J. H. Bae and C. B. Petrick, *Soc. Pet. Eng. J.* *21*: 573 (1981).
71. W. J. Benton, R. Hwan, C. A. Miller, and T. Fort, Jr., Structure of Solution of Synthetic Petroleum Sulfonates Before and After Addition of Oil, Third ERDA Symposium on Enhanced Oil and Gas Recovery and Improved Driving Methods, Tulsa, 1977, vol. 1, paper A-9.
72. R. Hwan, C. A. Miller, and T. Fort, Jr., *J. Colloid Interface Sci.* *68*: 221 (1979).
73. S. Qutubuddin, C. A. Miller, G. Berry, T. Fort, Jr., and A. Hussam, in *Surfactants in Solution*, K. L. Mittal and B. Lindman, eds., Plenum, New York, 1984, vol. 3, p. 1693.

74. C. A. Miller, R. Hwan, W. J. Benton, and T. Fort, Jr., *J. Colloid Interface Sci. 61*: 554 (1977).
75. J. W. Cahn and J. C. Hilliard, *J. Chem. Phys. 28*: 258 (1958).
76. J. S. Huang and M. W. Kim, *Soc. Pet. Eng. J. 24*: 197 (1984).
77. P. D. Fleming III and J. E. Vinatieri, *J. Chem. Phys. 66*: 3147 (1977).
78. P. D. Fleming III and J. E. Vinatieri, *AIChE J. 24*: 493 (1979).
79. A. M. Cazabat and D. Langevin, *J. Chem. Phys. 74*: 3148 (1981).
80. A. M. Cazabat, D. Langevin, J. Meunier, and A. Pouchelon, *Adv. Colloid Interface Sci. 16*: 175 (1982).
81. A. M. Cazabat, D. Chatenay, P. Guering, D. Langevin, J. Meunier, O. Sorba, J. Lang, R. Zana, and M. Paillette, in *Surfactants in Solution*, Vol. 3, K. L. Mittal and B. Lindman, eds., Plenum, New York, 1984, p. 1737.
82. R. Dorshaw and D. F. Nicoli, in *Measurement of Suspended Particles by Quasi-Elastic Light Scattering*, B. Dahneke, ed., Wiley, New York, 1983, p. 529.
83. Y. Talmon and S. Prager, *J. Chem. Phys. 69*: 2984 (1978).
84. L. E. Scriven, *Nature 267*: 333 (1976).
85. J. Jouffroy, P. Levinson, and P. de Gennes, *J. Phys. 43*: 1241 (1982).
86. B. Widom, *J. Chem. Phys. 81*: 1030 (1984).
87. M. Kahlweit and E. Lessner, in *Surfactants in Solution*, K. L. Mittal and B. Lindman, eds., Plenum, New York, 1984, vol. 1, p. 23.
88. B. M. Knickerbocker, C. V. Presheck, L. E. Scriven, and H. T. Davis, *J. Phys. Chem. 83*: 1984 (1979).
89. C. J. Glover, M. C. Puerto, J. M. Maerker, and E. L. Sandvik, *Soc. Pet. Eng. J. 19*: 183 (1979).
90. G. J. Hirasaki, *Soc. Pet. Eng. J. 22*: 181 (1982).
91. S. P. Gupta, *Soc. Pet. Eng. J. 24*: 38 (1984).
92. R. C. Nelson, in *Surface Phenomena in Enhanced Oil Recovery*, D. O. Shah, ed., Plenum, New York, 1981, p. 73.
93. M. C. Puerto and R. L. Reed, *Soc. Pet. Eng. J. 23*: 669 (1983).
94. Y. Barakat, L. N. Fortney, C. Lahanne-Casson, R. S. Schechter, W. H. Wade, U. Weerasoorija, and S. Yiv, *Soc. Pet. Eng. J. 23*: 913 (1983).
95. Y. Barakat, L. N. Fortney, R. S. Schechter, W. H. Wade, and S. Yiv, *J. Colloid Interface Sci. 92*: 561 (1983).
96. J. K. Jacobson, J. C. Morgan, R. S. Schechter, and W. H. Wade, *Soc. Pet. Eng. J. 17*: 122 (1977).
97. W. H. Wade, J. C. Morgan, R. S. Schechter, J. K. Jacobson, and J. L. Salager, *Soc. Pet. Eng. J. 18*: 242 (1978).
98. P. H. Doe, M. M. El-Emary, W. H. Wade, and R. S. Schechter, in *Chemistry of Oil Recovery*, R. T. Johansen and R. L. Bey,

eds., ACS Symp. Ser. 91, American Chemical Society, Washington, D.C., 1979, p. 17.
99. J. C. Morgan, R. S. Schechter, and W. H. Wade, in *Improved Oil Recovery by Surfactant and Polymer Flooding*, D. O. Shah and R. S. Schechter, eds., Academic Press, New York, 1977, p. 101.
100. M. Baviere, W. H. Wade, and R. S. Schechter, in *Surface Phenomena in Enhanced Oil Recovery*, D. O. Shah, ed., Plenum, New York, 1981, p. 117.
101. G. J. Hirasaki, *Soc. Pet. Eng. J. 22*: 971 (1982).
102. A. F. Chan and V. J. Kremesec, Jr., *Soc. Pet. Eng. J. 25*: 580 (1985).
103. O. Ghosh and C. A. Miller, *J. Colloid Interface Sci. 100*: 444 (1984).
104. D. Roux and A. M. Bellocq, *Chem. Phys. Lett. 94*: 156 (1983).
105. J. E. Puig, L. E. Scriven, H. T. Davis, and W. G. Miller, in Ref. [41].
106. J. L. Cayias, R. S. Schechter, and W. H. Wade, *J. Colloid Interface Sci. 59*: 31 (1977).
107. K. S. Chan and D. O. Shah, *J. Dispersion Sci. Tech. 1*: 55 (1980).
108. K. S. Chan and D. O. Shah in *Surface Phenomena in Enhanced Oil Recovery*, D. O. Shah, ed., Plenum, New York, 1981, p. 53.
109. M. C. Puerto and W. W. Gale, *Soc. Pet. Eng. J. 17*: 193 (1977).
110. C. Lelanne-Cassou, I. Carmona, L. Fortney, A. Samii, R. S. Schechter, W. H. Wade, U. Weerasooriya, V. Weerasooriya, and S. Yiv, Binary Surfactant Mixtures for Minimizing Alcohol Cosolvent Requirements, SPE Preprint 12055, presented at 58th Annual Meeting of APE-AIME, San Francisco, 1983.
111. K. Shinoda and H. Kuneida, *J. Colloid Interface Sci. 42*: 381 (1973).
112. H. Kuneida and K. Shinoda, *Bull. Chem. Soc. Jpn. 55*: 1777 (1982).
113. K. Shinoda, *Prog. Colloid Polym. Sci. 68*: 1 (1983).
114. M. Bourrel, Ch. Koukounis, R. S. Schechter, and W. H. Wade, *J. Dispersion Sci. Tech. 1*: 13 (1980).
115. M. Bourrel, J. L. Salager, R. S. Schechter, and W. H. Wade, *J. Colloid Interface Sci. 75*: 451 (1980).
116. H. L. Chang, U.S. Patent 4,008,769 (February 22, 1977).
117. W. H. Baldwin and G. W. Neal, in *Chemistry of Oil Recovery*, R. T. Johansen and R. L. Bey, eds., ACS Symp. Ser. 91, American Chemical Society, Washington, D.C., 1979.

118. L. J. Magid, *J. Colloid Interface Sci. 87*: 447 (1982).
119. H. R. Kraft and G. Pusch, Mobilization and Banking of Residual Oil in High Salinity Reservoir Systems with the Use of Aqueous Surfactant Solutions of Ethoxylated Carboxymethylates, SPE/DOE Preprint 10714, presented at the Third Symposium on Enhanced Oil Recovery, Tulsa, 1982.
120. D. Balzer, Carboxymethylated Ethoxylates as EOR Surfactants, paper presented at the Second European Symposium on Enhanced Oil Recovery, Paris, 1982.
121. C. A. Miller and S. Qutubuddin, U.S. Patent Application 394,418 (1982).
122. V. K. Bansal and D. O. Shah, *J. Colloid Interface Sci. 65*: 451 (1978).
123. V. K. Bansal and D. O. Shah, *J. Am. Oil Chem. Soc. 55*: 367 (1978).
124. M. E. Hayes, M. Bourrel, M. M. Emary, R. S. Schechter, and W. H. Wade, *Soc. Pet. Eng. J. 19*: 349 (1979).
125. J. L. Salager, M. Bourrel, R. S. Schechter, and W. H. Wade, *Soc. Pet. Eng. J. 19*: 271 (1979).
126. L. J. Magid, *J. Colloid Interface Sci. 87*: 460 (1982).
127. J. C. Noronha and E. J. Derderian, Microemulsion Compositions Exhibiting Optimum Properties at Very High Salinities, paper presented at the AIChE National Meeting, Houston, March 1983.
128. S. Qutubuddin, C. A. Miller, T. Fort, Jr., and W. J. Benton, in *Macro- and Microemulsions: Theory and Applications*, D. O. Shah, ed., ACS Symp. Ser. #272 (1985), p. 223.
129. S. Thach and S. J. Salter, Characterization of Broad Equivalent Weight Petroleum Sulfonates Used in a Field Test, SPE/DOE Preprint 10721, presented at the Third Symposium on Enhanced Oil Recovery, Tulsa, 1982.
130. C. Koukounis, W. H. Wade, and R. S. Schechter, *Soc. Pet. Eng. J. 23*: 301 (1983).
131. G. Gillberg and L. Eriksson, *Ind. Eng. Chem. Prod. Res. Dev. 19*: 304 (1980).
132 S. Qutubuddin, C. A. Miller, and T. Fort, Jr., *J. Colloid Interface Sci. 101*: 46 (1984).
133. R. C. Nelson, J. B. Lawson, D. R. Thigpen, and G. L. Stegemeier, Cosurfactant-Enhanced Alkaline Flooding, SPE/DOE Preprint 12672, presented at the Fourth Symposium on Enhanced Oil Recovery, Tulsa, 1984.
134. T. P. Castor, W. H. Somerton, and J. F. Kelley, in *Surface Phenomena in Enhanced Oil Recovery*, D. O. Shah, ed., Plenum, New York, 1981.
135. S. C. Jones and K. D. Dreher, *Soc. Pet. Eng. J. 16*: 161 (1976).

136. S. J. Salter, The Influence of Type and Amount of Alcohol on Surfactant-Oil-Brine Phase Behavior and Properties, SPE 6843, presented at the Fall SPE Meeting, Denver, October 1977.
137. D. O. Shah, *Dev. Pet. Sci. 13*: 1 (1981).
138. C. E. Blevins, G. P. Willhite, and M. J. Michnick, *Soc. Pet. Eng. J. 21*: 581 (1981).
139. M. Bourrel and C. Chambu, *Soc. Pet. Eng. J. 23*: 327 (1983).
140. D. H. Smith and S. A. Templeton, *J. Colloid Interface Sci. 68*: 59 (1979).
141. L. Cash, J. L. Cayias, G. Fournier, D. McAllister, T. Schery, R. S. Schechter, and W. H. Wade, *J. Colloid Interface Sci. 59*: 39 (1977).
142. G. R. Glinsmann, Surfactant Flooding with Microemulsions Formed in Situ—Effect of Oil Characteristics, SPE Preprint 8326, presented at the SPE Annual Technical Conference and Exhibition, Las Vegas, 1979.
143. M. K. Tham and P. B. Lorenz, The EACN of a Crude Oil: Variation with Cosurfactant and Water Oil Ratio, paper presented at the European Symposium on Enhanced Oil Recovery, Bournemouth, England, 1981.
144. S. Mukherjee, C. A. Miller, and T. Fort, Jr., *J. Colloid Interface Sci. 91*: 223 (1983).
145. R. C. Nelson, *Soc. Pet. Eng. J. 23*: 501 (1983).
146. R. J. Good, Effect of Pressure, Time and Composition on Oil-Water-Surfactant Systems for Tertiary Oil Recovery, First Annual Report, U.S. Dept. of Energy Contract DE-AS19-80 BC10326, 1982.
147. J. P. O'Connell, Temperature, Pressure and Composition Effects in the Phase Behavior of Some Concentrated Systems, presented at the Symposium on Chemistry of Enhanced Oil Recovery, American Chemical Society, Atlanta, 1981.
148. A. Skauge and P. Fotland, The Effect of Pressure and Temperature on the Phase Behavior of Microemulsions, SPE/DOE Preprint 14932, presented at the Fifth Symposium on Enhanced Oil Recovery, Tulsa, 1986.
149. S. L. Wellington, Soc. Pet. Eng. J. 23: 901 (1983).
150. R. G. Bauer and D. F. Klemmenson, A New Polymer for Enhanced Oil Recovery, SPE/DOE Preprint 10711, presented at the Third Symposium on Enhanced Oil Recovery, Tulsa, 1982.
151. R. K. Prud'homme, *Soc. Pet. Eng. J. 24*: 431 (1984).
152. C. L. McCormick, R. D. Hester, H. H. Neidlinger, and G. C. Wildman, in *Surface Phenomena in Enhanced Oil Recovery*, D. O. Shah, ed., Plenum, New York, 1981, p. 751.
153. R. Dawson and R. B. Lantz, *Soc. Pet. Eng. J. 12*: 448 (1972).
154. S. P. Trushenski, D. L. Dauben, and D. R. Parrish, *Soc. Pet. Eng. J. 14*: 633 (1974).
155. S. P. Trushenski, in *Improved Oil Recovery by Surfactant and Polymer Flooding*, D. O. Shah and R. S. Schechter, eds., Academic Press, New York, 1977.

156. M. T. Szabo, *Soc. Pet. Eng. J. 19*: 1 (1979).
157. F. Th. Hesselink and M. J. Faber, in *Surface Phenomena in Enhanced Oil Recovery*, D. O. Shah, ed., Plenum, New York, 1981, p. 861.
158. G. A. Pope, K. Tsaur, R. S. Schechter, and B. Wang, *Soc. Pet. Eng. J. 22*: 816 (1982).
159. N. W. Desai and D. O. Shah, *Polym. Prepr. 22*: 39 (1981).
160. S. Qutubuddin, W. J. Benton, C. A. Miller, and T. Fort, Jr., *Polym. Prepr. 22*: 41 (1981).
161. D. B. Siano and J. Bock, *Polym. Prepr. 22*: 61 (1981).
162. S. P. Gupta, Micellar Fluid/Polymer Phase Effects in Micellar Flooding, SPE 10355, in press.
163. C. Huh, *J. Colloid Interface Sci. 97*: 201 (1984).
164. Y. Barakat, L. N. Fortney, R. S. Schechter, S. H. Yiv, and W. H. Wade, Alpha-Olefin Sulfonates for Enhanced Oil Recovery, *Proc. Second European Symp. Enhanced Oil Recovery*, Paris, 1982.
165. A. Oswald, H. Huang, J. Huang, and P. Valent, Jr., U.S. Patent 4,434,062 (1984).
166. A. Calje, W. Agterof, and A. Vrij, in *Micellization, Solubilization, and Microemulsions*, K. L. Mittal, ed., Plenum, New York, 1977, vol. 2, p. 779.
167. D. Roux, A. M. Bellocq, and P. Bothorel, in *Surfactants in Solution*, K. L. Mittal and B. Lindman, eds., Plenum, New York, 1984, vol. 3, p. 1843.
168. L. E. Scriven, in *Micellization, Solubilization, and Microemulsions*, K. L. Mittal, ed., Plenum, New York, 1977, vol. 2.
169. H. E. Gilliland and F. R. Conley, *Oil Gas J.*, p. 43 (January 19, 1976).
170. J. H. Bae and C. B. Petrick, Glenn Pool Surfactant Flood Pilot Data: Comparison of Laboratory and Observation Well Data, SPE/DOE Preprint 12694, presented at the Fourth Symposium on Enhanced Oil Recovery, Tulsa, 1984.
171. R. C. Nelson, *Soc. Pet. Eng. J. 22*: 259 (1982).
172. G. J. Hirasaki, H. R. Van Domselaar, and R. C. Nelson, *Soc. Pet. Eng. J. 23*: 486 (1983).
173. R. L. Reed and C. W. Carpenter, U.S. Patent 4,337,159 (1982).
174. B. Lindman, P. Stilbs, and M. E. Moseley, *J. Colloid Interface Sci. 83*: 569 (1981).
175. A. C. Lam, R. S. Schechter, and W. H. Wade, *Soc. Pet. Eng. J. 23*: 781 (1983).
176. G. J. Hirasaki, Scaling of Non-Equilibrium Phenomena in Surfactant Flooding, SPE/DOE Preprint 8841, presented at Symposium on Enhanced Oil Recovery, Tulsa, 1980.
177. D. T. Wasan, F. S. Milos, and P. E. DiNardo, *AIChE Symp. Ser. 212*(78): 105 (1982).

178. C. W. Arnold, Jr., M. S. thesis, Univ. of Texas at Austin (1978).
179. W. J. Benton, C. A. Miller, and T. Fort, Jr., *J. Dispersion Sci. Tech. 3*: 1 (1982).
180. K. H. Raney, W. J. Benton, and C. A. Miller, in *Macro- and Microemulsions: Theory and Applications*, D. O. Shah, ed., ACS Symp. Ser. #272, p. 193 (1985).
181. K. J. Ruschak and C. A. Miller, *Ind. Eng. Chem. Fundam. 11*: 534 (1972).
182. J. F. Scamehorn, R. S. Schechter, and W. H. Wade, *J. Colloid Interface Sci. 85*: 463 (1982).
183. H. S. Hanna and P. Somasundaran, in *Improved Oil Recovery by Surfactant and Polymer Flooding*, D. O. Shah and R. S. Schechter, eds., Academic Press, New York, 1977, p. 253.
184. J. H. Harwell, Ph.D. dissertation, Univ. of Texas at Austin (1983).
185. J. E. Lawson and R. E. Dilgrin, *Soc. Pet. Eng. J. 18*: 75 (1978).
186. F. J. Trogus, R. S. Schechter, and W. H. Wade, *J. Colloid Interface Sci. 70*: 293 (1979).
187. S. Somasundaran and H. S. Hanna, *Soc. Pet. Eng. J. 19*: 221 (1979).
188. J. H. Bae and C. B. Petrick, *Soc. Pet. Eng. J. 17*: 358 (1977).
189. K. V. Viswanathan and P. Somasundaran, Abstraction of Sulfonates and Dodecane by Alumina, SPE Preprint 10602, presented at Symposium on Oilfield and Geothermal Chemistry, Dallas, 1982.
190. K. O. Meyers and S. J. Salter, *Soc. Pet. Eng. J. 21*: 500 (1981).
191. E. D. Manev, M. S. Celik, K. P. Ananthapadmanabhan, and P. Somasundaran, *Soc. Pet. Eng. J. 24*: 667 (1984).
192. S. L. Wellington, private communication.
193. S. P. Gupta, *Soc. Pet. Eng. J. 24*: 38 (1984).
194. P. Somasundaran, M. Celik, A. Goyal, and E. Manev, *Soc. Pet. Eng. J. 24*: 233 (1984).
195. L. W. Holm and S. D. Robertson, *J. Pet. Technol. 33*: 161 (1981).
196. S. A. Hong, J. H. Bae, and G. R. Lewis, An Evaluation of Lignosulfonate as a Sacrificial Adsorbate in Surfactant Flooding, SPE/DOE Preprint 12699, presented at the Fourth Symposium on Enhanced Oil Recovery, Tulsa, 1984.
197. R. G. Larson and G. J. Hirasaki, *Soc. Pet. Eng. J. 18*: 42 (1978).
198. G. J. Hirasaki, *Soc. Pet. Eng. J. 21*: 191 (1981).
199. R. G. Larson, *Soc. Pet. Eng. J. 19*: 411 (1979).
200. G. A. Pope and R. C. Nelson, *Soc. Pet. Eng. J. 18*: 339 (1978).

201. G. A. Pope, B. Wang, and K. Tsaui, *Soc. Pet. Eng. J. 19*: 357 (1979).
202. N. Van Quy and J. Labrid, *Soc. Pet. Eng. J. 23*: 461 (1983).
203. M. R. Todd, J. K. Dietrich, A. Goldburg, and R. G. Larson, Numerical Simulation of Computing Chemical Flood Designs, SPE Preprint 7077, presented at Symposium on Improved Oil Recovery, Tulsa, 1978.
204. P. D. Fleming, C. P. Thomas, and W. K. Winter, *Soc. Pet. Eng. J. 21*: 63 (1981).
205. M. J. Tham, R. C. Nelson, and G. J. Hirasaki, *Soc. Pet. Eng. J. 23*: 746 (1983).
206. S. P. Gupta, *Soc. Pet. Eng. J. 22*: 481 (1982).
207. S. A. Pursley, R. N. Healy, and E. I. Sandvik, *J. Pet. Technol. 25*: 793 (1973).
208. M. S. French, G. W. Keys, G. L. Stegemeier, R. C. Ueber, A. Abrams, and H. J. Hill, *J. Pet. Technol. 30*: 195 (1973).
209. Commercial Scale Demonstration (Robinson Field), Enhanced Oil Recovery by Micellar Polymer Flooding, U.S. Dept. of Energy Annual Report DOE/ETY/13077-23.
210. J. R. Bragg, W. W. Gale, W. A. McElhannan, O. W. Davenport, M. D. Petrichmak, and T. L. Ashcraft, Loudon Surfactant Flood Pilot Test, SPE/DOE Preprint 10862, presented at the Third Symposium on Enhanced Oil Recovery, Tulsa, 1982.
211. Big Muddy Field Low-Tension Flood Demonstration Project, U.S. Dept. of Energy Annual Reports DOE/SF/01424-13,-26, -39,-45, 1979–1982.

5

Interfacial Catalysis by Microphases in Apolar Media

CHARMIAN J. O'CONNOR Department of Chemistry, University of Auckland, Auckland, New Zealand

I. INTRODUCTION

Microenvironments encountered in bulk solvents rarely resemble those encountered in living systems. More usually, interfacial character and anisotropic interactions among constituents are hallmarks of biochemical environments, particularly those associated with biological membranes. But the catalyses frequently encountered in quaternary liquid mixtures, i.e., substrate, water, apolar solvent, and surfactant, suggest that these mixtures may serve as important model systems for the study of enzymic and membrane interactions.

Water can readily be dispersed in an organic medium such as *n*-heptane using sodium bis(2-ethylhexyl)sulfosuccinate, Aerosol OT, as a dispersant. The driving force for the structure formed, Fig. 1, is a decrease in the overall free energy of the system. Such structures are generally known as *reversed* or *inverted* micelles, although they are sometimes referred to as *microemulsion droplets*. They contain an aqueous core, or water pool, surrounded by a charged spherical interface, with the hydrocarbon tails directed into the organic solvent. Because of the high ionic strength of the aqueous phase, it is expected that the sodium counterions will be closely

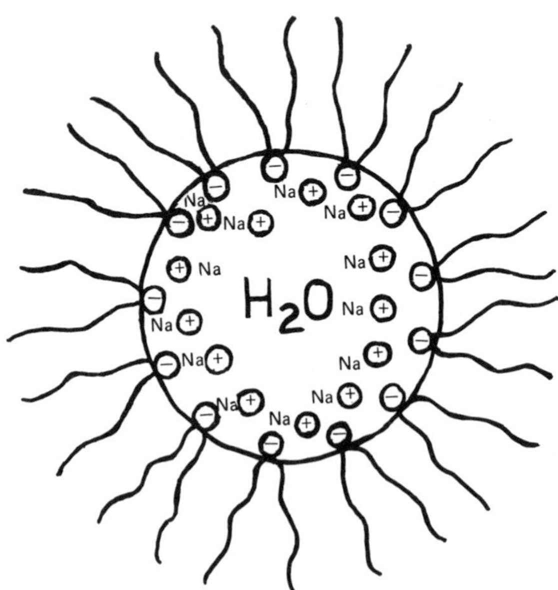

FIG. 1 Structure of an Aerosol OT-stabilized reversed micelle (microemulsion droplet).

associated with the sulfonate groups, forming a well-defined double-layer structure. Aerosol OT micelles in "dry" aprotic solvents are roughly spherical and monodisperse [1] and are characterized by distinct critical micelle concentrations (CMCs). (The CMC is regarded as the saturation concentration of free amphiphiles in a solution relative to the micellar pseudophase.) For example, the CMC and aggregation number of Aerosol OT in CCl_4 are 6×10^{-4} mol dm^{-3} and 17, respectively [2,3].

Many of the studies involving reversed micelles have used alkylammonium carboxylates [or alkanediaminebis(carboxylate)s] as the surfactants. These zwitterionic surfactants are stabilized by dipole-dipole and ion-pair interactions, by intermolecular hydrogen bonds, and by interactions of the hydrocarbon chains with the solvent. Aggregation numbers are typically less than 10 (compared with up to 100 for aqueous micelles) and there is evidence for "operational" CMCs of about 10^{-2} to 10^{-3} mol dm^{-3} [4]. Dodecylammonium propionate (DAP) is commonly chosen as the representative for this class of surfactants.

These aggregates are capable of solubilizing considerable quantities of water in their hydrophilic cavities. Furthermore, the presence of a minimum concentration of cosolubilized water is regarded as a necessary criterion for aggregation to occur at all [5]. The water molecules in the micellar core may be divided into two categories [6]. Type I water represents molecules residing near the micellar interface and is tightly bound to the ionic head groups of the surfactant [7–9]. As the core size expands, those water molecules which are not directly bound to the surfactant become available to form type II water and constitute the deepest core of the water pool (Fig. 2). This kind of water is regarded as being almost identical to bulk water [9]. Naturally, the distinction is not as discrete as the illustrations suggest, because spherically and ellipsoidally shaped aggregates have been described, as well as rodlike micelles—which at higher concentrations show typical features of lyotropic liquid crystalline mesophases. Due to the fluctuation of the monomers within the aggregate [10], the micelle can be considered to adopt (with a certain lifetime) a particular geometric structure. But the interpretation of the growing micellar aggregates by means of micellar association processes agrees very satisfactorily with a molecular model based on dipole-dipole interactions between the surfactant molecules and a spherical assumed aqueous pseudophase [11]. Probe molecules are distributed in the two aqueous "phases" depending on either electrostatic attraction or repulsion with the micellar interface.

The enclosed water droplet is about 1–5 nm in diameter and discussions of reversed micelles are generally restricted to a ratio R of $[H_2O]/[\text{surfactant}] < 10$. Reversed micelles are thus limiting examples of the ternary system of surfactant/water/apolar solvent which

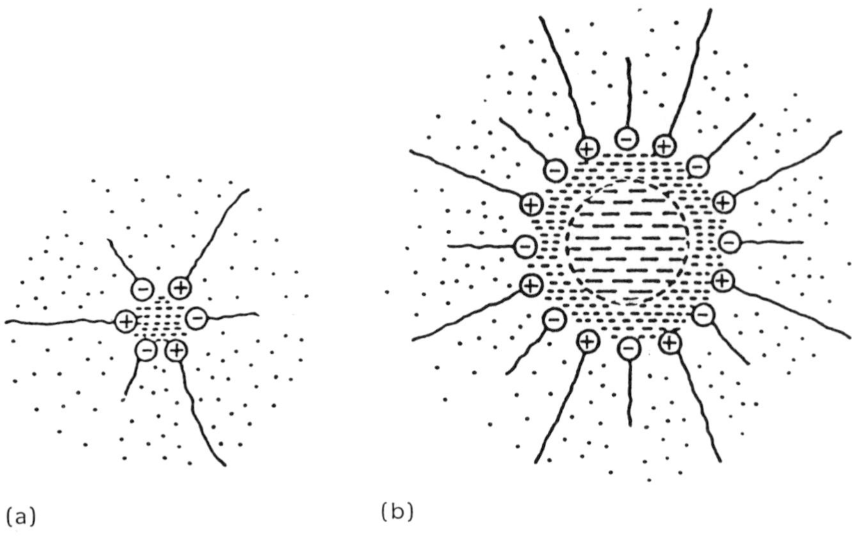

FIG. 2 Binary "phase" model for water in the interior of an alkyl-ammonium carboxylate reversed micelle, e.g., DAP, ⊕ , $\overset{+}{N}H_3$, ⊖ , ^-OOC. (a) At low water concentrations, $[H_2O]/[\text{surfactant}] \lesssim 1$, most water molecules are bound to the surfactant head group (type I water). (b) As the core size is expanded, type II water molecules, which are not bound directly to the surfactant, become available, and these constitute the softer core of the water pool [8].

forms an aggregate called a W/O (water-in-oil) microemulsion [12,13]. The earliest examples of catalysis in liquid crystalline phases are probably those of Friberg and Ahmed [14,15], who studied the rate of hydrolysis of *p*-nitrophenyl dodecanoate in the lamellar liquid crystalline phase of the system water/hexadecyltrimethylammonium bromide/hexanol and found that a pronounced increase in reaction rate accompanied changes in water mobility and counterion binding. However, this solvent system is more typical of a microemulsion than of the more common conception of a reversed micelle in an apolar solvent. Microemulsions which contain an ionic surfactant and oil have to consist of at least four components, the extra component being an uncharged cosurfactant. The cosurfactant is usually an alcohol, although amines can also act as cosurfactants. Whether microemulsions are true emulsions or solutions containing swollen micelles is a subject of controversy. The emulsion notion is supported by the relatively large size of 10 to 50 nm diameter, which is large enough for them to be regarded as droplets; the swollen micelle notion is supported by the stability, reproducibility, and reversible behavior of microemulsions, which are uncharacteristic of emulsions in general. But the issue is largely one of semantics, because they can both change reaction rates, and the factors that control chemical reactivity in reversed micelles will also be at work in W/O microemulsions. The submicrosocpic aggregates can bring reactants together or keep them apart, and they can also exert a medium effect. These factors have been analyzed in detail for reversed micellar solutions and are discussed in the following sections. The discussion will, however, be restricted to interfacial catalysis in microphases of reversed micelles in apolar solvents and chloroform and will not include a discussion of catalysis in aggregates which contain a cosurfactant. Nor will the account include examples taken from industrial, or potentially industrial, applications of interfacial catalysis in the microphases of vesicles (spherical or ellipsoidal single- or multicompartment closed bilayer structures, composed of phospholipids or synthetic surfactants), but the principles outlined below should remain equally applicable in these systems.

Reversed micelles in apolar solvents can influence considerably the kinetics of reactions involving organic and inorganic molecules and ions. Rate enhancements as large as 10^7-fold have been observed. The micelles can play two roles. First, they can raise or lower the energy of the transition state relative to the ground state, thereby retarding or enhancing the reaction rate. Second, the net result of the dipole-dipole interactions of the polar head groups with each other (or of a head group with its counterion) and with the solvent, and of the hydrocarbon chain with the solvent, is to form an interface between the bulk apolar solvent and the polar core. Through solubilization, the effective concentrations of reactants are increased in the restricted reaction field provided by this aqueous core. In particular, concentrations are increased at the interface, which is where many of the reactions probably occur, and it is this aspect which, in part, lends credibility to the notion that reversed micelles are effective mimics of enzymes. Enzymes are usually actively involved in their catalytic pathways, and in many of the reactions described below the head groups of the surfactants participate chemically. In some cases they seem to mimic the lock-and-key mechanisms which are believed to be responsible for the stereospecificity and other special properties of enzymes. While apolar solvents generally denature enzymes, solutions containing surfactants in apolar solvents can significantly increase enzymic activity.

Contributions to the activation energy are closely related to medium effects in general and are beginning to be understood. Important factors include differences in reactant environment, such as the changes in microscopic polarity and microviscosity of the reaction field as the size of the water pool increases, the dielectric constant at the interface, and the effects of large electric fields on bond strength. Entropies of activation for catalyses arising from reactivity between esters and zwitterionic surfactants, about −120 to −190 J K^{-1} mol^{-1} [16–20], are remarkably decreased from that for hydrolysis of an ester in an aqueous surfactant, −56 J K^{-1} mol^{-1}, and very similar to the value of −150 J K^{-1} mol^{-1} found for the deactylation of acetyl-enzyme by α-chymotrypsin [17]. These low values reflect the loss of translational (and rotational) degrees of freedom suffered by the substrate on its encapsulation in the restricted field.

The presence of the interface allows quantitative treatments to be applied to the micellar catalytic effects. Often, the observed trends can be explained by using concepts such as partition of solutes between micelles and solvent.

Micellar structure can be discussed in terms of either of two models, both of which are based, in the final analysis, on the assumptions of a thermodynamic equilibrium between monomers and micellar aggregates. The pseudophase model considers only the overall equilibrium between monomers and fully formed micelles and can be regarded as one simple and useful perspective on the process of micellization. The advantage of this treatment is a remarkable reduction in arbitrary parameters, but the other treatment is more readily justified theoretically. This alternative treatment describes the buildup of the micelle, in terms of a series of equilibria between monomeric and aggregated surfactant which would lead to a distribution of micellar size about a mean, and is often termed the mass action model or the multiple-equilibrium model. In this chapter the kinetic implications and applications of both models are surveyed.

No attempt has been made to cover the literature exhaustively. Instead, more exemplary reactions have been selected to illustrate the various important aspects of interfacial catalysis in microphases in apolar media and to stress the principles behind this phenomenon. Possible future perspectives are discussed along with existing results. A recent critical review should be consulted for a discussion of the forces governing aggregation, the models of association and the concept of the CMC, evidence for submicellar aggregates, and acid-base interactions within these micellar aggregates [21]. In addition, there are several excellent earlier reviews which give detailed accounts and examples of micellar catalysis and concepts in apolar media [4,13,22–26].

II. SATURATION EFFECTS IN MICELLAR CATALYSIS

Catalysis by microphases in apolar media shows that the micelles provide media which may favor or inhibit reactions. These micellar effects are true for both unimolecular and bimolecular reactions.

In unimolecular reactions, the kinetic rate-surfactant concentration profiles are of markedly different types, even for reactions carried out in a common solvent and with a common surfactant.

The simplest profile possible is that which mimics the saturation-type kinetics observed for many micellar catalyzed reactions in aqueous solution, i.e., a sigmoidal rate enhancement followed by a plateau. Figure 3 shows the pseudo-first-order rate constant for the mutarotation of 2,3,4,6-tetramethyl-α-D-glucose in DAP solutions in cyclohexane and benzene at 297.6 K [27]. The rate constants in the presence of DAP, dodecylammonium butanoate (DAB), and dodecylammonium benzoate (DABz) were enhanced by factors of 280, 688, and 457, respectively, relative to the rate in benzene. The corresponding rate enhancement by DAP in cyclohexane was 863. More favorable partitioning of the substrate between the micellar phase and the bulk apolar solvent in cyclohexane than in benzene was thought to account for this latter result. The catalytic efficiency of the surfactants paralleled the extent of binding of the carbohydrate ether to the micellar aggregate.

It was tempting to assume bifunctional catalysis involving the ammonium ion as a general acid and the carboxylate ions as a general base, following the mechanism which had been postulated for bifunctional catalysis in homogeneous nonaqueous systems [28].

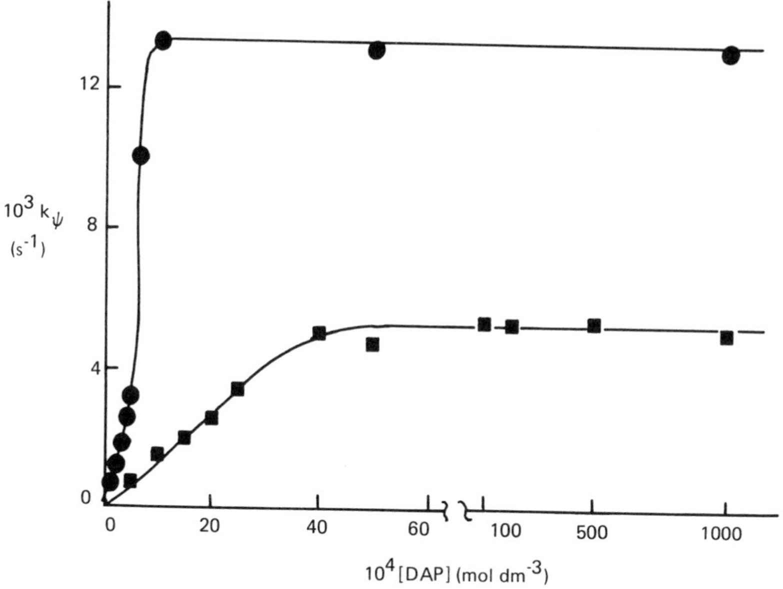

FIG. 3 Variation of the pseudo-first-order rate constants for the mutarotation of 2,3,4,6-tetramethyl-α-D-glucose (1.7×10^{-2} mol dm^{-3}) with concentration of DAP in cyclohexane (•) and in benzene (■) at 297.6 K [27].

Profiles similar to that shown in Fig. 3 have been observed for the trans-cis isomerization of the bis(oxalato)diaquochromate(III) anion, $[Cr(C_2O_4)_2(H_2O)_2]^-$, in the polar cavities of reversed alkylammonium carboxylate micelles in benzene at 297.5 K [29]. Compared to the isomerization in pure water, the rate constants in the plateau region were increased by a factor of up to 63-fold.

In some cases the plateau becomes rather short and increasing surfactant concentration then decreases the observed rate (i.e., a rate maximum is observed). Figure 4 shows this situation for the decomposition of the Meisenheimer complex ion, 1,1-dimethoxy-2,4,6-trinitrocyclohexadienylide ion, in the presence of DABz aggregates in benzene at 297.5 K [30]. The rate constant was greater by factors of 62,900 and 1,800 than in pure benzene and water, respectively, and DAP and DAB were equally effective catalysts. Decomposition of the methoxy adduct of 1-methoxy-2,4-dinitronaphthalene at high surfactant and low substrate concentrations (conditions under which maximum rate enhancement is expected) were too fast for measurement under the experimental conditions. It was speculated that this more bulky substrate was held more rigidly in the micellar

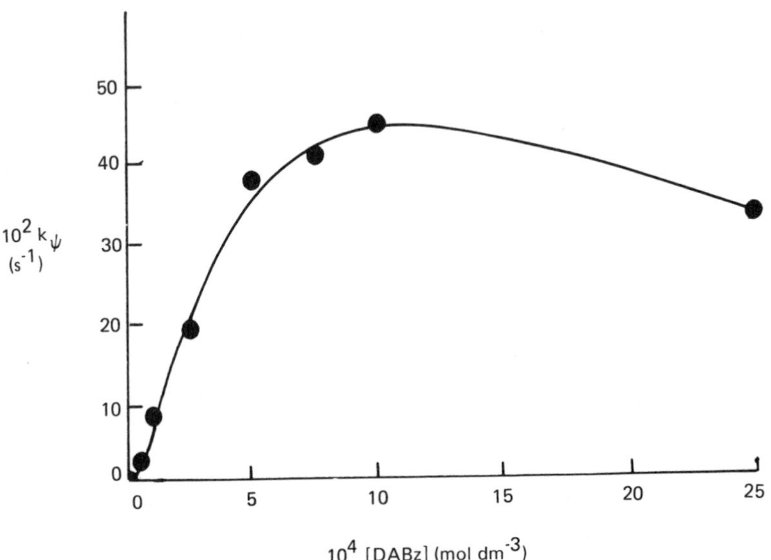

FIG. 4 Variation of the pseudo-first-order rate constants for decomposition of 1,1-dimethoxy-2,4,6-trinitrocyclohexadienylide ion (5×10^{-5} mol dm^{-3}) with concentration of DABz in benzene at 297.6 K [30].

cavity than the monocyclic substrate. A mechanism was suggested, similar to that for the reactivity of tetramethyl glucose, in which the polar substrate was solubilized in the hydrophilic cavity of the reversed micelle, where hydrogen bond formation, proton transfer, and enhanced water activity facilitated product formation. Additional factors, such as the geometry of the substrate and head groups on the surfactant, the size of the micelles and of the cavity, and the polarity of the bulk solvent and of the substrate, were considered likely to influence the extent and mechanism of the catalysis. Phosphatidylethanolamine is analogous to dodecylammonium carboxylates in that it can transfer protons from the ammonium ion, while L-α-lecithin-β-γ-diacylphosphorylcholine resembles hexadecyltrimethylammonium butanoate (CTABu) in that it cannot transfer protons in this manner. While CTABu did not catalyze the decomposition of the Meisenheimer complexes, catalysis was observed in the presence of both the phospholipid micelles in benzene, showing that proton transfer was not the only mode of catalysis. Enhanced water activity in the micellar cavity was invoked as an additional factor contributing to the observed rate enhancement.

A. Pseudophase Model for Unimolecular Reactions

Most of the observations of micellar effects on reaction rates in apolar solvents have shown that it is useful to consider the solvent and the micelles as if they were separate phases and to consider the overall reaction rate, R_{total}, to be the sum of the reaction rates occurring in the bulk solvent, R_0, and in the pseudophase provided by the aggregate, R_M:

$$R_{total} = R_0 + R_M \tag{1}$$

Menger and Portnoy [31] were the first to develop a quantitative treatment for aqueous micellar reactions, and this treatment is readily adapted to provide a satisfactory explanation for many reactions in reversed micelles. This account will consider only those processes which occur on time scales slower than the dissolution or dissociation of aggregates. The mean lifetime of a surfactant moelcule in reversed micelles and microemulsions [9,32] is of the order of microseconds, and the restriction imposed thus excludes ultrafast photophysical events ($t_{1/2} < 10^{-7}$ sec), for on this time scale the reversed micelle can be considered to be frozen. Electron and proton transfer reactions ($t_{1/2} \sim 10^{-5}$–10^{-7} sec) will also be excluded, for these can go to completion faster than the exchange of water pools among neighboring aggregates, and the events are confined to defined reversed micelles. Intervesicle exchange of surfactant-solubilized water pools, at least for Aerosol OT, occurs on the millisecond time scale [9,29]. The rates of most chemical processes ($t_{1/2} > 10^{-4}$ sec) are governed by statistical distributions and intervesicle exhange of water pools and reactants.

The phase separation approach implies that the micellization process follows a monomer ⇌ n-mer equilibrium:

$$nm \underset{}{\overset{K_n}{\rightleftharpoons}} M_n \tag{2}$$

where K_n is the micelle association constant between n molecules of monomeric surfactant, m, and the aggregate, M_n.

This model postulates that micellization is a phase transition and its description follows that of a two-phase equilibrium. In its simplest form it overemphasizes the cooperativity of the aggregation for it does not contain a size-limiting step, and therefore it is of little value in accounting for the formation of the small aggregates seen in apolar media. More recent theories of interfacial catalysis by microphases in apolar media have taken these into account, and these will be discussed in Sec. III.

Application of conservation of mass to Eq. (2) gives:

$$\frac{[m]}{[S]} + n\left(\frac{[m]}{[S]}\right)^n [S]^{n-1} K_n = 1 \tag{3}$$

where [S] is the total molar concentration of surfactant. If one assumes that K_n is the product of n−1 individual and equal mass action constants, Eq. (4) is obtained and the concentration of micellized surfactant, $[M_n]$, is then obtained from Eq. (5):

$$\frac{[m]}{[S]} + n\frac{[M_n]}{[S]} = 1 \tag{4}$$

$$[M_n] = [S] - CMC \tag{5}$$

It is assumed that the concentration of monomeric surfactant is given by the critical micelle concentration, CMC.

Although consideration should really be given to the principle that binding of the reactant molecules to the micelles physically involves a partition of the substrate between the micellar and apolar phases rather than association in the stoichiometric proportions 1:1, this refinement is not normally included in analysis of unimolecular reactions.

For these reactions, it is assumed that the substrate does not perturb the equilibria of the system and that the reactions in the aggregates are described by scheme 1.

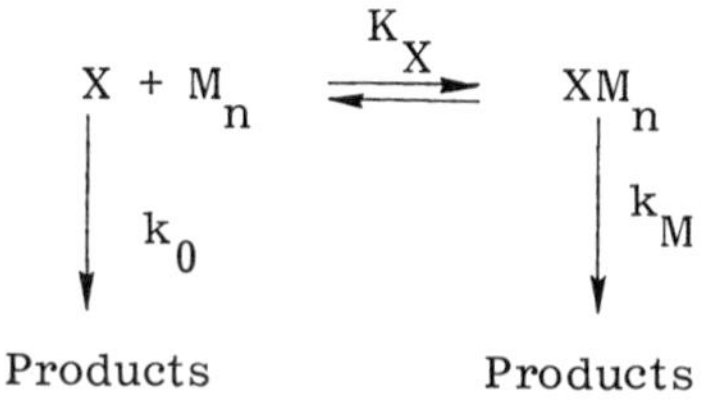

$$X + M_n \underset{}{\overset{K_X}{\rightleftharpoons}} XM_n$$

$$X + M_n \xrightarrow{k_0} \text{Products} \qquad XM_n \xrightarrow{k_M} \text{Products}$$

Scheme 1

The substrate X forms a complex XM_n with the micelle, and k_0 and k_M are the first-order rate constants for reaction in the solvent and micellar pseudophases, respectively. The binding constant K_X is given by:

$$K_X = \frac{[XM_n]}{[X][M_n]} \tag{6}$$

The observed first-order rate constant for the reaction, k_ψ, is given by:

$$k_\psi = \frac{k_0 + k_M K_X [M_n]}{1 + K_X [M_n]} \tag{7}$$

(The binding constant K_X is given by K/N in a number of earlier references, where N is the aggregation number of the micelle.) It is further assumed that the substrate does not complex with the monomeric surfactant.

It is easy to recognize the analogy between Eq. (7) and the Michaelis-Menten equation for enzyme-catalyzed reactions, although, in the above treatment, the surfactant is in large excess over the substrate, whereas in enzymic reactions the substrate is usually in excess over the enzyme. Rearrangement of Eq. (7) to the reciprocal form, Eq. (8):

$$\frac{1}{k_0 - k_\psi} = \frac{1}{k_0 - k_M} + \frac{1}{(k_0 - k_M) K_X [M_n]} \tag{8}$$

gives an equation which is similar to the Lineweaver-Burk equation of enzyme kinetics, and which allows calculation of the rate and equilibrium constants k_M and K_X from plots of $(k_0 - k_\psi)^{-1}$ against $[M_n]^{-1}$. In cases when k_ψ is significantly different from k_M it is often more convenient to use an alternative form, Eq. (9):

$$\frac{k_\psi - k_0}{k_M - k_\psi} = K_X M_n \tag{9}$$

Application of Eq. (8) or (9) is obviously restricted to a surfactant concentration in which [S] > [CMC], and is further restricted to the region of sigmoidal increase in the rate-surfactant concentration profile, before the onset of saturation, inhibition, or counterion catalysis. Figure 5 shows a typical plot of k_ψ^{-1} against $[M_n]^{-1}$, that for the esterolysis of *p*-nitrophenyl acetate (PNPA) in solutions of DAP dissolved in benzene at 298 K [18]. (In apolar solvents k_0 is frequently equal to zero.)

Small extensions of this treatment have been given. Thus, in the reaction between PNPA and DAP, El Seoud et al. [33] interpreted k_0 as being the rate constant for the reaction of the ester with the surfactant monomers and they assumed this to be negligible above the CMC. Equation (8) then reduced to Eq. (10):

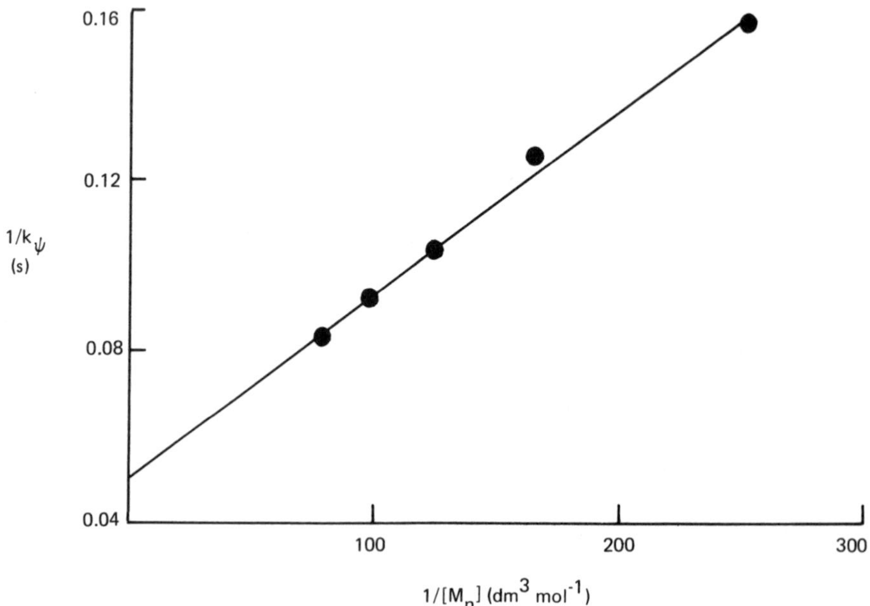

FIG. 5 Micelle-substrate association (binding) constant plot for *p*-nitrophenyl acetate (5×10^{-5} mol dm^{-3}) and DAP in benzene according to Eq. (8) [18].

$$\frac{1}{k_\psi} = \frac{1}{k_M} + \frac{1}{k_M K_M [M_n]} \tag{10}$$

However, the treatment of Kon-no et al. [34] considered the reaction between PNPA and DAP monomers to be governed by Eq. (11):

$$k_{\psi m} = \frac{k_0 + k_m K_M [S]}{1 + K_m [S]} \tag{11}$$

where $k_{\psi m}$ is the pseudo-first-order rate constant observed in the monomer solution (and $k_{\psi M}$ the corresponding rate constant in micellar solution above the CMC), k_m is the rate constant in the monomer solution, and K_m is the association constant of the substrate with the monomers (and K_M the corresponding association constant in the micelle). The derived values of k_M and K_X will obviously depend on the value substituted for the CMC in Eq. (5), but for some surfactants this value is well documented. Independent measurements of the CMC for DAP give values of $3-7 \times 10^{-3}$ mol dm^{-3} and 4.5×10^{-3} mol dm^{-3} by ^1H NMR spectroscopy [35] and vapor pressure osmometry [34], respectively.

Under similar experimental conditions, the value of the micelle-substrate binding constant has been determined as 33 [18], 28 [34], and 11 [33] mol^{-1} dm^3 and the rate constant in the micellar phase as 2.04 [18], 2.33 [34], and 4.3 [33] sec^{-1}.

In a study of functional micelles in aqueous solution, Fornasier and Tonellato [36] used a more critical extension of the phase separation approach which included contributions for the reaction occurring in the micellar pseudophase of fractional volume $\bar{V}$ and that in the solvent pseudophase of fractional volume $1 - \bar{V}$. Provided $k_0 \ll k_\psi$, their treatment leads to Eq. (12):

$$\frac{[M_n]}{k_\psi} = \frac{1}{k^{app}} + \frac{K_X^{app}}{k^{app}} [M_n] \tag{12}$$

in which k^{app} and K_X^{app} refer to the apparent catalytic rate constant and substrate binding constant, respectively. They are further related by Eq. (13):

$$k^{app} = \frac{K_X^{app} k_{2,M}}{\bar{V}} \tag{13}$$

where $k_{2,M}$ refers to the second-order rate constant in the micellar pseudophase and $\bar{V}$ is the molar volume of the micelle (i.e., the reciprocal of the surfactant concentration in the micellar pseudophase).

If the assumptions involved in the derivation and application of Eqs. (8) and (12) to reactions of surfactants in apolar solvents were reasonably valid, then values of K_X and K_X^{app} and of $k_{2,M}/\bar{V}$ and k_M should be comparable. Excellent agreement was obtained for catalysis in aqueous micellar systems [37], but there were large discrepancies between the values obtained by these two analyses for the rate data for the reaction of PNPA in a series of α,ω-alkyldiamine bis(dodecanoate) surfactants [38]. This is but one of the pieces of evidence which suggests that the simple pseudophase separation model may not adequately explain all of the kinetic phenomena observed in reversed micellar solutions. Further evidence is outlined in Sec. III.

B. General Acid-General Base Catalysis

Rate profiles of the general shape described by Figs. 3 and 4 were found in early investigations of interfacial catalysis in reversed micellar solutions. It is probably true to say that, in spite of the explosion of interest in such systems in more recent years, there have been very few examples of these simple saturation-type kinetics other than those [18,27–30,33,34] mentioned above. It was the early expectation that these interfacial systems would mimic more precisely enzymic catalysts than did aqueous micelles, but only recently has attention been given to more extensive interpretation of the causes of the variations in the rate-surfactant concentration profiles, which cannot be accounted for by the theories developed for catalysis (or inhibition) in aqueous micellar systems. Some of the variations in these profiles are shown in Figs. 6–10.

Fiture 6 shows the speudo-first-order rate constant for decomposition of 2,4-dinitrophenyl sulfate to form the corresponding phenol in DAP in benzene at 312.8 K [39]. Increasing concentrations of surfactant initially increased the rate constant in a sigmoidal fashion in the region of the CMC, after which k_ψ continued to increase linearly. The data indicated that, in addition to micellar catalysis, components of DAP also enhanced the decomposition rate. The observed rate constant k_ψ at a given DAP concentration, or more generally at a given alkylammonium carboxylate concentration, can be described by Eq. (14):

$$k_\psi = k_M + (k_{RNH_3^+} + k_{^-O_2CR'})[RNH_3^{+-}O_2CR'] \tag{14}$$

where k_M is the rate constant for micellar catalysis and $k_{RNH_3^+}$ and $k_{-O_2CR'}$ represent rate constants due to the alkylammonium and

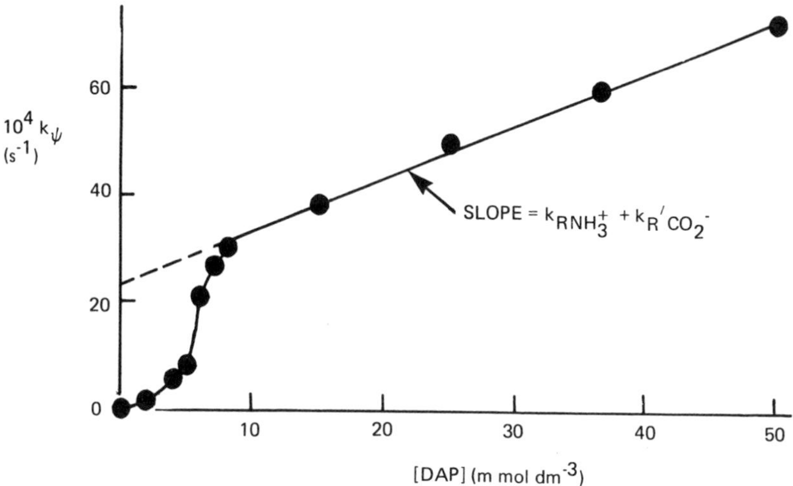

FIG. 6 Variation of the pseudo-first-order rate constants for decomposition of 2,4-dinitrophenyl sulfate (10^{-5} mol dm^{-3}) with concentration of DAP in benzene at 312.8 K [39].

carboxylate groups of the surfactant, respectively. Below the CMC, the reaction was held to be due entirely to $k_{RNH_3^+} + k_{^-O_2CR'}$. Above the CMC, however, Eq. (14) was obeyed over a sufficiently large concentration range that k_M values could be calculated from the intercept of the linear portion extrapolated back to the y axis and $k_{RNH_3^+} + k_{^-O_2CR'}$ values could be obtained from the slopes. Throughout the concentration range, the alkylammonium and carboxylate ions act as a general acid and general base. Superimposed on these effects, above the CMC, it was considered that enhanced activity of the substrate and solubilized water in the micellar microenvironment, relative to that in water, as well as proton transfer contributed to the observed micellar catalysis of up to 70-fold that obtained for hydrolysis of the sulfate ester in water.

Similar qualitative explanations were given to account for the decomposition of PNPA in benzene solutions of a series of alkylammonium propionates, $CH_3(CH_2)_nNH_3^{+-}O_2CCH_2CH_3$ (where n = 3, 5, 7, 9, or 11), and a series of dodecylammonium carboxylates, $CH_3(CH_2)_{11}NH_3^{+-}O_2C(CH_2)_nCH_3$ (where n = 1, 2, 6, or 7) [40], and also for the decomposition of PNPA in benzene solutions of a series of alkane-α,ω-diamine bis(dodecanoate)s, $CH_3(CH_2)_{10}COO^{-+}NH_2(CH_2)_nH_2N^{+-}OOC(CH_2)_{10}CH_3$ (where n = 2–7, 9, 10, and 12) [38]. Typically, the rate-surfactant concentration profiles showed an initial sigmoidal dependence of rate on surfactant concentration followed by a region of increasing rate (which was generally linear) with increase in [S].

In the cases described above, for the decomposition of sulfate and acetate esters, the onset of the linear increase in rate with increasing surfactant concentration occurred immediately at the end of the sigmoidal rise. But in the decomposition of *p*-nitrophenylurea solubilized in the interior of DAP micelles in benzene at 323 K, to form *p*-nitrophenyl isocyanate, the rate constant initially reached a plateau value corresponding to k_M, and then, at $[S] > 20 \times 10^{-3}$ mol dm^{-3}, increased linearly with increasing [S], thus showing again the ability of the components of the surfactant to act as general acid and base catalysts [41]. Such catalysis was not surprising in view of the general acid and base catalysis found in aqueous buffer solutions [42]. The rate-surfactant concentration profile is given in Fig. 7. The plateau value of the rate constant was 3000-fold faster than the rate constant for hydrolysis of *p*-nitrophenylurea in aqueous solution at pH 6.7, but addition of water so that the ratio $[H_2O]/[DAP]$ was about unity suppressed the decomposition to a negligible

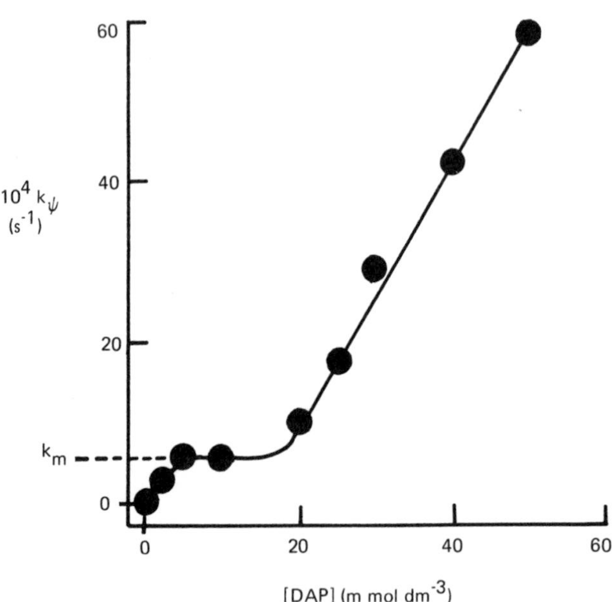

FIG. 7 Variation of the pseudo-first-order rate constants for decomposition of *p*-nitrophenylurea (1.6×10^{-4} mol dm^{-3}) with concentration of DAP in benzene at 323 K [41].

rate. At this ratio, the surfactant head groups will be fully hydrated (see Sec. V,C) and this result indicated that proton transfer to inner core solubilized water competed with the initial preequilibrium proton transfer necessary for decomposition.

Some of the rate-surfactant concentration profiles for decomposition of PNPA in the alkane-α-ω-diamine bis(dodecanoate) salts [38], particularly those for the ethane and pentane derivatives, showed steps similar to but less pronounced than those in Fig. 7.

Changing the solvent from benzene to either toluene or cyclohexane resulted in the rate-surfactant concentration profiles for decomposition of PNPA in DAP micelles exhibiting convex curvature [43] (Fig. 8). A plot of k_ψ against $[DAP]^2$ was linear with a nonzero intercept, suggesting that the rate law be given by:

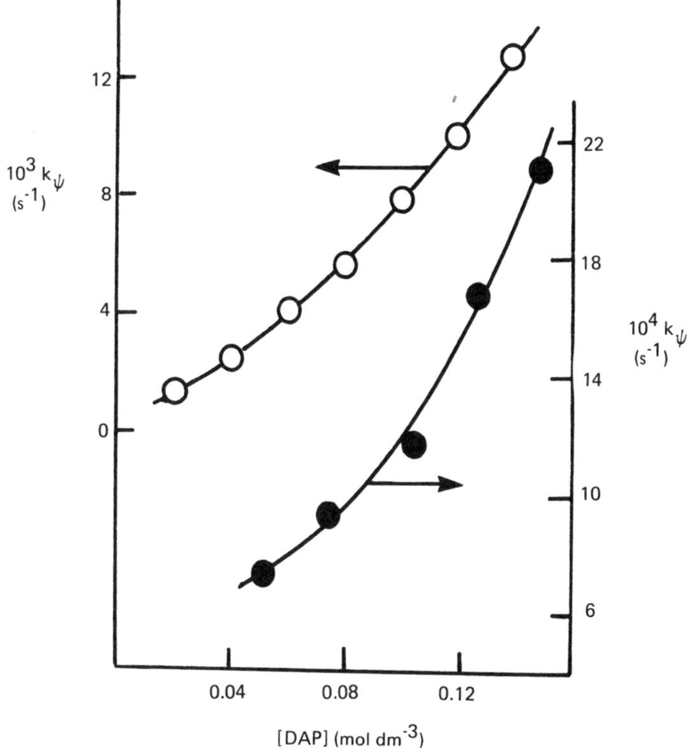

FIG. 8 Variation of the pseudo-first-order rate constants for decomposition of p-nitrophenyl acetate (5×10^{-3} mol dm^{-3}) with concentration of DAP in toluene at 298 K (○), and in cyclohexane at 303 K (●) [43].

$$\text{rate} = k_1[\text{PNPA}] + k_3[\text{PNPA}][\text{DAP}]^2 \tag{15}$$

Kon-no et al. [34] had previously reported a qualitatively similarly shaped profile for the reaction of PNPA with DAP or DAB in cyclohexane, and Kondo et al. [16] found that the rate constants for decomposition of *p*-nitrophenyl carbonates in DAP aggregates in benzene, hexane, or carbon tetrachloride always fitted the rate equation

$$\text{rate} = k_3[\text{ester}][\text{DAP}]^2 \tag{16}$$

As the solvent was changed from benzene to toluene and then to cyclohexane, the rate of decomposition of PNPA in DAP aggregates increased [43]. The solvent polarity constants $E_T(30)$ for these three solvents are 144.3, 141.8, and 130.5 kJ mol^{-1}, respectively. Thus the increase in rate might be attributed to decreased solvation of the attacking nucleophile in the less polar solvent. The observed pseudo-first-order rate constant for the hydrolysis of methyl *p*-nitrophenyl carbonate, in the presence of DAP and water, also increased with a decrease in the $E_T(30)$ value of the solvent [39]. But consideration of solvent polarity is not sufficient to provide an explanation for the altered dependence on concentration of DAP in the different solvents. The CMC values of DAP in benzene and cyclohexane are similar [44], and it is reasonable to assume that the aggregation behavior in toluene will also be similar.

Consideration of the head group participation as a factor in interfacial catalysis in apolar media is not restricted to studies in these alkylamine carboxylate surfactants. Thus, the reversible hydration/dehydration reactions of 1,3-dichloroacetone [45]:

$$(ClCH_2)_2C{=}O + H_2O \rightleftharpoons (ClCH_2)_2C(OH)_2 \tag{17}$$

are both increased as a function of increasing Aerosol OT concentration, probably because in the equilibrium, Eq. (17), the sulfonate group (acting as a general base) can form a stronger hydrogen bond with the hydrate (pK_a 11.6) [46] than with the attacking water.

Changing the surfactant, for studies of the interfacial catalysis of the esterolysis reaction of PNPA in benzene or toluene, to a series of dodecylammonium 4-substituted phenoxide salts, $CH_3(CH_2)_{11}NH_3^+$-$^-OC_6H_4Y$-*p*, (where Y = H, Cl, Br, Me, or MeO) also resulted in rate-surfactant concentration profiles [19] similar to those shown in Fig. 8. The rate data again precisely fitted Eq. (15) and plots of $k_\psi/[S]$ against [S] were linear. This behavior was similar to that for aminolysis of PNPA by dodecylamine [19].

The association between *p*-nitrophenol and a base as strong as triethylamine has been shown by spectroscopic studies [47] to be described by a tautomeric equilibrium between a simple hydrogen-bonded complex and a proton-transferred species, Eq. (18):

$$RNH_2\cdots H{-}OAr \rightleftharpoons R\overset{+}{N}H_3\cdots\overset{-}{O}Ar \qquad (18)$$

Having assumed that the position of Eq. (18) favored the unprotonated amine, it was then concluded that the reaction of PNPA with the phenoxide surfactants was aminolysis of the ester [19].

The reaction between esters and amines in aprotic solvents has been extensively investigated [48] and shown to obey a two-term rate law, Eq. (19):

$$k_\psi = k_1[\text{amine}] + k_2'[\text{amine}]^2 \qquad (19)$$

in which k_1 is the rate constant for reaction due to one amine molecule and k_2' that due to two amine molecules.

In surfactant solutions there is an analogous equation, Eq. (20):

$$k_\psi = k_1[\text{amine}] + k_2[\text{amine}][S] \qquad (20)$$

in which k_2 is the corresponding rate constant for attack by one molecule of amine and one of surfactant [48].

El Seoud et al. [33] showed that a combination of these rate constants was necessary to account for the reaction of PNPA with added dodecylamine in the presence of DAP, for which Eq. (21) was found to hold:

$$k_\psi = k_1[\text{amine}] + k_2'[\text{amine}]^2 + k_2[\text{amine}][M_n] \qquad (21)$$

Menger and Smith [48] have proposed a unifying mechanism, scheme 2, for ester aminolysis in aprotic solvents.

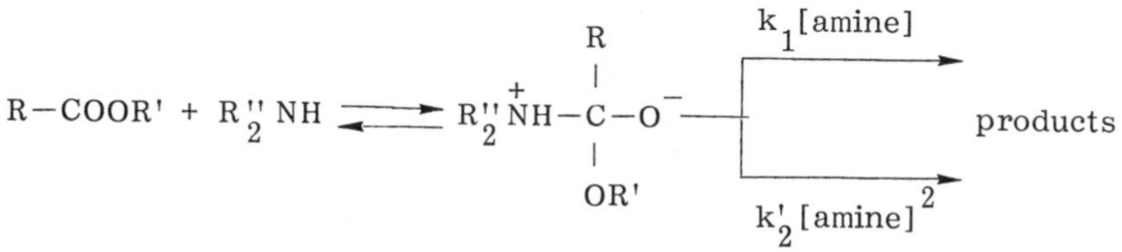

$$R{-}COOR' + R''_2NH \rightleftharpoons R''_2\overset{+}{N}H{-}\underset{OR'}{\overset{R}{\underset{|}{\overset{|}{C}}}}{-}O^- \;\begin{cases} \xrightarrow{k_1[\text{amine}]} \\ \xrightarrow[k_2'[\text{amine}]^2]{} \end{cases}\; \text{products}$$

Scheme 2

Reversible formation of a tetrahedral zwitterionic intermediate was followed by rate-determining collapse of the intermediate to products by one of two paths. It was suggested that the second amine molecule in the k_2' step functioned by removing a proton from the tetrahedral intermediate, thereby avoiding the formation of a high-energy N-protonated amide when the intermediate collapses.

It may be possible to account for the shape of the rate-surfactant concentration profiles for the reaction of alkylamine carboxylate surfactants with PNPA by postulating a change in the rate-determining step as the concentration of surfactant increases. For example, if the reaction of the amine with the ester were to proceed by formation of a tetrahedral intermediate, as in scheme 2, and the slow breakdown of the intermediate were rate-determining and subject to general acid-general base catalysis, then it would be possible to imagine a situation in which breakdown of the intermediate becomes faster than its formation, and the rate-determining step is the attack of the amine on the ester. The kinetic evidence for such a change in rate-determining step would be revealed by a change from a second-order dependence on surfactant concentration to a pseudo-first-order dependence as the concentration of the surfactant was increased, i.e., as the rate-surfactant concentration profile climbed through the initial sharp, curved increase and then followed a linear increase, in a manner similar to that shown in Fig. 6. Such a change was seen in other alkylamine carboxylate catalyzed reactions [26,38,43,49,50].

This explanation is attractive, but Jencks [51] has warned against assuming a change in the rate-determining step before excluding the possibility of changes in concentrations of the reactive species. Although the possibility of such a change in mechanism cannot be rigorously excluded, consideration of the physical properties of alkylamine carboxylates in apolar solvents allows an equally valid explanation which can qualitatively explain the changes in slope seen in the plots of k_ψ against [S]. This explanation is included in the discussion of the consequences of the continuous aggregation model in Sec. III.B.

C. Aquation of Inorganic Complexes

In stark contrast to the theme of the results exemplified by the data shown in Fig. 8, for which increasing surfactant concentration resulted in a second-order dependence, i.e., $k_\psi \propto [S]^2$, is that demonstrated by the data shown in Fig. 9. These data, for the aquation of the tris(oxalato)chromate(III) anion, $[Cr(C_2O_4)_3]^{3-}$, in water pools trapped in the polar cavity of octylammonium tetradecanoate (OAT) reversed micelles in benzene at 297.5 K, show an exponential decrease in k_ψ with increasing concentration of surfactant, at any given water concentration [52]. Logarithmic plots of k_ψ against stoichiometric surfactant concentration [S] were linear.

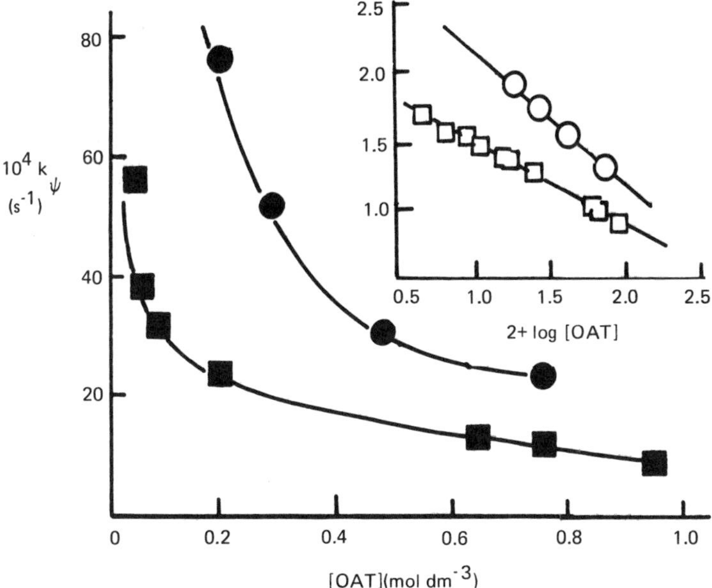

FIG. 9 Variation of the pseudo-first-order rate constants for aquation of tris(oxalato)chromate(III) anion ($\sim 10^{-4}$ mol dm^{-3}) by surfactant-solubilized water with concentration of OAT in benzene at 297.5 K: $[H_2O] = 16 \times 10^{-2}$ mol dm^{-3} (●); 5.5×10^{-2} mol dm^{-3} (■). Inset: logarithmic plots of rate constants for hydrolysis of the $[Cr(C_2O_4)_3]^{3-}$ ion against concentration of OAT in benzene: (○, □): concentrations as above [52].

Similar trends were found for the aquation of $[Cr(C_2O_4)_3]^{3-}$, of $[Co(C_2O_4)_3]^{3-}$, and of *cis*-$[Co(C_2O_4)_2(H_2O)_2]^-$ in water pools solubilized by DAP micelles in benzene. The concentrations of surfactants were not taken low enough in this study to achieve maximum rate enhancement, but, even so, the results obtained were dramatic. The observed maximum rate enhancements for aquation of $[Cr(C_2O_4)_3]^{3-}$ by OAT and DAP reversed micelles were 5×10^6-fold and 1×10^6-fold that of the aquation rate in aqueous solution, and about 100-fold faster than in aqueous perchloric acid solution. A later study [53], in which the concentration of OAT was decreased, found that there were two rate maxima (for discussion of this phenomenon see Sec. III) and that $k_{\psi(max)}$ was 5×10^7-fold greater than that in aqueous solution. Organized assemblies, such as reversed micelles, are being actively explored as membrane-mimetic reagents, for it has always been the hope that the unique environment provided by the polar interior of these surfactant aggregates in apolar solvents would prove to be closely analogous to that of an enzyme, but this expectation has rarely resulted in the identification of rate enhancements of the magnitude that we have come to expect from enzymic catalysis.

However, very few kinetic, as distinct from equilibrium, studies have been made using these unique interfacial catalytic systems as media for studying inorganic reaction mechanisms. This area is certainly one which warrants more extensive investigation.

In the present reactions, the origin of the catalysis by the reversed micelles was clearly not favorable partitioning, since the rate of aquation in both pure benzene and water is negligible. Favorable partitioning was invoked to account for the rate enhancement described for imidazole-catalyzed hydrolysis of PNPA in water pools solubilized by OAT in octane [54]. It was suggested that in the aquation of these very polar inorganic substrates, micellar catalysis was probably the consequence of hydrogen bonding between the oxygen atoms of the substrate and the ammonium ion of the surfactant, thereby facilitating proton transfer and increasing the concentration of the one-ended dissociated intermediate in the preequilibrium steps. The decrease in k_ψ values with increasing surfactant concentration is then the consequence of a decreasing effective water concentration per micelle, subsequent to optimum saturation of the polar cavity of the reversed micelle by water.

III. MASS ACTION OR MULTIPLE-EQUILIBRIUM MODEL

It was noted above that, at low concentrations of DAP and OAT micelles in benzene, the rate-surfactant concentration profiles for aquation of $[Cr(C_2O_4)_3]^{3-}$ exhibited two maxima [53]. At the time, no explanation was offered for this phenomenon, but it is now thought that these complex rate profiles are an indication that the pseudophase model may not give a completely satisfactory definition of the aggregation process occurring in apolar media. In aqueous solution, similar "bumpy" profiles have been observed for the hydrolysis of trimethyl orthobenzoate [55] in solutions of a variety of bile salt micelles. The rate-concentration profiles were characterized by peaks and troughs which coincided with the discontinuities in the slopes of the surface tension-concentration profiles [56]. It has been well documented that bile salt micelles do not always conform to the pseudophase model. Various alternative models, including a complex pattern of dimers and one or more higher oligomers [57], a continuous association similar to stacking [58], and a stepwise association process [59], have been formulated.

Such experimental observations, showing a concentration-dependent growth of the aggregates, were in keeping with a multiequilibrium model, corresponding to the type of stepwise aggregation shown in Eq. (22):

$$\text{monomer} \underset{}{\overset{K_{12}}{\rightleftharpoons}} \text{dimer} \underset{}{\overset{K_{23}}{\rightleftharpoons}} \text{trimer} \underset{}{\overset{K_{ij}}{\rightleftharpoons}} \text{n-mer} \qquad (22)$$

In the discussion of the observed discontinuities in the surface tension studies and the supporting rate data, it was suggested [56, 60] that micelle formation of bile salts passes through several distinct stages. The association process is one of multiple association of monomers or small oligomers, rather than a monomer $\rightleftharpoons$ n-mer equilibrium, and each small aggregate possesses its own characteristic substrate binding constant and its own kinetic identity as well as its own unique ability to modify the surface tension properties of the solvent.

When the decomposition of PNPA was studied at low concentrations of alkanediamine bis(dodecanoate) salts in benzene at 341 K, the rate data did not fit on continuous smooth curves, but on profiles which bore a close resemblance to roller coasters. Such a profile is shown in Fig. 10 for the reaction in the presence of ethane-1,2-diamine bis(dodecanoate) (EtDB) [38,50].

These data provide further evidence that the pseudophase model is inadequate to account for the interfacial catalytic effects in apolar media and suggest that a model which allows for several small aggregates may be more applicable. Kertes [61] emphasized that when ionic surfactants undergo stepwise aggregation in hydrocarbon solvents, the dipole-dipole interactions which promote aggregation give rise to the characteristic stability of aggregates with low numbers of monomers and aggregates at very low concentrations of surfactants (typified, for example, by alkylamine carboxylate systems). The data in Fig. 10 provide some kinetic evidence for a stepwise aggregation, and even in more concentrated surfactant solutions it has been shown that good agreement between theory and kinetic results is obtained [49].

The multiple-equilibrium model, Eq. (22), implies the coexistence of monomers, dimers, trimers, and additional higher aggregates. Such aggregates, some present in undetectably small concentrations, are not to be regarded as persistent long-lived entities of well-defined aggregation number [62].

Lo et al. [44] made a detailed study of the state of aggregation of DAP in benzene and cyclohexane, using vapor pressure osmometry. Plots of MW_1/MW_{na} against concentration (where MW_1 is the molecular weight of the monomer and MW_{na} that of the number average) were used to test for the following types of association:

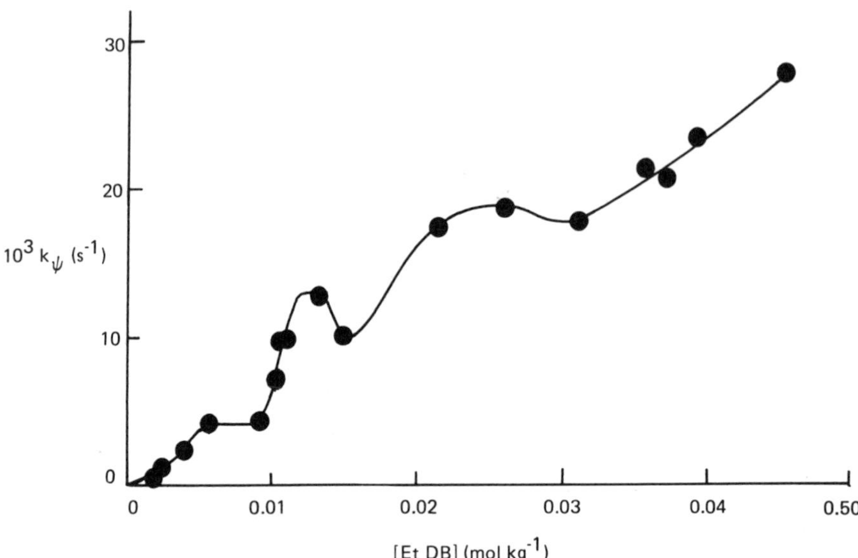

FIG. 10 Variation of the pseudo-first-order rate constants for decomposition of *p*-nitrophenyl acetate ($\sim 10^{-5}$ mol kg^{-1}) with concentration of EtDB in benzene at 341 K [38,50].

1. Monomer $\underset{}{\overset{K_n}{\rightleftharpoons}}$ n-mer association $\quad n = 2,3,\ldots$ (23)
2. Indefinite self-association

$$\text{monomer} \overset{K_{12}}{\rightleftharpoons} \text{dimer} \overset{K_{23}}{\rightleftharpoons} \text{trimer} \overset{K_{34}}{\rightleftharpoons} \text{tetramer} \cdots \quad (24)$$

Four types were considered.

Type I: sequential indefinite—all association species are assumed to be present and all molar equilibrium constants K_{ij} are assumed to be equal.

Type II: even indefinite—trimers, pentamers, and all other odd species are absent, $K_{12} = K_{24} = K_{26} = \cdots = K$.

Type III: sequential indefinite, $K_{12} \neq K_{23} = K_{34} = \cdots K$.

Type IV: even indefinite, $K_{12} \neq K_{24} = K_{26} = \cdots K$.

3. Monomer–m-mer association, e.g., (1,n,m) = (1,2,4), (1,2,6), (1,2,8)
$nm \rightleftharpoons M_n \qquad mm \rightleftharpoons M_m$

Vapor pressure osmometric data for AOT in benzene were consistent with either type I or type II (monomer ⇌ 6-mer ⇌ 14-mer) association [63], but the aggregation of dodecylpyridinium iodide in benzene was best described by the equilibrium monomer ⇌ 7-mer and only the monomer and micelle concentrations were significant [64].

Kertes and Gutman [62] argued that the experimental data for aggregation in apolar media can be satisfactorily explained only in terms of a stepwise aggregation process. They distinguished between the *extent* (the number of aggregated units) and the *degree* (the size of aggregated units) of aggregation. According to this argument, the concept of a CMC will no longer be appropriate, since this implies an equilibrium between monomer and aggregates of size defined by the aggregation number, and does not provide for the possible existence of clusters of intermediate size smaller than this micellar aggregate.

This conclusion may be a little drastic, for there is much evidence to demonstrate the occurrence of at least an "operational CMC," as defined by Fendler [24]. The sharp breaks in the plots of the intensity of the long-lived component in the positron lifetime spectrum against concentration of DAP [65] and the abrupt changes in measurements of electric field effects for AOT [38] were not reconcilable with a multiple-equilibrium model. Eicke and Denss [66] suggested that the search for evidence for a CMC in apolar media was not meaningful unless the aggregation data could be fitted to the pseudophase model.

The controversy concerning the nature of association of surfactant molecules in apolar solvents and the applicability of the CMC concept remains unresolved. It seems unlikely that any one theory will serve to explain the aggregation behavior of nonionic, cationic, and anionic surfactants in all apolar solvents. Nevertheless, physical evidence is accumulating that the sequential self-association model, as described by Lo et al. [44], best fits many of the experimental data. If all the values of the equilibrium constants K_{12}, K_{23}, . . ., K_{ij}, in Eq. (22) are assumed to be equal, then the weight fraction of the monomer, f, is related to the stoichiometric concentration of surfactant, [S], by Eq. (25):

$$\frac{(1 - f)}{f}^{1/2} = K_{ij}[S] \qquad (25)$$

Although Prigogine and Defay [67] developed Eq. (25) in 1954, it was not until 1983 that an attempt was made to fit kinetic data with a sequential self-association model [49].

A. Decomposition of Aryl Esters

Cox and Jencks [68] have suggested that the methoxyaminolysis of phenylacetates (in aqueous solution) might proceed through a preassociation mechanism. In this mechanism the reactants and catalyst come together in an encounter complex prior to addition of the amine.

Their results and those of Cordes et al. [69] for the methoxyaminolysis of PNPA indicated that general acid catalysis by alkylammonium ions (or carboxylic acid) should give a Brönsted plot with slope α equal to 0.16. O'Connor and Lomax [38] showed that, for the reaction of PNPA with a homologous series of alkane-α,ω-diamine bis(dodecanoate)s in benzene, α varied from a value of 0.085 at low surfactant concentrations to 0.17 at higher concentrations. These results were consistent with Cox and Jencks's postulate that with strong acids the rate-limiting attack of the amine on the ester occurred with the catalyzing acid being present in a position in which it could rapidly protonate the oxygen atom of the addition intermediate. In the alkylamine carboxylate/apolar solvent systems, the interactions between the ester and the surfactant are likely to be strong at low surfactant concentrations. The highly polar ammonium ion will be destabilized in benzene compared with its stability in water, whereas the diffuse charge on the protonated ester will be more stable. This strong interaction between the ester and the general acid resulted in a lessened sensitivity of the reaction to the pK_a of the general acid. The Brönsted slope therefore became close to zero at very low surfactant concentrations.

At high surfactant concentrations hydrogen bonding between the component species of the surfactant should cause an effective decrease in acidity. This postulate was confirmed, for the reaction became more sensitive to acidity as the interaction between the ester and the surfactant became weaker, and the resultant Brönsted slope increased.

Independent evidence has been produced to show that the reaction proceeded by aminolysis of the unprotonated ester, thus supporting the concept of a preassociation mechanism [70]. Calculations were made for the ρ value, approximately +1, for the esterolysis reactions of an extensive series of 4'-nitrophenyl 4-substituted benzoates, in DAP and butane-1,4-diamine in benzene, by substituting the kinetic rate constants into Hammett σ and Tsuno-Yukawa linear free energy relationships.

Carbon-13 NMR product analysis for the reaction between ^{13}C-enriched *p*-nitrophenyl propionate (PNPP) and DAP showed the existence of an exclusive pathway, i.e., aminolysis to form *N*-propylpropionamide [20]. Within 1% accuracy, pathways representing partial hydrolysis or carboxylate ion attack were excluded.

This unambiguous result thus solved the problem of the identity of the attacking nucleophile in this amine-carboxylate-salt system when the ratio of ester to DAP was 1:50. However, it was emphasized that it did not exclude the possibility of carboxylate ion attack at very low surfactant concentrations, i.e., approximately a 30-fold dilution in the low-concentration region of the rate-surfactant concentration profiles. The possibility that hydrolysis might also occur in the presence of large amounts of added water was also not excluded. Such a possibility was indicated by the work of Kon-no et al. [71] and Wallerberg et al. [72].

El Seoud et al. also reported [73] that in the reaction between PNPA and DAP only one amide product, CH_3CONHR, where R represents the dodecyl group, was produced. The carboxylate ion therefore acts as a general base, because if it had participated as a nucleophile, leading to formation of $CH_3COOCOC_2H_5$ as an intermediate followed by rapid amine attack, then two amides, namely CH_3CONHR and C_2H_5CONHR, would have been identified.

The reversed micelles of neither didodecyldimethylammonium bromide (DDAB) nor Aerosol OT catalyze the decomposition of PNPA by themselves. However, these reversed micelles increase the action of imidazole in the decomposition of PNPA by solubilizing it. This interesting carrier effect of the reversed micelle on imidazole was investigated [71], and while aminolysis occurred in the anhydrous systems, addition of water to the reaction system of imidazole-PNPA-surfactant in CCl_4, so that $[H_2O]/[S] \sim 1$, resulted in hydrolysis. Water pooled in the reversed micelle accelerated the catalytic effect more markedly than free water.

Nevertheless, the magnitude and sign of ρ, approximately +1 [38], the ^{13}C product analysis, which showed that the amine was the reactive nucleophile [20], and the kinetic isotope effect (k_H/k_D) value of 1.2 for the reaction of PNPA in 0.2 mol dm^{-3} DAP in benzene at 298 K [33] reflect enforced association.

B. Role of Individual Oligomers

The changing dependence of rate on surfactant concentration is more easily seen if it is plotted in the form $k_\psi/[S]$. Figure 11 shows such a plot for the decomposition of PNPA in solutions of DAP in benzene at 341 K [20], and the inset in Fig. 11 shows a plot of the fraction of surfactant which exists in the monomeric form, calculated for the indefinite sequential self-association model, K_{ij} = 300 [49].

Plots of [oligomer]/[S] should mimic the changes in $k_\psi/[S]$ if the rate is proportional to [oligomer]. It is possible to describe the concentrations of oligomers in a simple aggregating system, the indefinite sequential self-association model, by specifying the association constant.

The data in Fig. 11 show a remarkable resemblance between the shapes of the profile for fraction of monomer present and the values of $k_\psi/[S]$, up to a surfactant concentration of 1 mol kg^{-1}, suggesting that the concentration of the monomer in this system is of paramount importance in determining the magnitude of the reaction rate.

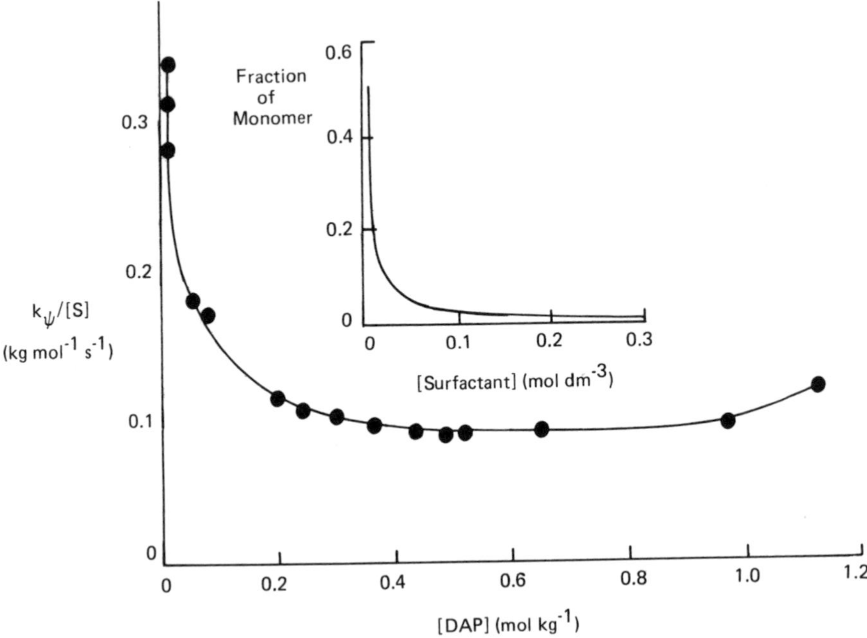

FIG. 11 Variation of the values of $k_\psi/[S]$ for decomposition of *p*-nitrophenyl acetate ($\sim 10^{-5}$ mol kg^{-1}) with concentration of dodecylammonium propionate in benzene at 341 K [20]. Inset: fraction of surfactant which exists in the monomeric form calculated for the sequential indefinite self-association model, K_{ij} = 300 [49].

In the range of surfactant concentrations investigated, the weight fraction of monomer, f, is related to the stoichiometric concentration of surfactant by Eq. (25). The constant K_{ij} is inversely proportional to the dielectric constant of the medium, which in turn is a function of both the dielectric constant of the solvent and that of the surfactant. As the mole fraction of surfactant increases, the dielectric constant of the medium and therefore f increase. At [S] > 1 mol kg^{-1}, $k_\psi/[S]$ increases, a result consistent with the postulate that, if the contribution of the monomer to the overall reaction is equal to or greater than that of other oligomeric species, then at high mole fraction of surfactant, where f increases, the dependence of k_ψ on [S] once again approaches second order. Other occurrences of second-order dependence at high surfactant concentrations (e.g., [16]) have been noted. However, at moderate [S], formation of oligomers tends to remove monomer from the solution and values of k_ψ, which become more nearly dependent on [S], reflect this association.

In the series of alkanediamine bis(dodecanoate) surfactants, plots of $k_\psi/[S]$ against [S] were generally of the shape shown in Fig. 12. This figure shows the data for decomposition of PNPA in solutions of butane-1,4-diamine bis(dodecanoate) (BuDB) in benzene at 341 K [20]. Similar distorted bell-shaped profiles followed by a plateau were obtained for each of the diamine bis(dodecanoate) surfactant salts [38,49,50], but the shape, height, and position of the maximum in the profile, which occurs at small [S], were very dependent on the length of the alkanediamine chain. Increasing carbon chain length decreased the maximum value of $k_\psi/[S]$ but increased the value of [S] at which this occurred.

But there was close similarity between these plots and the fractional plot of [dimer]/[S] against [S]. This may be seen by comparing Fig. 12 with Fig. 13. The latter figure shows the variation in the fraction of various oligomers present on increasing [S], calculated for the sequential indefinite self-association model, K_{ij} = 100 [49].

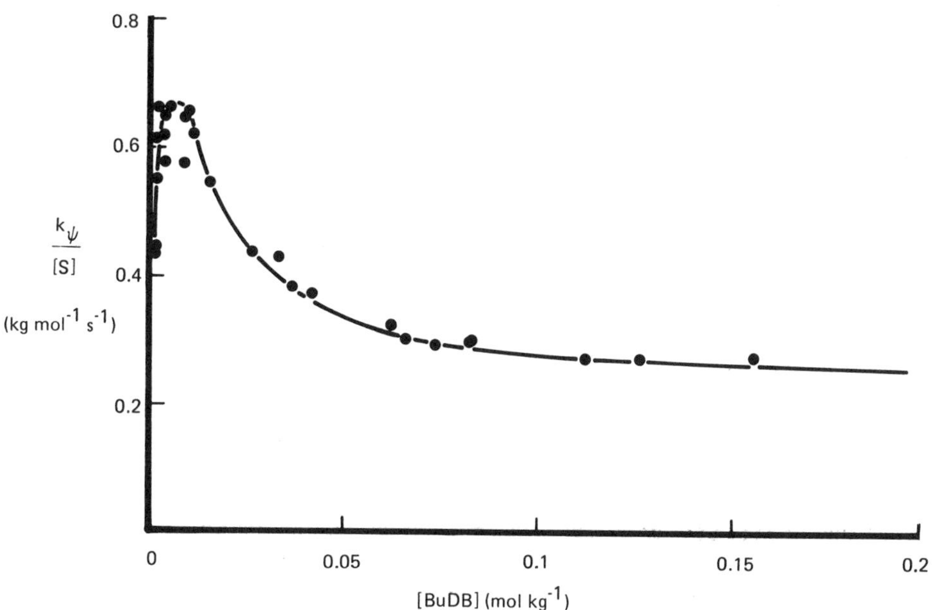

FIG. 12 Variation of the value of $k_\psi/[S]$ for decomposition of *p*-nitrophenyl acetate ($\sim 10^{-5}$ mol kg^{-1}) with concentration of BuBD in benzene at 341 K [20].

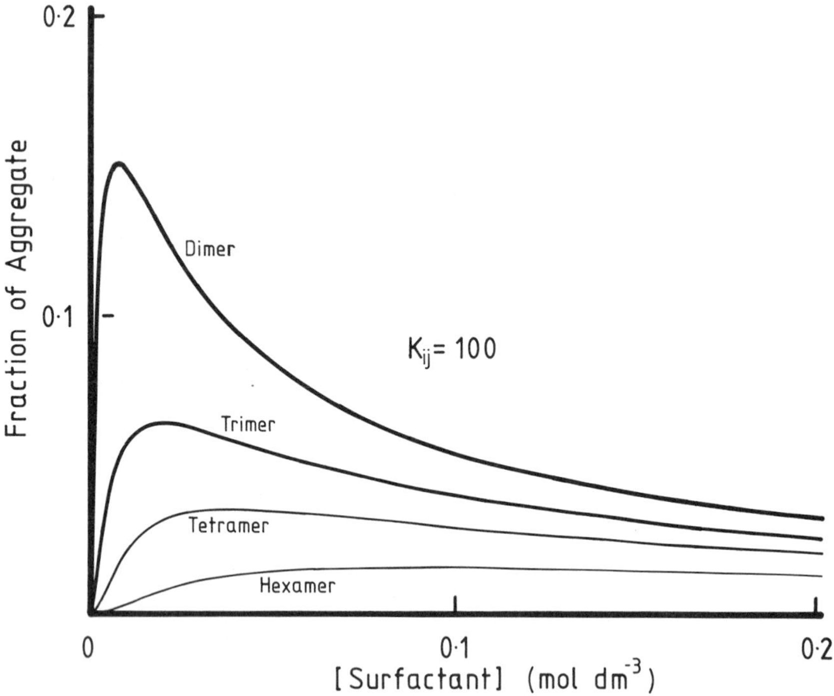

FIG. 13 Variation of the fraction of surfactant which exists as various oligomers upon surfactant concentration, calculated for the sequential indefinite self-association model, K_{ij} = 100 [49].

However, it was then formulated that more than one oligomer catalyzed the reaction, and that the catalysis could be treated on the basis of the total number of species in solution. The observed rate data were best fitted by Eq. (26):

$$k_\psi = k_1[\text{monomer}]^2 + k_2[\text{dimer}] + k_3[\text{trimer}] + \cdots + k_n[n\text{-mer}] \tag{26}$$

in which the overall rate constant is the sum of a number of individual rate constants multiplied by the concentrations of their relative oligomers, and includes a contribution from the monomer.

The analysis was further refined by testing the application of a model in which the values of K_{ij} were allowed to be markedly different. The curves of fraction of [total aggregate]/[S] against [S], shown in Fig. 14, were calculated by using a sequential indefinite

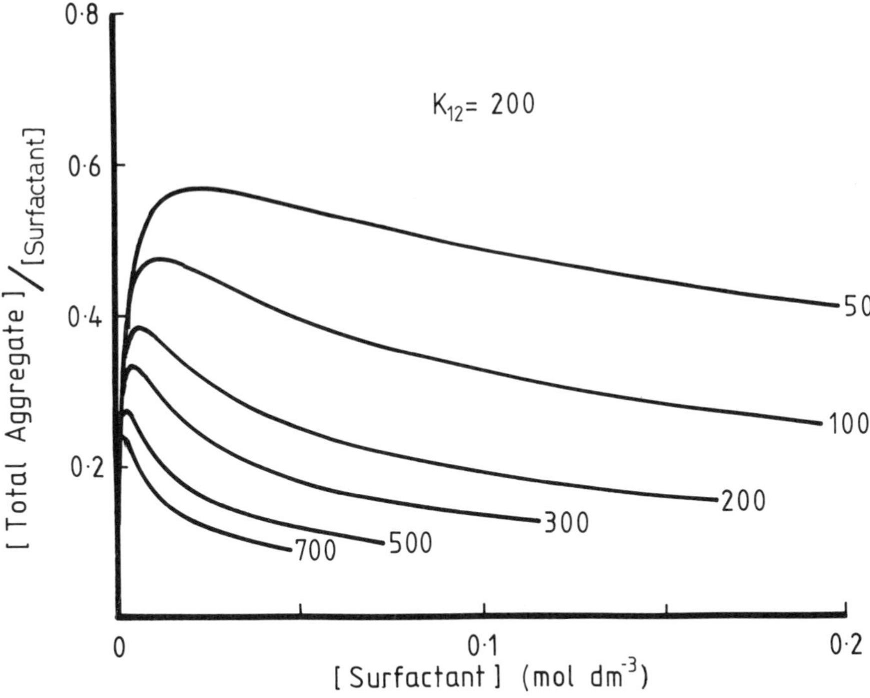

FIG. 14 Variation of the fraction of total aggregate upon surfactant concentration, calculated for the sequential indefinite self-association model, K_{12} = 200, $K_{23} = \cdots = K_{ij}$ varying as shown [49].

self-association model in which the first step association constant, K_{12}, was set equal to 200, and all other K_{ij} were assumed equal but were varied over values between 50 and 700 as shown. The curve for $K_{12} = K_{ij}$ = 200 resembles Fig. 12 more closely than any of the others [49].

It was concluded [49] that while aggregation affected the rate of reaction, it might have an inhibitory effect on the rate at moderate concentrations of surfactant, for such a process reduces the total number of discrete species in solution.

IV. DISTRIBUTION MODEL FOR BIMOLECULAR REACTIONS

In Sec. II it was shown that the simple distribution model, scheme 1, is adequate for at least qualitative treatment of micellar effects on unimolecular reactions because these effects can be treated in terms of the distribution of only *one* reactant between the apolar and micellar pseudophases.

But this model fails for micellar catalyzed bimolecular reactions, whose rate constants generally pass through maxima with increasing surfactant concentrations. Figure 15 shows a typical example for the apparent rate constants of formation (anation), k_2^{app}, and decomposition (aquation), k_{-2}^{app}, of the glycine adduct of vitamin B_{12a} in solutions of DAP in benzene at 298 K. Similarly shaped profiles were obtained for the imidazole and sodium azide adducts in solutions of DAP in benzene and also for all three adducts in solutions of Aerosol OT in benzene [74].

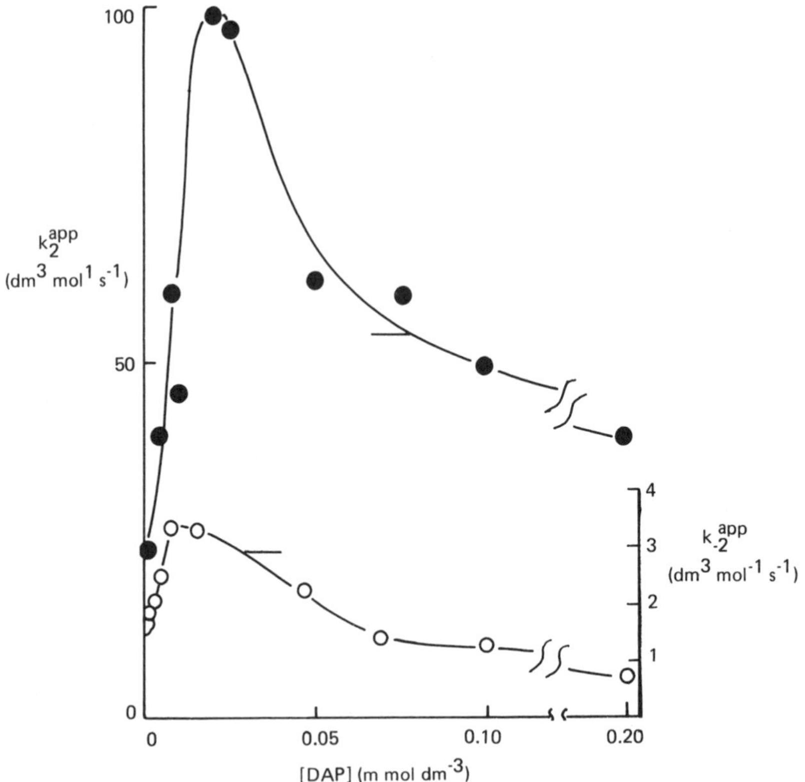

FIG. 15 Variation of the second-order rate constants for formation, k_2^{app} (•), and aquation, k_{-2}^{app} (○), of the vitamin B_{12a} adduct with glycine in benzene in the presence of 0.01 mol dm^{-3} solubilized water at 298 K [74].

A. Micellar Incorporation of Solutes

At low surfactant concentrations the rate-surfactant concentration profiles are similar to those found for unimolecular reactions, i.e., the rate constants increase as the substrates are taken up by the micelles, but instead of reaching a constant value they go through maxima, indicating the presence of opposing effects. If it were possible to predict these profiles by estimating the concentrations of both reactants in the micellar pseudophase, then it should be possible, in principle, to estimate the true second-order rate constants in the micellar pseudophase and thereafter decide whether the catalysis is due merely to an increase of reactant concentration in the micellar pseudophase or to a favorable environmental effect on the reaction.

It is easier to ask this question than to answer it, because any comparison of second-order rate constants depends on the units in which the reactant concentrations are measured. Typically, solution kineticists measure concentrations in molarity, or in rarer cases, where the experimental conditions make it more convenient to measure the mass of the reactants and their solvents rather than their volumes, in molalities [20,38,70]. One approach arising from the former method is to estimate the volume of the micellar pseudophase and use this as the volume element of the reaction [75]. But this approach may have its problems because some substrates may be solubilized, at least partially, by hydrophobic interactions with the tails of the surfactant molecules rather than entirely within the polar core, so there is uncertainty as to whether one should use the volume of the complete micellar pseudophase or only that of the polar core. The choice of the volume element is arbitrary and will affect the numerical value of the rate constants, and the method usually adopted is to quote apparent second-order rate constants, k_2^{app}, calculated from the molar concentrations of the reactant species.

B. Rate-Surfactant Profiles in Bimolecular Reactions

A rate maximum in the interfacial catalysis of bimolecular reactions in apolar media has also been observed for the acylation of benzimidazole with 2,4-dinitrophenyl acetate in octane solutions containing Aerosol OT at 299 K [76], but generally the phenomenon has not been as well documented as it has for the corresponding micellar catalyzed bimolecular reactions in aqueous media.

This last-mentioned study provided the first example of a quantitative analysis of the general physicochemical principles responsible for the influence of reversed micelles on the kinetics of chemical reactions in organic solvents. The main aim of the work was to check whether it was possible, within the framework of pseudophase theory, to describe the kinetic laws of chemical reactions even in the presence of inverted surface-active agents in organic solvents.

The solution of the problem is conceptually simple, for the pseudophase distribution model predicts that the reaction rates will initially increase as the two reactants are brought together in the micellar pseudophase, and then decrease as the reactants eventually become diluted in that pseudophase as the surfactant concentration is increased.

The quantitative treatment of bimolecular micellar catalyzed reactions requires estimation of the extents of micellar incorporation of both reactants. This problem may not be soluble in all cases, because either substrate may affect the incorporation of the other, and may be additionally regulated by varying the amount of solubilized water, for this factor leads to substantial changes in kinetic parameters [77].

The kinetic treatment which has been applied, and which was modeled on the quantitative analysis of aqueous micellar catalyzed reactions, assumes that the micelles can be treated as if they are a separate phase.

C. Generalized Treatment of Interfacial Catalysis of Bimolecular Reactions

The overall observed first-order rate constant k_ψ is given by Eq. (7), and the individual first-order rate constants k_0 and k_M can be written in terms of the appropriate second-order rate constants $k_{2,0}$ and $k_{2,M}$. Scheme 3 illustrates the reaction between a substrate X and a nucleophile N occurring in the organic solvent and in the micelle.

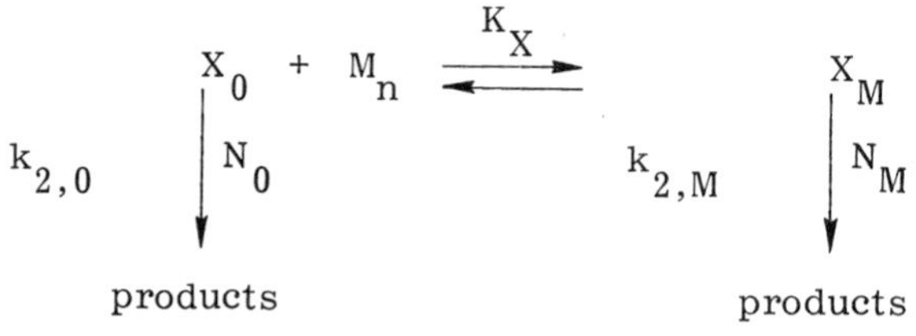

$$
\begin{array}{ccccc}
 & X_0 + M_n & \underset{}{\overset{K_X}{\rightleftharpoons}} & & X_M \\
k_{2,0} & \downarrow N_0 & & k_{2,M} & \downarrow N_M \\
 & \text{products} & & & \text{products}
\end{array}
$$

Scheme 3

(The subscripts 0 and M denote the solvent and micellar pseudophases.)

The overall rate, described by Eq. (1), is modified to

$$R_{total} = k_{2,M}[X]_M[N]_M[M_n]\bar{V} + k_{2,0}[X]_0[N]_0(1 - [M_n]\bar{V}) \qquad (27)$$

where $\bar{V}$ is the molar volume of the surfactant aggregate and the concentrations of reagents X and N are given by the equations of mass balance:

$$[X]_{total} = [X]_M[M_n]\bar{V} + [X]_0(1 - [M_n]\bar{V}) \tag{28}$$

$$[N]_{total} = [X]_M[M_n]\bar{V} + [X]_0(1 - [M_n]\bar{V}) \tag{29}$$

If the chemical reactions described by scheme 3 do not affect partition equilibria:

$$[X]_0 \underset{}{\overset{P_X}{\rightleftarrows}} [X]_M \tag{30}$$

and

$$[N]_0 \underset{}{\overset{P_N}{\rightleftarrows}} [N]_M \tag{31}$$

where P_X and P_N represent the partition coefficients of X and N, respectively. The observed second-order rate constant for reaction in the presence of aggregate is given by [75,77,78]:

$$k_2^{app} = \frac{k_M P_X P_N [M_n]\bar{V} + k_0(1 - [M_n]\bar{V})}{[1 + (P_X - 1)[M_n]\bar{V}][1 + (P_N - 1)[M_n]\bar{V}} \tag{32}$$

If the volume fraction of the micellar phase is small ($[M_n]\bar{V} \ll 1$), i.e., the surfactant solution is dilute, and if both X and N bind strongly to the host ($P_X \gg 1$ and $P_N \gg 1$), then Eq. (32) simplifies to Eq. (33):

$$k_2^{app} = \frac{(k_M/\bar{V})K_X K_N [M_n] + k_0(1 - [M_n]\bar{V})}{(1 + K_X[M_n])(1 + K_N[M_n])} \tag{33}$$

where the binding constants are expressed by

$$K_X = (P_X - 1)\bar{V} \tag{34}$$

$$K_N = (P_N - 1)\bar{V} \tag{35}$$

The experimental data on the dependence of k_2^{app} on $[M_n]$ can be used to calculate k_M, K_X, and K_N. For this purpose Eq. (33) is transformed to

$$\frac{[M_n]}{k_2^{app} - k_0} = \alpha + \beta[M_n]\frac{k_2^{app}}{k_2^{app} - k_0} + \gamma[M_n]^2\frac{k_2^{app}}{k_2^{app} - k_0} \tag{36}$$

where

$$\alpha = \frac{\bar{V}}{k_M K_X K_N} \tag{37}$$

$$\beta = \alpha(K_X + K_N + k_0\bar{V}) \tag{38}$$

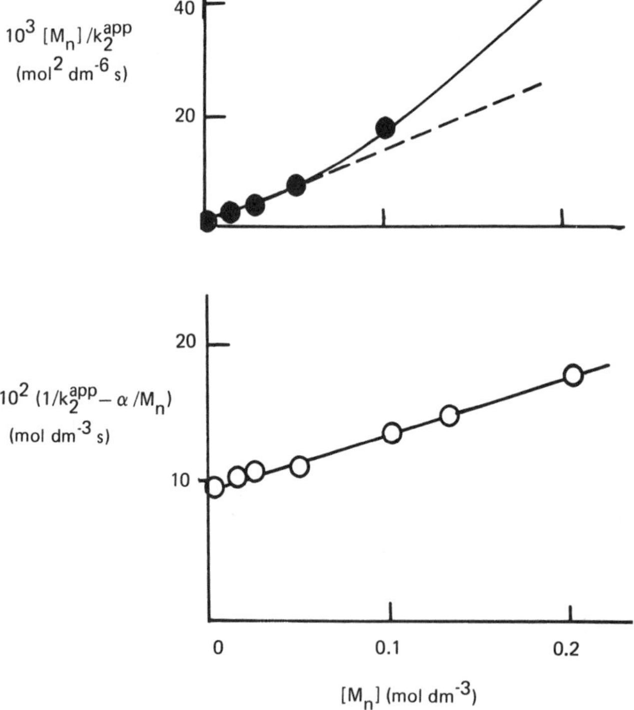

FIG. 16 Graphical analysis of the experimental second-order rate constant for the acylation of benzimidazole with 2,4-dinitrophenyl acetate in Aerosol OT solutions in octane at 299 K. The experimental results shown are given in the coordinates (a) of Eq. (36) and (b) of Eq. (40) [76].

$$\gamma = \alpha K_X K_N \tag{39}$$

A plot of the data according to Eq. (36) gives a value for the intercept α. This allows further analysis in terms of the rearranged equation:

$$\left(\frac{1}{k_2^{app}} - \frac{\alpha}{[M_n]}\right)\left(1 - \frac{k_0}{k_2^{app}}\right) = \beta + \gamma[M_n] \tag{40}$$

which in turn provides numerical values for β and γ and hence for k_M, K_X, and K_N.

Figure 16 shows a graphical analysis of the curve of k_2^{app} against $[M_n]$ for acylation of benzimidazole with 2,4-dinitrophenyl acetate in solutions of Aerosol OT in octane [76]. Independent determination of the binding constants of the reactants with the micelles, from results on the influence of the solubilities of the reactants, were extremely close to those calculated from the kinetic results. The validity of the pseudophase model was confirmed by calculation of the theoretical curves from Eq. (33) using these experimental values of the binding constants. In all cases the experimental and theoretical results were superimposable.

Thus it was seen that the pseudophase model could be applied successfully to the quantitative analysis of the observed kinetic laws. This model provides the possibility of predicting the optimum conditions for performance of micellar catalysis [78]. Fendler [79] has stressed that a proper understanding of these kinetic processes is highly relevant for the meaningful design of appropriate systems for lubrication and corrosion control.

V. ROLE OF COSOLUBILIZED WATER

The addition of controlled amounts of surfactant-entrapped water in apolar solvents provides a unique medium for interactions and reactions of polar substrates. The term *water pool* was originally coined by Menger et al. [54] to describe reversed micelles containing large amounts of water, but since then usage of the term has been extended to describe the interior of aggregates formed in apolar solvents, because these contain some water even after careful drying. Eicke and Christen [5] suggest that this trace amount of water may be necessary for aggregation to occur at all. Since DAP can be prepared water-free, the contamination of water in "dry" benzene due to solvent has been estimated as 0.002 mol dm^{-3} [80] or 0.03% w/w [81]. Seno et al. [82] found that, at 308 K, 5 ml of a 0.2 mol dm^{-3} DAP solution in benzene could solubilize 230 μl of water, i.e., a stoichiometric concentration of 2.55 mol dm^{-3}. The presence of water has a profound effect on the size of the aggregate. Addition of 0 to 7 molecules of water per molecule of dodecylamine butanoate in benzene changed the aggregation number from 4.3 to 34.5 molecules per aggregate [83]. In the Aerosol OT/octane system it has been estimated that 2–4 water molecules are used in hydration of the cation, 10 water molecules in hydration of the anion, and further water molecules form an independent microphase in the polar interior of the aggregate [84]. The dependence of the apparent CMC on water concentration (i.e., relatively independent at low, but not at high, water concentration) results in loss of significance of the concept of a CMC [5]. Another complication to appear is that the apparent acidities of the polar cores change with increasing surfactant concentration [85] and the ability of the DAP/cyclohexane system to act as a buffer depends on water concentration [86].

A. Molecular Interactions

In the light of the critical presence of this third component in the structure of reversed micelles, it is not surprising that many of the interfacial catalyses observed in apolar media arise in aquation or similar reactions and result in the net consumption of water molecules.

Crystal violet underwent a hydrolysis reaction in DAP/cyclohexane-solubilized water which resulted in formation of crystal violet carbinol [87]. The rate of this fading reaction was strongly dependent on the solubilized water concentration.

Examples, which have been discussed above, include the aquation of the tris(oxalato)chromate(III) anion, which showed a linear dependence on increasing concentration of solubilized water [52], and the reversible hydration of 1,3-dichloroacetone in Aerosol OT in CCl_4 [45], for which catalysis was explained in terms of the replacement, in the transition state, of a surfactant anionic head group by a water molecule.

Martinek et al. [88], however, believed that the increase in the rate of hydrolysis of picryl chloride in Aerosol OT dissolved in octane was due not to the increased concentration of the reactants in the surfactant aggregate, but to the increased basicity of water arising from specific interactions of the molecules of water with the metal cation. The values of k_ψ increased up to an R value ($[H_2O]/[surfactant]$) equal to three and then remained almost constant. The size of the water pool was also found not to affect the rate of oxygen atom exchange between the carboxylate oxygen atoms of DAP and solubilized $H_2{}^{18}O$, but at a constant concentration of $HClO_4$, representing 75% conversion of DAP to DAPH, the rate of exchange was termolecular with respect to added water as solubilizate [89]. (Other studies [90,91] have indicated that there may be preferred complexes, besides the obvious 1:1 complex, between amines and carboxylic acids. These complexes were described as "orientationally ill-defined 3:1 aggregates" containing three carboxylic acid groups to one amine molecule.)

The oxygen atom exchange reaction between the propionate ions of DAP and cosolubilized water and the trans-cis isomerization of the bis(oxalato)diaquochromate(III) anion do not consume water molecules, but they do involve active participation of the cosolubilizate during the course of the reaction. In this latter system also, independence of added water concentration was observed, following an initial sharp rise as the number of water molecules per micelle was increased up to an R value of unity. The plateau value was interpreted as the consequence of saturation kinetic behavior arising from an optimum saturation of the polar cavity of the reversed micelle by water, in excess of the solvation requirements of the substrate. The initial linear rate increase with increasing water concentration implied the involvement of water in the rate-determining step. (See Sec. V,C for further interpretation of the significance of an R value of unity.)

The aquation of cobalt chloride has been investigated in solutions of dodecylpyridinium chloride solubilized in $CHCl_3$ [92]. Spectral evidence showed that, while in the absence of water the solubilized cobalt(II) ion was present as the tetrahedral $CoCl_4^{2-}$ complex, addition of water converted this tetrahedral species to an octahedral one. Reaction scheme 4 was formulated:

$$CoCl_4^{2-} + 4H_2O \rightleftharpoons CoCl_2(H_2O)_4 + 2Cl^-$$

$$CoCl_2(H_2O)_4 + 2H_2O \rightleftharpoons Co(H_2O)_6^{2+} + 2Cl^-$$

Scheme 4

It was concluded from temperature jump measurements that the rate-determining step was conversion of $CoCl_3(H_2O)_3^-$ to $CoCl_2(H_2O)_4$. This study found that the equilibria were best described on the basis of microscopic concentrations (i.e, the localized concentration around the metal ion in the micellar core), but that the reaction rates depended on both macroscopic and microscopic concentrations. The dependence of reaction rate on macroscopic concentrations was supported by the observation of Eicke et al. [93] that there was rapid exchange of water and electrolyte solutions between the aggregates present in solutions of Aerosol OT in octane. This observation was based on measurements of the fluorescence spectrum of the probe Tb^{3+}.

Robinson et al. [94] have also shown that there appears to be virtually no energy barrier to the communication of material between water pools in the Aerosol OT/H_2O/heptane system. The kinetic results for complexation of nickel(II) solubilized in murexide-containing micelles supported a mechanism involving rapid exchange of reactants between aqueous pools followed by rate-limiting loss of a solvated water molecule from the metal ion in a water pool. Rate constants for ligand release and water exchange in heterogeneous water were similar to those in bulk water and, to a first approximation, were independent of the size of the pool.

Aquation studies on surfactant-solubilized cobalt(II) salts in $CHCl_3$ also showed that addition of water converted the tetrahedral CoX_4^{2-} species, initially present, to the octahedral $[CoX_4(H_2O)_2]^{2-}$ species. This equilibrium was temperature-dependent [7]. Similar solvochromism/thermochromism effects have been reported for surfactant-solubilized copper(II) halides in $CHCl_3$ [95,96].

Many of these aquation reactions in reversed micelles are extremely rapid. The aquation of $FeCl_3$ in dodecylpyridinium chloride-solubilized water in $CHCl_3$ has been studied by the temperature jump method [97] and scheme 5 was formulated for the reaction

$$\mathrm{DP{-}FeCl_3} + \mathrm{H_2O} \underset{}{\overset{K}{\rightleftharpoons}} \mathrm{DP{-}FeCl_3(H_2O)}$$

$$\mathrm{DP{-}FeCl_3(H_2O)} + \mathrm{H_2O} \underset{k_{-1}}{\overset{k_1}{\rightleftharpoons}} \mathrm{DP{-}FeCl_3(H_2O)_2}$$

Scheme 5

The slow relaxation time T_2 was inversely dependent on $[H_2O]^2$ according to Eq. (41):

$$\frac{1}{T_2} = k_1K[H_2O]^2 + k_2 \qquad (41)$$

and this result was interpreted as meaning that the k_1 pathway represented the rate-determining step.

Formation of a complex with strong bonding between copper(II) and norleucine *p*-nitrophenyl ester in Aerosol OT in $CHCl_3$ was postulated by Sunamoto et al. [98,99] to explain the observed saturation-type kinetics—scheme 6.

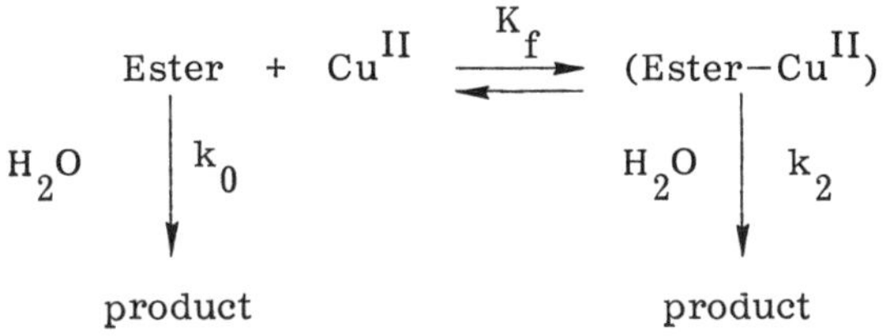

$$
\begin{array}{ccc}
\text{Ester} + Cu^{II} & \xrightleftharpoons{K_f} & (\text{Ester}-Cu^{II}) \\
H_2O \Big\downarrow k_0 & & H_2O \Big\downarrow k_2 \\
\text{product} & & \text{product}
\end{array}
$$

Scheme 6

Nucleophilic attack by water was enhanced by hydrogen bond formation between the water and the anionic surfactant, but addition of excess water resulted in aquation of the "naked" metal ion and a decrease in the metal-ester complex formation.

B. Polar Environment of the Reaction Field

Consideration of the aquation/hydration/hydrolysis reactions outlined above indicates that, generally, when these occur at the interface of the microphase in apolar media and cause a change in the net number of water molecules present in the water pool, either permanently or temporarily (because the water molecules are active participants in the reaction mechanims), there may be a positive dependence on added water concentration when the R value is small ($R \leqslant 3$), but at higher concentrations the substrate is saturated with respect to water, and the reaction becomes pseudo-first-order and independent of water concentration.

In some of these reactions, however, increasing water concentration causes a decrease in the rate of reaction. Thus, the changes in reaction rates for hydrolysis of sucrose in dodecylbenzenesulfonic acid in dioxane/water with respect to changes in substrate concentration and cosolubilized water were found to show behavior parallel to that of the stability of the micelle [100,101], and increasing water concentration resulted in a decrease in the rate of reaction when compared with the second-order rate constant for hydrolysis in aqueous hydrochloric acid solution.

The spontaneous hydration of acetaldehyde in solutions of the nonionic surfactant Triton X-100 in CCl_4 containing H_2O or D_2O also showed a decrease in the specific acid-catalyzed hydration with increasing concentration of solubilized water [102].

In the bimolecular ligand exchange reactions of vitamin B_{12a} in benzene [74], the aggregates of DAP were much larger than the aggregates of alkylammonium carboxylates normally found in apolar solvents [4]. From the dependence of its solubility on the concentration of DAP, it was deduced that the large vitamin B_{12a} molecule, contained in the restricted water pools, was surrounded by some 300 surfactant molecules and was therefore effectively shielded from the bulk apolar solvent.

A linear correlation existed between the absorbance maximum of vitamin B_{12a} in various solvents and the solvent polarity $E_T(30)$. A similar correlation was demonstrated for the solubility of glycine. The microscopic polarities of the environments of vitamin B_{12a} and glycine, solubilized in benzene by DAP, were estimated from these correlations and were shown to depend on the concentrations of water and of surfactant. The environment of DAP resembled that of highly structured water, while that of glycine approximated the polarity of alcohols.

Rate constants for the formation, k_2^{app}, and decomposition, k_{-2}^{app}, of the adducts of vitamin B_{12a}, Bzm—Co—OH_2, according to Eq. (42):

$$\text{Bzm—Co—OH}_2 + \text{Y} \underset{k_{-2}}{\overset{k_2}{\rightleftharpoons}} \text{Bzm—Co—Y} + \text{H}_2\text{O} \qquad (42)$$

(Y = glycine, imidazole, or sodium azide) were determined and k_2^{app} in the presence of glycine was 6.6×10^4-fold greater than that in pure water, while k_{-2}^{app} was approximately 445-fold greater than in water.

The maximum values of k_2^{app} and k_{-2}^{app} decreased with increasing water concentration, in a manner parallel to the decrease in the apparent microviscosity of vitamin B_{12a}. It was speculated, therefore, that the observed rate enhancements were largely due to tightening of the solvation shell around the substrates.

Results such as these, which relate the considerable specificity of reaction mechanisms to restricted polar environments, should help to unravel the binding mechanisms of complex amino acids and peptides and help to elucidate transport processes in vivo.

Stimulated by the similarity of the picture of a water droplet, surrounded by its monolayer of surfactant molecules with their polar heads toward the water phase core, to that of the water pockets in bioaggregates such as mitochondrial membranes, there have been several efforts to utilize the systems for mimicking enzymic functions. Following this approach, Seno et al. [103] studied the hydrolysis of adenosine 5'-triphosphate (ATP) solubilized by DAP in *n*-hexane, Eq. (43):

$$\text{ATP} + \text{H}_2\text{O} \xrightarrow{k^{app}} \text{ADP} + \text{H}_3\text{PO}_4 \qquad (43)$$

The pseudo-first-order rate constant k^{app} was proportional to the surfactant concentration and inversely proportional to the water concentration, according to Eq. (44):

$$k^{app}[\mathrm{ATP}] = k\frac{[\mathrm{DAP}][\mathrm{ATP}]}{[\mathrm{H_2O}]} \tag{44}$$

Somewhat disappointingly, k^{app} was only 1- to 5-fold greater than in aqueous solution, but the rate enhancement was increased 10- to 100-fold when Mg^{2+} or Ca^{2+} ions were added to the system.

The rate enhancement was attributed to a lowering of the polarity of the water phase in the core of the reversed micelle, and the [DAP]/ $[H_2O]$ term in Eq. (44) was taken as a representation of the degree of lowering of the polarity.

Divalent metal cations, especially those of Mg^{2+} and Ca^{2+}, are known to activate ATPase in living systems. Both cations are hard Lewis acids and have, therefore, a stronger interaction with the harder base, phosphate, than with the adenine ring. The additional rate enhancement caused by the divalent metal ions, Mg^{2+} being more effective than Ca^{2+}, was considered to be due to neutralization of the negative charges on the phosphate part of ATP, thus facilitating its hydrolysis by nucleophilic attack.

The rate equation for the reversed micellar catalysis reaction of alkyl *p*-nitrophenyl carbonates by DAP in benzene and hexane, Eq. (45) [16]:

$$\text{rate} = k_2\frac{[\mathrm{DAP}]^2}{[\mathrm{H_2O}]}\ \text{substrate} \tag{45}$$

was very similar to Eq. (44) and the findings were in accordance with these previous results.

When the substrate is anchored very close to the catalyst, the catalyst will work more effectively. In addition, dehydration from the ammonium ion and carboxylate ion head groups (the hydrophobic ion pair) [104,105] will render them more powerful catalysts than those in bulk aqueous media. The entrapment of substrates in the rigid interior core of reversed micelles brings about a convenient rigidity effect, which then causes them to become anchored at the reaction site in a manner very similar to that of the binding between an enzyme and a substrate.

In a further study aimed at demonstrating an enzyme model, Seno et al. [106] investigated the coordination reactions of ketones in *n*-hexane in the presence of DAP. Scheme 7 was proposed for the reaction mechanism:

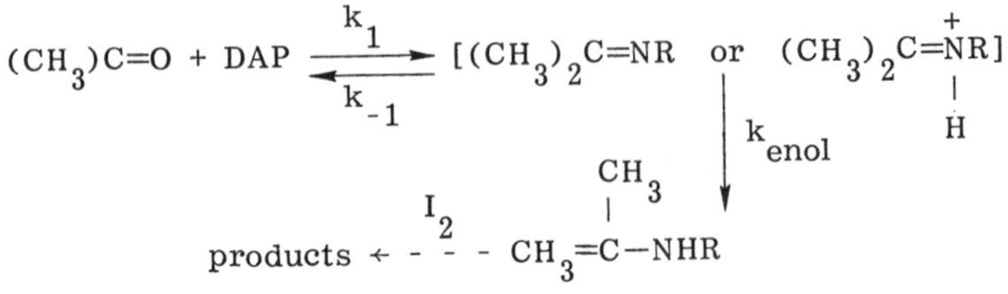

$$(CH_3)C{=}O + DAP \underset{k_{-1}}{\overset{k_1}{\rightleftharpoons}} [(CH_3)_2C{=}NR \text{ or } (CH_3)_2C{=}\overset{+}{N}(H)R] \xrightarrow{k_{enol}} CH_2{=}C(CH_3){-}NHR \xrightarrow{I_2} \text{products}$$

Scheme 7

in which k_1, k_{-1}, and k_{enol} are the rate constants for formation, decomposition, and enolization, respectively, of ketimine. The second-order rate constant of iodination of acetone was 10^6- to 10^7-fold greater in the presence of DAP than in its absence and depended in a complex fashion on the concentration of water, being nearly independent at low but becoming inversely dependent as the concentration of water increased. However, the value of k_{enol} was nearly proportional to the $[DAP]/[H_2O]$ ratio [cf. Eq. (44)], suggesting by analogy with the results of this earlier system [103] that the enolization of the ketimine occurred in the cores of the reversed micelles to form enamine, to which iodine added rapidly.

The rates of iodination of the higher alkyl ketones were not enhanced to the same extent by DAP in *n*-hexane. This selectivity of iodination would be caused by a steric factor rather than by the electronic influence of the more hydrophobic substrates. The partition of the substrate and the intermediate enamine between the microscopic heterogeneous phase, i.e., bulk hexane, and the reversed micellar phase with core water will play an important role in governing the selectivity of reactivity, and results such as these are of interest in relation to the specificity of enzymes.

The importance of substrate distribution between the bulk solvent and the micellar water pool is also illustrated in a study of the kinetics of the aminolysis of *p*-nitrophenyl carboxylates in the presence of DAP and Aerosol OT aggregates in benzene [33,107]. As was found for the iodination reaction of ketones [106], increasing the alkyl chain length of the *N*-alkylimidazole and/or the ester decreased the reaction rate. Addition of water decreased the observed rates of aminolysis in DAP, due to hydration of the head groups inhibiting their ability to act as acid-base catalysts, but in the presence of Aerosol OT the rates of methylimidazole-catalyzed ester hydrolysis increased as a function of added water. In this surfactant, substrate-micelle interactions are weak due to the absence of hydrogen bonding. This weakness, in turn, leads to decreased reagent concentrations in the micellar core, and simple partitioning of the reactants plays an important role. An increase in water concentration should affect the solubilized ester and imidazole in opposite directions, but the determining factor in the increasing rates was apparently the increased concentration of the diazole in the water pool.

C. Two-Step Hydration Mechanism

The last two reactions considered above, i.e., iodination of ketones and aminolysis of esters, are both bimolecular reactions, and neither involves the net consumption of water molecules. More important, however, neither involves the participation of water molecules directly in the reaction mechanism, and so the complex role observed for the water molecules may be compared with their role in the many pseudo-first-order unimolecular reactions which involve neither their net consumption nor active involvement.

Typical examples of this complexity are shown in Fig. 17. The variation of the pseudo-first-order rate constants for decarboxylation of 6-nitro-1,2-benzisoxazole-3-carboxylate, k_D, scheme 8 [108]:

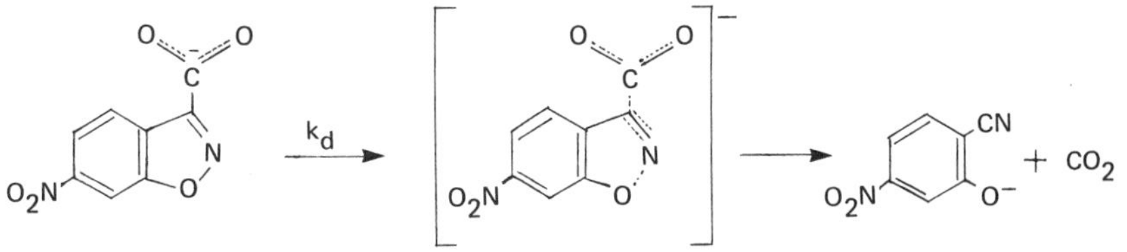

Scheme 8

in hexadecyltrimethylammonium chloride in $CHCl_3$ shows concave curvature with increasing values of $R = [H_2O]/[S]$, while that of the pseudo-first-order rate constants for decomposition of PNPA in butane-1,4-diamine bis(dodecanoate) in benzene [20] shows convex curvature.

Within these two patterns fall most of the curves showing dependence of first-order rate constants on R values, e.g., decomposition of PNPA in benzene solutions of dodecylammonium phenoxides [19] and of alkylammonium and alkanediamine carboxylates [18,20,43].

Within these same two patterns fall the values for many parameters evaluated in investigations of the state and nature of water interacting with surfactants in apolar media. Thus, Fig. 18 shows the convex curves for variation of the O—H vibrational energy of water molecules affected by the head groups of the surfactant DAP dissolved in $CHCl_3$, in the near-infrared spectrum [109], and the concave curve for variation of the microviscosity, estimated from the ^{13}C spin lattice relaxation times of glycine solubilized in DAP reversed micelles [110], with R values.

Other parameters which fit these patterns are the chemical shift and line width of the water proton signal in the NMR and the fluorescence intensity of terbium chloride [93], the microscopic polarity

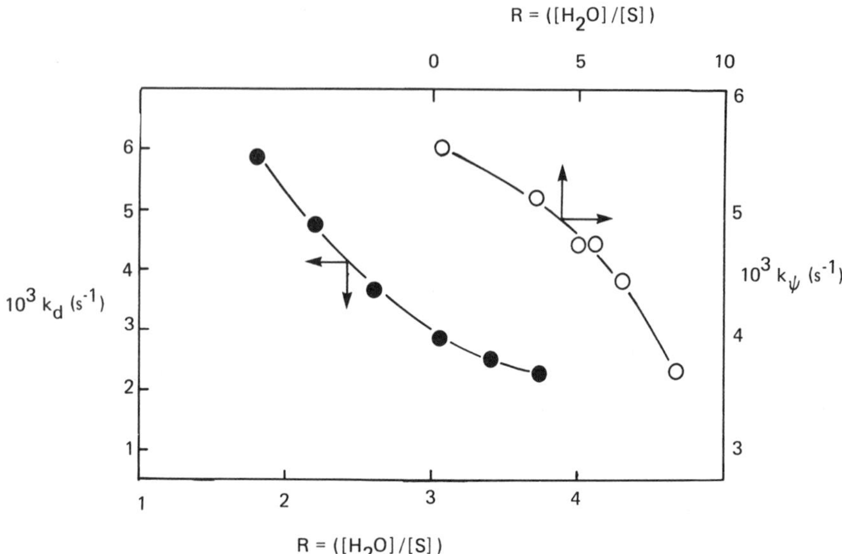

FIG. 17 Variation of pseudo-first-order rate constants with R value for the decomposition of 6-nitro-1,2-benzisoxazole-3-carboxylate in the system 0.20 mol dm^{-3} hexadecyltrimethylammonium chloride/ 7.4×10^{-3} mol dm^{-3} aqueous NaOH/$CHCl_3$ at 298 K (•) [108], and for the decomposition of *p*-nitrophenylacetate in the system 5.76×10^{-2} mol dm^{-3} butane-1,4-diamine bis(dodecanoate)/water/benzene at 341 K (○) [20].

[111,112], and the viscosity [110,112–114] of the specific and restricted field provided by reversed micelles. In all these investigations, an obvious inflection was always observed at the point where the mole ratio of water to surfactant is approximately unity.

The most plausible explanation involves a two-step hydration mechanism [6–9,99,110]. When water is first introduced into a system, it tends to interact with surfactants, irrespective of the presence of other polar or ionic solutes, to form water phase I. Upon completion of the first hydration of the surfactant, water molecules start to interact with themselves by hydrogen bonding, or to bind with other polar solutes cosolubilized in the interior core to form water phase II. During this second step there is formation of a water pool, whose properties approach that of bulk water as the concentration of water increases (see Fig. 2).

Reversed micelles can thus provide a "multiple field assistance" effect [114,115] for a wide variety of organic and inorganic reactions. This very restricted field formed by reversed micelles or surfactant-

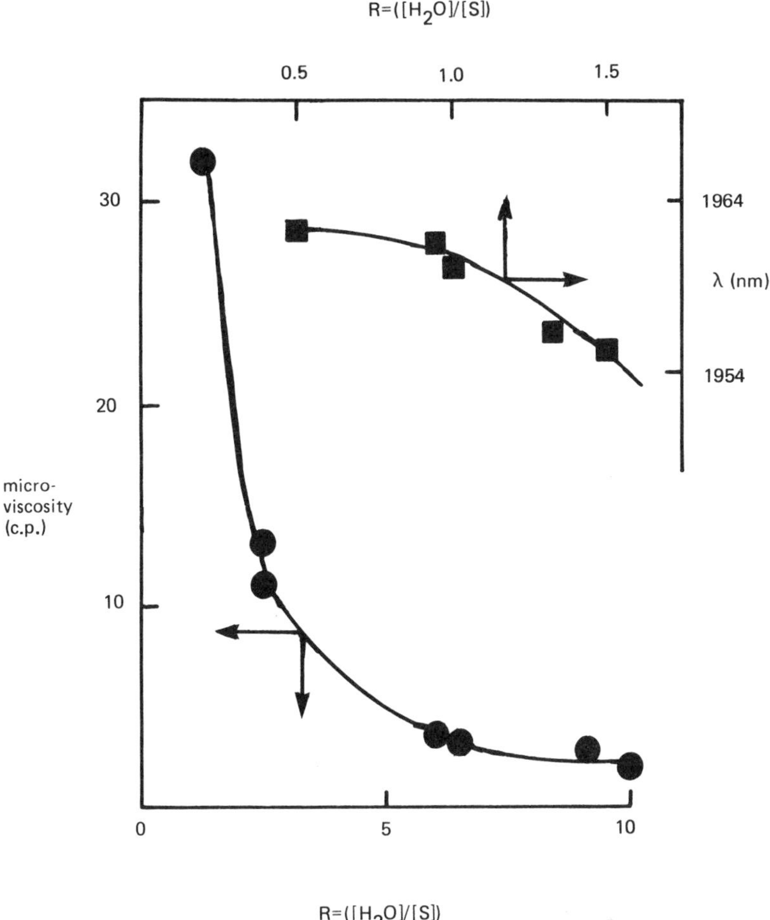

FIG. 18 Variation of the microscopic viscosity of the water pool solubilized in DAP reversed micelles with R (= $[H_2O]/[S]$) values (●) [110], and variation of the O—H vibrational mode of water in the near-infrared spectrum of DAP in chloroform at 298 K with R values (■) [109].

water aggregates provides, multiply and simultaneously, different effects such as a proximity, an electronic or structural strain, an anchoring effect, and so forth, which can specifically control chemical equilibria and reaction pathways or rates by *simultaneously* adjusting pH, polarity, viscosity, and activity in the microenvironment. The situation closely resembles the hydrophobic pocket of enzymes or biomembranes.

In general, an increase in the R value leads to an increase in the micropolarity and a decrease in the microviscosity of the water pool [111,116]. It also leads to restricted mobility of the reacting species [112,113]. The intramolecular cyclization of the Schiff base formed from histidine and pyridoxal, scheme 9:

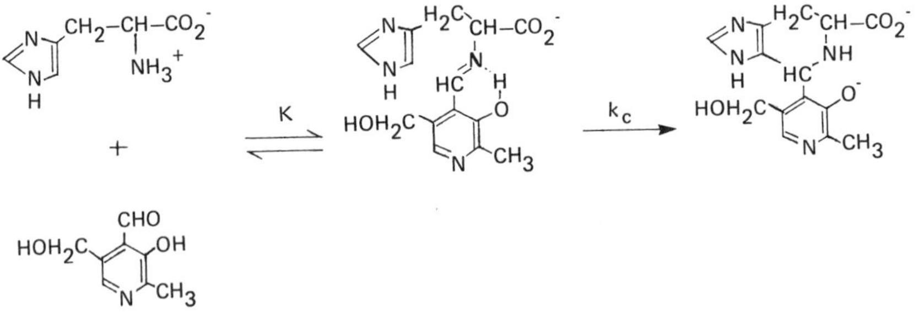

Scheme 9

has been investigated in both cationic hexadecyltrimethylammonium chloride/H_2O/$CHCl_3$ and anionic Aerosol OT/H_2O/heptane reversed micelles [114]. The rate of formation of Schiff based from pyridoxal and amino acids is considerably enhanced in reversed micelles [117] because of the local concentration effects on the bimolecular reaction. The equilibrium constant K decreases with an increase in the R value of the micelles because the reactants become diluted as the water, produced as the reaction proceeds, expands the pool size.

The cyclization reaction rate constant, k_c in Scheme 9, decreases with a decrease in the R value and thus is effectively retarded by a decrease in the pool size. In addition, the value of k_c in the reversed micelles is much smaller than that in bulk water. As the core size expands, the micropolarity in the interior core approaches that of pure water [112], and this is unfavorable for Schiff base formation but favorable for cyclization. The restriction effect on the mobility of the substrate in the reaction field contributes to these results. It is known that the cyclization product is an inhibitor of histidine decarboxylase [118], and this investigation points the way to an enormous vista of both organic and inorganic reactions whose reactivity, in the restricted reaction field provided by water pools solubilized in reversed micelles, will assist our understanding of the origin and extent of the activity of enzyme active sites.

VI. HYDROPHOBIC ENVIRONMENT

Several studies have been made on the effect of substitution of a longer-chain homolog for a shorter-chain component of the surfactant. The first [34] used PNPA as a kinetic probe and measured its rate of decomposition in the presence of a number of dodecylammonium carboxylates (the carbon chain length of the carboxylate moiety was varied from 2–6, 8, and 12). Analysis of the data according to the pseudophase model suggested that an increase in carbon chain length produced a maximum micellar rate constant (or a minimum monomer rate constant) for the four- and five-carbon-chain homologs. It was further shown that the rate of decomposition of PNPA increased as the number of polar groups on the surfactant molecule increased. The order of reactivity was DAP < propane-1,3-diamine bis(oleate) < ethane-1,1,2-triamine tris(dodecanoate). However, these experiments did not clarify whether an effect due to pK_a may have been involved, because the set of surfactants chosen contained several variables and the variation observed may have been a statistical one. However, the study did reveal that the number of labile hydrogen atoms per reactive head group had a direct bearing on the reaction rate. The order of catalytic activity in cyclohexane was DAP > didodecylammonium propionate > hexadecyldimethylammonium propionate. This trend was explained in terms of catalysis accomplished by transfer of protons liberated from ammonium ions to the ester substrate and is thus similar to the mechanism suggested by Satchell and Secemski [119] for ester aminolysis reactions in apolar media. El Seoud et al. [33,45,107] later reported on the aminolysis reaction of PNPA with various amines in DAP solutions in benzene, and found that the maximum rate occurred for octylamine.

The decomposition of PNPA has also been measured in benzene solutions of alkylammonium propionates (alkyl chain length 4, 6, 8, 10, or 12) and dodecylammonium carboxylates (alkyl chain length of the carboxylate 2, 3, 7, or 8) [18,26,40,43] and in benzene solutions of alkanediamine bis(dodecanoate)s (alkyl chain length of the amine 2–7, 9, 10, or 12) [38]. The rate data were treated according to Eq. (15).

The micellar rate constant k_M increased as the alkyl chain length in the alkylammonium head group lengthened [40], and this behavior was parallel to the increasing aggregation tendency, i.e., decreasing CMC values, of the surfactants, as determined by ^1H NMR spectroscopy [4]. The values of k_M in the presence of the diamine surfactants fell into two distinct groups, each one dependent on whether the carbon chain length of the parent was even or odd, but both sets showed decreasing rates with increasing length [38].

The bimolecular rate constant k_2 was affected by both the chain length (and whether the number of carbon atoms was even or odd) and the acidity of the amine (or diamine) head groups [38,40], but was unaffected by carboxyl carbon chain length [38]. This finding supported the postulate of acid-catalyzed amine attack and formation of an amide as product [20]. Moreover, the reaction rate was then expected to be proportional to the concentration of monomer at small surfactant concentrations, in keeping with the experimental findings [49].

Similar trends were found for the decomposition of 2,4-dinitrophenyl sulfate in alkylammonium carboxylate surfactant solutions in benzene [39]. The logarithms of the rate constants k_M and k_2 increased linearly with the number of carbon atoms in both the carboxyl and amine groups of the surfactant. Changes in carbon chain length of both groups affected k_M to the same extent, but, as might have been expected for a reaction subject to general acid-general base catalysis, the value of $k_{RNH_3^+}$ depended on chain length to a greater extent than did $k_{^-O_2CR'}$.

Exactly the opposite trend was found in the rate of trans-cis isomerization of the bis(oxalato)diaquochromate(III) anion in the presence of these same surfactants in benzene [29]. An increase in the alkyl chain length of both the alkylammonium and carboxylate ions of the surfactant caused an increase in the rate of isomerization; however, changes in the latter had a greater effect than in the former. The log of the rate constant decreased linearly with ΔpK_a (the difference between the pK_a values of the relevant amine and carboxylic acid). Since ΔpK_a reflects the tightness of the ion-pairs and consequently the electron density, and therefore acidity, of the alkylammonium ion, these results indicated that proton transfer from the ammonium headgroups to the complex anion was a primary factor in the rate enhancement.

The polarity in the interior of the aggregates was also important in determining substrate reactivity. An increase in the oxyethylene chain length of the surfactant in imidazole/surfactant/CCl_4, systems, or an increase in the molar ratio of the surfactant, caused the catalytic action of imidazole on PNPA to be increased [120].

VII. MICELLAR ENZYMOLOGY

One of the most exciting developments in interfacial catalysis in microphases in apolar media in recent years has been the investigation of the reactivity of enzymes solubilized in reversed micelles

[121,149]. Aerosol-OT and phospholipids are most frequently used as surfactants because of their ability to solubilize large water pools in hydrocarbon solvents.

It is possible to solubilize up to 1 mg/ml of large proteins (M.W. > 5×10^5) in these solutions and most proteins investigated, to date, are completely surrounded by the surfactants. In contrast to the expansion in aggregate size induced by solubilizing vitamin $B_{12}a$ in DAP solutions in benzene [103], the entrapped proteins do not seem to render significant changes in the size of the reversed micelle. Deviations were observed only when the inner cavity of the reversed micelle was smaller than the effective size of the solubilized protein molecule [122].

Figure 19 shows schematic representations of reversed micelles containing hydrophilic, surface-active, and membrane hydrophobic proteins [123,124].

Organic solvents, on their own, usually denature enzymes with a dramatic or total loss of catalytic activity, but in these surfactant solutions conditions can be chosen so that the enzyme does not lose any active centers, the reaction rate constants obey Michaelis-Menten kinetics, and the solutions remain stable for several months [125,126].

A. Pseudophase Model for Enzymic Reactions

Martinek et al. [123,127] have reviewed this field of reactivity, for which they have coined the phrase "micellar enzymology," and have

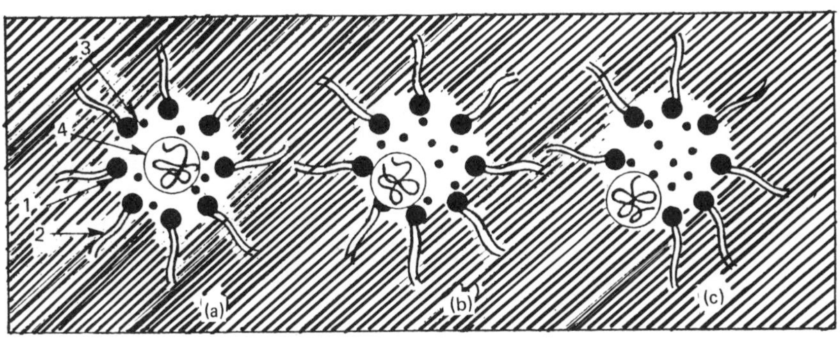

FIG. 19 Schematic representation of reversed micelles containing (a) hydrophilic, (b) surface-active, and (c) membrane hydrophobic proteins; 1, polar head group; 2, hydrocarbon tail of the surfactant molecule; 3, counterion and/or water molecule; and 4, protein molecule [123,124].

cited many examples of these enzymic reactions occurring in micelle-solvent pseudobiphasic systems. They have discussed the application of the pseudophase model for the regularities of catalysis by enzymes solubilized by reversed surfactant micelles in organic solvents. This model assumes a uniform distribution of reactants over the entire volume of the hydrated micelles, but it was pointed out that the model should actually be more involved and allow for the microheterogeneity of the inner cavity of the reversed micelles; in particular, it should allow for the distribution of the substrate between the inner surface layer and the water core.

Consider the kinetics of the reaction between an enzyme E and substrate X in an organic solvent/surfactant system, Eq. (46):

$$E + X \underset{}{\overset{K_m}{\rightleftharpoons}} EX \xrightarrow{k_{cat}} E + \text{Products} \tag{46}$$

At equilibrium the substrate is distributed between the bulkier organic phase and the phase of hydrated micelles, Eq. (47):

$$X_0 \rightleftharpoons X_M \tag{47}$$

and the partition coefficient may be represented as in Eq. (48):

$$P_X = \frac{[X]_M}{[X]_0} \tag{48}$$

(The subscripts M and 0 denote micellar and bulk phases, respectively.)

Let the reaction rate in the micellar phase obey the Michaelis-Menten equation. Then, at the beginning of reaction, the concentration of product is very much smaller than that of the substrate, which in turn is much greater than that of the enzyme. The steady-state rate of reaction, R, referred to the total volume of the system may be expressed by Eq. (49):

$$R = \frac{k_{cat,M}[E]_{init,M}[X]_{init,M}}{K_{m,M} + [X]_{init,M}} \nu \tag{49}$$

where ν is the volume fraction of the micellar phase, "init" denotes initial concentrations, K_m is the Michaelis-Menten constant, and k_{cat} is the rate constant for the enzyme-catalyzed reaction, Eq. (46).

It is assumed that the exchange of substrate molecules between the phases proceeds quickly, i.e., the course of the reaction in Eq. (46) does not distort the equilibrium, Eq. (47).

The concentrations of reactants are then determined from Eq. (48) and the equations of material balance:

$$[X]_{init,total} = [X]_{init,M}\nu + [X]_{init,0}(1 - \nu) \tag{50}$$

$$[E]_{init,total} = [E]_{init,M}\nu \tag{51}$$

Equation (48) holds only for dilute solutions in which $[X] \ll [S]$. Substitution of Eqs. (47), (50), and (51) into Eq. (49) gives Eq. (52):

$$R = \frac{k_{cat}^{app}[E]_{init,total}[X]_{init,total}}{K_m^{app} + [X]_{init,total}} \tag{52}$$

where

$$k_{cat}^{app} = k_{cat,M} \tag{53}$$

and

$$K_m^{app} = K_{m,M}\frac{1 - \nu(P_X - 1)}{P_X} \tag{54}$$

If the reaction involves a charged substrate, Eq. (54) can be simplified by assuming that the substrate molecules are present only in the water-micellar phase and hence both $P_X \gg 1$ and $P_X\nu \gg 1$. Then:

$$K_m^{app} = K_{m,M}\nu \tag{55}$$

Equation (52) is usually used in the form of its reciprocal, Eq. (56):

$$\frac{E_{init,total}}{R} = \frac{K_m^{app}}{k_{cat}^{app}}\frac{1}{[X]_{init,total}} + \frac{1}{k_{cat}^{app}} \tag{56}$$

Equation (56) is known as the Lineweaver-Burke equation and plots of $1/R$ against $1/[X]_{init,total}$ then lead to the evaluation of k_{cat}^{app} and K_m^{app}.

Plots of these apparent values against ν, the ratio of the volume of water to the total reaction volume, give values of $k_{cat,M}$ and

$K_{m,M}$. As expected from the assumptions which have been made, the value of k_{cat}^{app} is independent of ν, and the slope of the linear plot, obtained by using Eq. (56), allows calculations of the true value of this constant for the reaction in the micellar phase.

Although the values of k_{cat}^{app} are lower than in aqueous systems, the concentration of reagents in the polar water pool may (but usually does not) lead to an overall rate enhancement compared with the aqueous system. The charge on the surfactant and substrate can play a major role in determining the rate of reaction. Like charges on substrate and surfactant do not produce great differences in k_{cat} and K_m between the reversed micelle and aqueous systems, but unlike charges produce a pronounced decrease in the total second-order rate constant k_{cat}/K_m.

The predictions of this theory were confirmed in a study of the substrate specificity of horse liver alcohol dehydrogenase for oxidation of normal aliphatic alcohols as a function of hydrocarbon chain length [124]. An increase in the alcohol hydrophobicity shifted the alcohol partition between the micellar water (polar) phase and the organic phase in favor of the latter and hence increased the local concentration of the substrate around the enzyme. This accounted for the rise in K_m^{app}.

Such microheterogeneous media should prove to be good models for the microenvironment of an enzyme located on or inside biological membranes, for effects similar to those so strongly pronounced in this model system should undoubtedly occur in vivo. Moreover, in vitro studies in reversed micellar media may prove to be better than those in aqueous solutions for yielding information about the possible catalytic activity and specificity that an enzyme displays in nature. Data furnished from aqueous solutions may be distorted and show essential discrepancies.

For example, Menger and Yamada [128] measured the rate of hydrolysis of *N*-acetyl-L-trytophan methyl ester catalyzed by α-chymotrypsin in the water pool solubilized by Aerosol OT in heptane. There was reduced activity due not to protein denaturation, but to a 1.5 pH unit shift to the right in the sigmoidal plot of enzyme activity against pH. With increasing pH of the water pool, $k_{cat,M}$ became larger than $k_{cat,water}$. The insensitivity of the enzyme activity to water pool size (and therefore aggregate size) was attributable to specific interactions between the enzyme and the surfactant giving rise to a solubilizing aggregate, rather than to the enzyme occupying existing fixed-size aggregates.

Others have noted substantial shifts in the position of maximum activity in activity-pH profiles on transferring the reaction from aqueous to reversed micellar solutions [127,129–133]. Martinek and co-workers [123] summarized the reasons for this as follows. First, use of ionogenic surfactants which form a charged (double electric) layer around the enzyme may cause a local pH shift (dependent on the charge, sign, and degree of surfactant ionization). This will account for a shift in pH, and hence in pK_a, of about 1–2 units. Second, the ionogenic groups of the solubilized enzyme may alter as a result of a change in the nature of the microenvironment of the enzyme, which may cause its partial dehydration and subsequently a change in pK_a of several units. Third, the ionogenic groups, including those controlling the enzymatic reaction, may cause conformational changes of the enzyme upon its solubilization, and a related shift in pK_a.

Although circular dichroism shows that the gross conformations of several enzymes in reversed micelles do not alter appreciably, increased ellipticities indicate increased helicites [129,130,134]. However, Luisi et al. [133] reported that both proteins and nucleic acids maintained rigid folding when they were solubilized into hydrocarbon solvents with the help of reversed micelles, and that the actual conformation was sometimes drastically changed with respect to water solutions. Generally there was an increase of conformational rigidity, and in the case of the high molecular weight DNA, evidence to indicate supercoiling.

Gierasch et al. [135] explored the way in which the microenvironment of a reversed micelle can affect the folding of a polypeptide chain by measuring the conformational impact of these systems on model peptides solubilized within the core. This property was found to be dominated by counterions. For example, water pools in samples containing 3% w/v Aerosol OT in *n*-heptane and 1% added water (v/v), a molar ratio of water to surfactant of eight, behave as though their effective sodium ion concentration were 5 mol dm^{-3}. In addition, the peptide solubilizate affects the micellar aggregate and accompanying interfacial water, for it alters distribution of the population of water molecules within a reversed micelle, causing a decrease in the number of the bulklike water pool.

(Kon-no et al. found that reversed micelles exhibited a characteristic enantioselectivity and high reactivity, similar to enzymic action, in the aminolysis reactions of optically active esters solubilized into the rigid core of the micelles [136]. The chiral environment of the reaction field was a significant factor in enantioselectivity in the reaction with optically active amine surfactants [137], and for effective enantioselectivity by the catalytic functional group $-\overset{+}{N}H_3{}^{-}OOC-$ the asymmetric center must be placed near the cationic head group [138]).

B. Hydration of the Micelles

One of the most striking effects observed in micellar enzymology is the dependence of the catalytic activity of solubilized enzymes on the degree of hydration of reversed micelles, w_0, i.e., the parameter defining the size and properties of the inner cavity of the micelles. (The ratio $[H_2O]/[S]$ was termed w_0 by Martinek et al. in their studies on micellar enzymology and it corresponds to the R values used by many authors in describing this ratio in reversed micellar kinetics.) This dependence has been well documented [127,130,131, 134,139–142]. It is usually evidenced by a bell-shaped curve, as shown in Fig. 20, for the dependence of k_{cat} for peroxidase oxidation of pyrogallol upon w_0 in the Aerosol OT/octane system [123]. This result suggests [139] that there exists an optimal value of w_0 at which the catalytic activity of the solubilized enzyme is maximal.

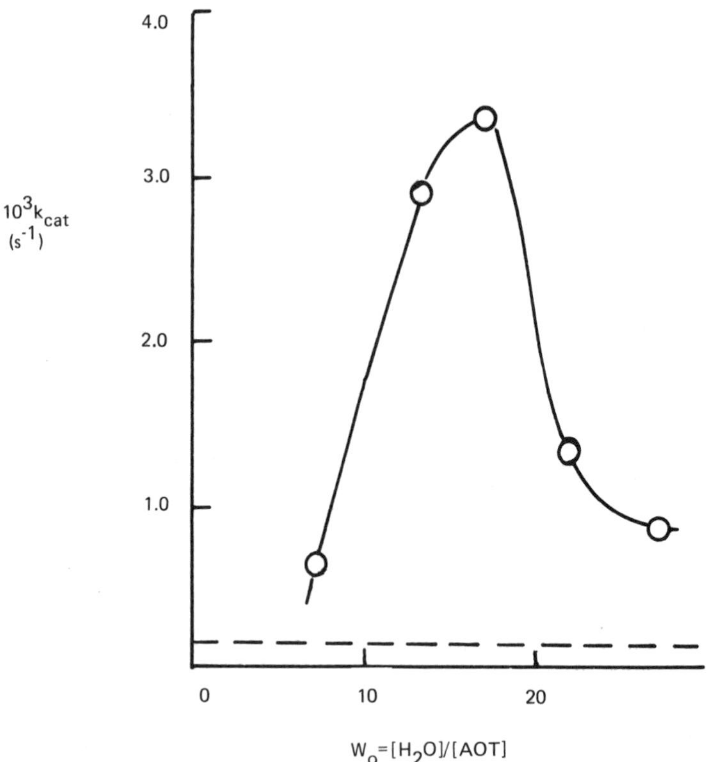

FIG. 20 Variation of the values of the first-order rate constant k_{cat} for peroxidase oxidation of pyrogallol with w_0 values in Aerosol OT (0.1 mol dm^{-3})-octane-water (0.025 mol dm^{-3} phosphate buffer, pH 7.0) at 299 K. Dashed line is the value of k_{cat} in aqueous solution [123].

At this value, the surfactant suffers a moderate degree of hydration and the microenvironment of the enzyme molecule inside the micelle differs greatly from that in aqueous solution. It seems that the phenomenon of superactivity [128,130] arises partially from the effects of the microenvironment of the enzyme molecule in the reversed micelle. It further follows from analysis of the data in Fig. 20 that the catalytic activity of the enzyme in aqueous solution is some hundreds of times smaller than its true reactivity [123].

Systems which use an enzyme, solubilized within the aqueous core of a reversed micelle, as a catalyst to promote the rapid establishment of chemical equilibrium, have potential as a means of introducing biocatalysis into preparative organic synthesis and may be treated as a very realistic model of biological membranes.

Furthermore, the properties of the solvent may confine substrate specificity, which is rightfully considered to be one of the most characteristic properties of enzymes. In a study of the oxidation of normal aliphatic alcohols, Eq. (57):

$$H(CH_2)_nOH + NAD^+ \rightleftharpoons H(CH_2)_{n-1}CHO + NADH + H^+ \qquad (57)$$

for which the carbon chain length n varied from 2 to 10 atoms, catalyzed by horse liver alcohol dehydrogenase in a colloidal solution of water solubilized by Aerosol OT in octane, a change in substrate specificity (compared with that in aqueous solution) was observed. The value of k_{cat} increased with increasing w_0 and became nearly equal to the value obtained in aqueous solution, but did not affect the substrate specificity. However, plots of the second-order rate constant k_{cat}/K_m^{ROH} against the length of the hydrocarbon chain showed bell-shaped curves with a maximum at n=4 in the organic solvent and n=8 in aqueous solution [124,143]. This unique (thus far) change was ascribed to the effect of the local concentration of the alcohol substrate around the enzyme in the micelle. This result is important, because it may show that we should expect substantial differences between the substrate specificity that an enzyme exhibits in nature and that found in vitro in studies in aqueous solution.

Although the field of micellar enzymology is still in its infancy, evidence is rapidly accumulating that it may successfully be applied to a wide range of problems and provide a deeper insight into the fundamental knowledge of enzymes and the wider possibilities of the application of enzymic catalysis. Reversed micelles can be exploited in the selective transport of enzymes, and they are potentially applicable to enzyme-mediated synthesis using hydrocarbon-soluble substrates, e.g., reduction of decanol to decanal [134], enzymic synthesis of peptides [144], and spontaneous polymerization of alanyl adenylate to form an oligopeptide containing 40 monomer units [145].

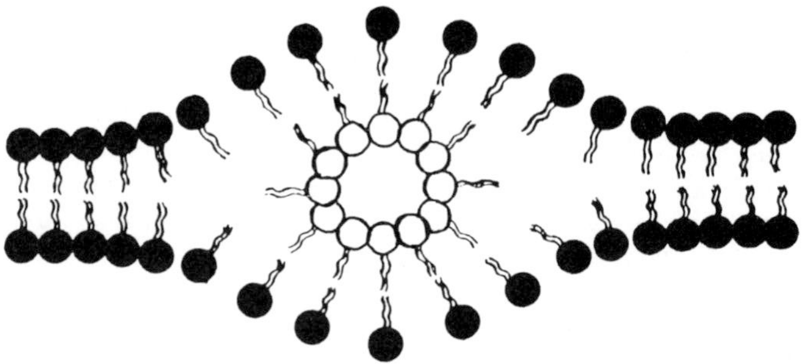

FIG. 21 Schematic representation of a lipidic particle [146].

Kruijff et al. [146] have discovered novel "lipidic particles," resembling reversed micelles, sandwiched between the monolayers of the lipid bilayer in biological membranes (Fig. 21), and these provide strong evidence that a reversed micelle, with a molecule of an enzyme entrapped in it, may be viewed as an elementary fragment of a biomembrane.

The field of cryoenzymology in aqueous-apolar media was pioneered in 1977 [147,148], but the necessary high concentration of organic cosolvent drastically altered the environment of the enzyme. The use of sufficiently large surfactant-solubilized water pools has alleviated this problem and provided convenient media for investigation of enzyme-catalyzed reactions at subzero temperatures. The supercooled water pools surrounding the enzyme are stabilized against freezing, due to heterogeneous nucleation by the surfactants, and temperatures as low as 235 K have been utilized for measurement of activation energies [132].

VIII. CONCLUDING REMARKS

Micellar catalysis of spontaneous, unimolecular reactions in the restricted reaction field of reversed micelles in apolar media depends on the properties of the micelle as a submicroscopic reaction medium. Factors which destabilize the initial state or stabilize the transition state (relative to their solutions in bulk solvent) increase the reaction rate, but extensive micellar incorporation simply means that the initial state has a lower free energy in the micelles than in the bulk organic phase. Therefore, catalysis of unimolecular reactions requires that the micelles stabilize the transition state more than the initial state, relative to the organic solvent.

However, the situation is more complex for bimolecular reactions where the two reactants come together in the transition state, because concentration of the reactants into the small volume in the restricted reaction field will itself increase the reaction rate. The micelle is often drawn as if it had a relatively smooth surface with little penetration of water beyond the head groups, but this model may be oversimplified and some of the methylene carbons of the hydrophobic surfactant tails may be in contact with water.

Comparison of true second-order rate constants in the apolar and micellar pseudophases inevitably depends on the choice of concentration units and the assumed magnitude of the volume element of reaction. Nevertheless, catalysis is apparently caused by concentration of the two reactants into the small volume of the aqueous core and inhibition arises from incorporation of one reactant into the micellar pseudophase and exclusion of the other from it.

The surfactants themselves may not always be chemically inert, and the polar head groups of some surfactants may participate in a bimolecular reaction. Thus far, functional head groups containing a protonated amine, a carboxylate ion, or a sulfate ion have been found to serve as reactive species in nucleophilic reactions. Such chemically active head groups may compete with a reactive ion such as H^+ or OH^- for a charged substrate in a pseudo-first-order reaction, and then the amount of reactive substrate in the polar core will depend on the extent of charge neutralization occurring at the interface. The kinetic treatment discussed earlier predicts that the reaction rates will then steadily increase to plateaus with increasing surfactant concentration as increasing amounts of substrate are taken up by the micelles.

In these descriptions of micellar reactions, the micelles are treated as if they were a separate phase, and the model is similar to those applied to other submicroscopic aggregates such as enzymes [51] and the many membrane-mimetic reagents whose chemistry has been reviewed by Fendler [13]. These descriptions seem to be generally satisfactory, but evidence is accumulating that submicellar aggregates, as well as fully formed micelles, may markedly affect reaction rates.

It may be useful in this context to distinguish between reactions which involve small hydrophilic ionic reactants and those which involve hydrophobic reactants. Submicellar aggregates of an ionic surfactant may well interact with hydrophilic substrates, but in apolar solvents they will probably not interact strongly with hydrophobic solutes. Thus one could predict that bimolecular reactions on submicellar aggregates could be important when *both* reagents are relatively hydrophilic, but not when one of them is hydrophobic.

Because it is relatively easy to obtain large rate enhancements by the use of surfactant aggregates in apolar solvents (there are examples of rate enhancements as high as 10^7), the potential of these systems is becoming more widely recognized. Fendler [79] summarized the roles of these systems in reactivity control, formation of catalytically efficient colloid particles, tertiary oil recovery, lubrication, corrosion inhibition, artificial photosynthesis, enzyme-mediated synthesis, cryoenzymology, macromolecular conformation, membrane fusion, and drug encapsulation. This wide variety of practical applications will ensure the future utilization of surfactant aggregates in apolar solvents, and additional research is clearly required in this highly relevant area of interfacial catalysis.

ACKNOWLEDGMENT

Some of the investigations reported here were supported financially by the Research Committees of the New Zealand Universities' Grants Committee and the University of Auckland. This support is gratefully acknowledged. The author also expresses her appreciation to Clifford Bunton and Janos Fendler for friendship and encouragement throughout her research career, to Karel Martinek and Junzo Sunamoto for making copies of their publications available, and to Terence Lomax and Robyn Ramage for their enthusiastic assistance as co-workers.

SYMBOLS

Chemicals

ADP	adenosine 5'-diphosphate
Aerosol OT	sodium bis(2-ethylhexyl)sulfosuccinate
ATP	adenosine 5'-triphosphate
BuDB	butane-1,4-diamine bis(dodecanoate)
CTABu	hexadecyltrimethylammonium butanoate
DAB	dodecylammonium butanoate
DABz	dodecylammonium benzoate
DAP	dodecylammonium propionate
DDAB	didodecyldimethylammonium bromide
DP	dodecylpyridinium chloride
E	enzyme
EtDB	ethane-1,2-diamine bis(dodecanoate)
n	number of alkyl units in surfactant chain
N	nucleophile
NAD^+	alcohol dehydrogenase

NADH	alcohol hydrogenase
OAT	octylammonium tetradecanoate
PNPA	*p*-nitrophenyl acetate
PNPP	*p*-nitrophenyl propionate
S	surfactant
X	substrate
Y	substituent

Kinetic Parameters

CMC	critical micelle concentration
$E_T(30)$	solvent polarity constant
f	weight fraction of monomer
k	rate constant
k_{cat}	enzyme-catalyzed rate constant
K	association constant
K_a	acid dissociation constant
K_m	Michaelis-Menten constant defined in Eq. (46)
K_n	micelle association constant defined in Eq. (2)
K_X	substrate-micelle binding constant defined in Eq. (6)
$K_{12} = \cdots = K_{ij}$	association constants between oligomers containing 1,2, ..., i,j monomers
m	surfactant monomer
M_n	surfactant aggregate
MW_1	molecular weight of monomer
MW_{na}	molecular weight of number average of aggregates
n	number of molecules of monomer
N	aggregation number of micelle
P	partition coefficient
R	rate of reaction
R	ratio of $[H_2O]$ to [surfactant]
T_2	relaxation time
$\bar{V}$	molar volume of micelle

w_0 hydration of surfactant according to ratio $[H_2O]/[S]$

W/O water-in-oil microemulsion

α parameter defined in Eq. (37)

β parameter defined in Eq. (38)

γ parameter defined in Eq. (39)

σ Hammett substituent constant

ρ sensitivity of reactivity to changing substituents

ν fractional volume of micellar pseudophase

Subscripts

D decarboxylation

m monomer

M micellar pseudophase

0 bulk apolar solvent

ψ pseudo first order

1 first order

2 second order

3 third order

Superscript

app apparent value calculated from observed rate

REFERENCES

1. J. B. Perl, *J. Colloid Interface Sci. 29*: 6 (1969).
2. S. Muto and K. Meguro, *Bull. Chem. Soc. Jpn. 46*: 1316 (1973).
3. K. Kon-no and A. Kitahara, *J. Colloid Interface Sci. 35*: 636 (1971).
4. J. H. Fendler and E. J. Fendler, *Catalysis in Micellar and Macromolecular Systems*, Academic Press, New York, 1975.
5. H.-F. Eicke and H. Christen, *Helv. Chim. Acta 61*: 2258 (1978).
6. P. E. Zinsli, *J. Phys. Chem. 83*: 3223 (1979).
7. J. Sunamoto and T. Hamada, *Bull. Chem. Soc. Jpn. 51*: 3130 (1978).
8. H. Kondo, I. Miwa, and J. Sunamoto, *J. Phys. Chem. 86*: 4826 (1982).
9. M. Zalauf and H.-F. Eicke, *J. Phys. Chem. 83*: 480 (1979).
10. H.-F. Eicke, *Chimia 31*: 265 (1977).

11. H.-F. Eicke and J. Rehak, *Helv. Chim. Acta 59*: 2883 (1976).
12. K. Kon-no and A. Kitahara, *J. Colloid Interface Sci. 35*: 409 (1971).
13. J. H. Fendler, *Membrane Mimetic Chemistry*, Wiley (Interscience), New York, 1982.
14. S. Friberg and S. I. Ahmad, *J. Phys. Chem. 75*: 2002 (1971).
15. S. I. Ahmad and S. Friberg, *J. Am. Chem. Soc. 91*: 5196 (1972).
16. H. Kondo, K. Fujiki, and J. Sunamoto, *J. Org. Chem. 43*: 3584 (1978).
17. K. Kon-no, K. Fujino, and A. Kitahara, *Yukagaku 30*: 239 (1981).
18. C. J. O'Connor and R. E. Ramage, *Aust. J. Chem. 33*: 757 (1980).
19. C. J. O'Connor and R. E. Ramage, *Aust. J. Chem. 33*: 757 (1980).
20. C. J. O'Connor and T. D. Lomax, *Aust. J. Chem. 36*: 895 (1983).
21. C. J. O'Connor, T. D. Lomax, and R. E. Ramage, *Adv. Colloid Interface Sci. 20*: 21 (1984).
22. H.-F. Eicke, *Top. Curr. Chem. 87*: 85 (1980).
23. A. Kitahara, *Adv. Colloid Interface Sci. 12*: 109 (1980).
24. J. H. Fendler, *Acc. Chem. Res. 9*: 153 (1976).
25. S. E. Friberg, *Colloids Surfaces 4*: 201 (1982).
26. C. J. O'Connor, T. D. Lomax, and R. E. Ramage, in *Solution Behavior of Surfactants—Theoretical and Applied Aspects, Vol. 2*, K. L. Mittal and E. J. Fendler, eds., Plenum, New York, 1982, p. 803.
27. J. H. Fendler, E. J. Fendler, R. T. Medary, and V. A. Woods, *J. Am. Chem. Soc. 94*: 7288 (1972).
28. C. G. Swain and J. F. Brown, *J. Am. Chem. Soc. 74*: 2538 (1952).
29. C. J. O'Connor, E. J. Fendler, and J. H. Fendler, *J. Am. Chem. Soc. 96*: 370 (1974).
30. J. H. Fendler, E. J. Fendler, and S. A. Chang, *J. Am. Chem. Soc. 95*:3273 (1973).
31. F. M. Menger and C. E. Portnoy, *J. Am. Chem. Soc. 89*: 4698 (1967).
32. M. Algrem, F. Grieser, and J. K. Thomas, *J. Am. Chem. Soc. 102*: 3188 (1980).
33. O. A. El Seoud, A. Martins, L. P. Barbur, M. J. de Silva, and V. Aldrigue, *J. Chem. Soc. Perkin Trans. II* 1674 (1977).
34. K. Kon-no, T. Matsuyama, H. Mizuno, and A. Kitahara, *Nippon Kagaku Kaishi 11*: 1857 (1975).
35. J. H. Fendler, E. J. Fendler, R. T. Medary, and O. A. El Seoud, *J. Chem. Soc. Faraday Trans. I 69*: 280 (1973).

36. R. Fornasier and U. Tonellato, *J. Chem. Soc. Faraday Trans. I 76*: 1311 (1980).
37. C. J. O'Connor and A. J. Porter, *Aust. J. Chem. 34*: 1603 (1981).
38. C. J. O'Connor and T. D. Lomax, *Aust. J. Chem. 36*: 906 (1983).
39. C. J. O'Connor, E. J. Fendler, and J. H. Fendler, *J. Org. Chem. 38*: 3371 (1971).
40. C. J. O'Connor and R. E. Ramage, *Aust. J. Chem. 33*: 779 (1980).
41. K. J. Mollett and C. J. O'Connor, *J. Chem. Soc. Perkin Trans. II* 369 (1976).
42. C. J. O'Connor and J. W. Barnett, *J. Chem. Soc. Perkin Trans. II* 1331 (1973).
43. C. J. O'Connor and R. E. Ramage, *Aust. J. Chem. 33*: 771 (1980).
44. F. Y.-F. Lo, B. M. Escott, E. J. Fendler, E. T. Adams, R. D. Larson, and P. W. Smith, *J. Phys. Chem. 79*: 2609 (1975).
45. O. A. El Seoud, M. J. de Silva, L. P. Barbur, and A. Martins, *J. Chem. Soc. Perkin Trans. II* 331 (1978).
46. J. E. Critchlow, *J. Chem. Soc. Faraday Trans. I* 1774 (1972).
47. C. L. Bell and G. M. Barrow, *J. Chem. Phys. 31*: 1158 (1959).
48. F. M. Menger and J. H. Smith, *J. Am. Chem. Soc. 94*: 3824 (1972).
49. C. J. O'Connor and T. D. Lomax, *J. Colloid Interface Sci. 95*: 204 (1983); *Tetrahedron Lett. 24*: 2917 (1983).
50. C. J. O'Connor and T. D. Lomax, in *Surfactants in Solution*, Vol. 3, K. L. Mittal and B. Lindman, eds., Plenum, New York, 1984, p. 1435.
51. W. P. Jencks, *Catalysis in Chemistry and Enzymology*, McGraw-Hill, New York, 1969.
52. C. J. O'Connor, E. J. Fendler, and J. H. Fendler, *J. Chem. Soc. Dalton Trans.* 625 (1974); *J. Am. Chem. Soc. 95*: 600 (1973).
53. C. J. O'Connor and R. E. Ramage, *Aust. J. Chem. 33*: 695 (1980).
54. F. M. Menger, J. A. Donohue, and R. F. Williams, *J. Am. Chem. Soc. 95*: 286 (1973).
55. C. J. O'Connor and B. T. Ch'Ng, *Bull. Chem. Soc. Jpn. 56*: 3021 (1983).
56. C. J. O'Connor, B. T. Ch'Ng, and R. G. Wallace, *J. Colloid Interface Sci. 95*: 410 (1983).
57. G. Kortum, *Lehrbuch der Electrochemie*, Verlag Chemie, Weiheim, 1966.
58. A. Norman, *Acta Chem. Scand. 14*: 1295 (1960).

59. N. A. Mayer, R. F. Kwasnick, M. C. Carey, and G. B. Benedek, in *Micellization, Solubilization and Microemulsions*, Vol. 1, K. L. Mittal, ed., Plenum, New York, 1977, p. 445.
60. C. J. O'Connor, B. T. Ch'Ng, and R. G. Wallace in *Surfactants in Solution*, Vol. 2, K. L. Mittal and B. Lindman, eds., Plenum, New York, 1984, p. 875.
61. A. S. Kertes, in *Micellization, Solubilization and Microemulsions*, Vol. 1, K. L. Mittal, ed., Plenum, New York, 1977, p. 445.
62. A. S. Kertes and H. Gutman, in *Surface and Colloid Science*, Vol. 8, E. Matijevic, ed., Wiley (Interscience), New York, 1975, p. 193.
63. K. Tamura and Z. A. Schelly, *J. Am. Chem. Soc. 103*: 1013, 1018 (1981).
64. S. Harada and Z. A. Schelly, *J. Phys. Chem. 86*: 2098 (1982).
65. Y.-C. Jean and H. J. Ache, *J. Am. Chem. Soc. 100*: 6320 (1978).
66. H.-F. Eicke and A. Denss, *J. Colloid Interface Sci. 64*: 386 (1978).
67. I. Prigogine and R. Defay, in *Chemical Thermodynamics*, D. H. Everett, translator, Longmans, London, 1954, Ch. 26.
68. M. M. Cox and W. P. Jencks, *J. Am. Chem. Soc. 103*: 572 (1981).
69. L. Do Amaral, K. Koehler, D. Bartenbach, T. Pletcher, and E. H. Cordes, *J. Am. Chem. Soc. 89*: 3537 (1967).
70. C. J. O'Connor and T. D. Lomax, *Aust. J. Chem. 36*: 917 (1983).
71. K. Kon-no, A. Kitahara, and M. Fujiwara, *Bull. Chem. Soc. Jpn. 51*: 3165 (1978).
72. G. Wallerberg, J. Boger, and P. Haake, *J. Am. Chem. Soc. 93*: 4938 (1971).
73. O. A. El Seoud, R. C. Vieira, M. I. El Seoud, J. P. S. Farah, M. T. Miranda, and P. P. Brotero, 185th Am. Chem. Soc. Meeting, Seattle, Washington, March 20–25, 1983, *Abstract Coll.* 45.
74. J. H. Fendler, F. Nome, and H. C. van Woert, *J. Am. Chem. Soc. 96*: 6745 (1974).
75. K. Martinek, A. K. Yatsimirski, A. V. Levashov, and I. V. Berezin, in *Micellization, Solubilization and Microemulsions*, Vol. 2, K. L. Mittal, ed., Plenum, New York, 1977, p. 489.
76. V. I. Pantin, A. V. Levashov, K. Martinek, and I. V. Berezin, *Dokl. Akad. Nauk SSSR 247*: 1194 (1979); English transl., p. 697.
77. I. V. Berezin, K. Martinek, and A. K. Yatsimirski, *Russ. Chem. Rev., Eng. Transl. 42*: 787 (1973).
78. L. S. Romsted, in *Micellization, Solubilization and Microemulsions*, Vol. 2, K. L. Mittal, ed., Plenum, New York, 1977, p. 509.

79. J. H. Fendler, in *Reverse Micelles: Biological and Technological Relevance of Reversed Micelles and Other Amphiphilic Structures in Apolar Media*, P. L. Luisi and B. E. Straub, eds., Plenum, New York, 1984, p. 305.
80. H.-F. Eicke and A. Denss, *Croat. Chem. Acta 52*: 105 (1979).
81. U. Hermann and Z. A. Schelly, *J. Am. Chem. Soc. 101*: 2665 (1979).
82. M. Seno, K. Araki, and S. Shiraishi, *Bull. Chem. Soc. Jpn. 49*: 899 (1976).
83. S. R. Palit and V. Venkateswarlu, *Proc. R. Soc. London Ser. A 208*: 542 (1951).
84. A. V. Levashov, V. I. Pantin, and K. Martinek, *Kolloid. Zh. 41*: 453 (1979).
85. F. Nome, S. A. Chang, and J. H. Fendler, *J. Chem. Soc. Faraday Trans. I 72*: 296 (1976).
86. N. Miyoshi and G. Tomita, *Z. Naturforsch. Teil B 35*: 736 (1980).
87. N. Miyoshi and G. Tomita, *Aust. J. Chem. 34*: 1545 (1981).
88. K. Martinek, A. V. Levashov, V. I. Pantin, and I. V. Berezin, *Dokl. Akad. Nauk. SSSR 238*: 626 (1978); English Transl., p. 107.
89. C. J. O'Connor and T. D. Lomax, *J. Am. Chem. Soc. 100*: 5910 (1978).
90. F. Kohler, H. Atrops, H. Kalali, E. Liebermann, E. Wilhelm, F. Ratkovics, and T. Salamon, *J. Phys. Chem. 85*: 2520 (1981).
91. F. Kohler, R. Gopel, G. Götze, H. Atrops, M. A. Demiriz, E. Liebermann, E. Wilhelm, F. Ratkovics, and B. Palagyi, *J. Phys. Chem. 85*:2524 (1981).
92. T. Masui, F. Watanabe, and A. Yamagishi, *J. Phys. Chem. 81*: 494 (1977).
93. H.-F. Eicke, J. C. W. Shepherd, and A. Steinemann, *J. Colloid Interface Sci. 56*: 168 (1976).
94. B. H. Robinson, D. C. Steytler, and R. D. Tack, *J. Chem. Soc. Faraday Trans. I 75*: 481 (1979).
95. J. Sunamoto, H. Kondo, S. Yamamoto, Y. Matsuda, and Y. Murakami, *Inorg. Chem. 19*: 3668 (1980).
96. U. K. A. Klein and D. J. Miller, *Ber. Bunsenges. Phys. Chem. 80*: 115 (1976).
97. A. Yamagishi, T. Masui, and F. Watanabe, *J. Phys. Chem. 84*: 34 (1980).
98. H. Kondo, T. Hamada, S. Yamamoto, and J. Sunamoto, *Chem. Lett.* 809 (1980).
99. J. Sunamoto, H. Kondo, and K. Akimaru, *Chem. Lett.* 821 (1978).
100. K. Arai, Y. Ogiwara, and K. Ebe, *Bull. Chem. Soc. Jpn. 49*: 1059 (1976).
101. K. Arai and Y. Ogiwara, *Bull. Chem. Soc. Jpn. 51*: 182 (1978).

102. O. A. El Seoud, *J. Chem. Soc. Perkin Trans. II* 1947 (1976).
103. M. Seno, S. Shiraishi, K. Araki, and H. Kise, *Bull. Chem. Soc. Jpn. 48*: 3678 (1975).
104. T. Kunitake, S. Shinkai, and Y. Okahata, *Bull. Chem. Soc. Jpn. 49*: 540 (1976).
105. S. Shinkai and T. Kunitake, *J. Chem. Soc. Perkin Trans. II 2*: 980 (1976).
106. M. Seno, K. Araki, and S. Shiraishi, *Bull. Chem. Soc. Jpn. 49*: 1901 (1976).
107. O. A. El Seoud, P. Pivêtta, J. P. S. Farah, and A. Martins, *J. Org. Chem. 44*: 4832 (1979).
108. J. Sunamoto, K. Iwamoto, S. Nagamatsu, and H. Kondo, *Bull. Chem. Soc. Jpn. 56*: 2469 (1983).
109. J. Sunamoto, T. Hamada, T. Seto, and S. Yamamoto, *Bull. Chem. Soc. Jpn. 53*: 583 (1980).
110. K. Tsujii, J. Sunamoto, and J. H. Fendler, *Bull. Chem. Soc. Jpn. 56*: 2889 (1983).
111. F. M. Menger and G. Saito, *J. Am. Chem. Soc. 100*: 4376 (1978).
112. H. Kondo, I. Miwa, and J. Sunamoto, *J. Phys. Chem. 86*: 4826 (1982).
113. J. Sunamoto, K. Iwamoto, M. Akutagawa, M. Nagase, and H. Kondo, *J. Am. Chem. Soc. 104*: 4904 (1982).
114. J. Sunamoto, H. Kondo, J. Kikuchi, H. Yoshinaga, and S. Takei, *J. Org. Chem. 48*: 2423 (1983).
115. J. Sunamoto, in *Solution Behavior of Surfactants: Theoretical and Applied Aspects*, Vol. 2, K. L. Mittal and E. J. Fendler, eds., Plenum, New York, 1982, p. 767.
116. J. H. Fendler, F. Nome, and H. C. van Woert, *J. Am. Chem. Soc. 96*: 6745 (1974).
117. H. Kondo, H. Yoshinaga, and J. Sunamoto, *Chem. Lett.* 973 (1980).
118. D. Mackay and D. Shepherd, *Biochim. Biophys. Acta 59*: 553 (1962).
119. D. P. N. Satchell and I. I. Secemski, *J. Chem. Soc. B* 130 (1969).
120. K. Kon-no, T. Inoue, T. Hanada, K. Nakamura, and A. Kitahara, *Yukagaku 29*: 670 (1980).
121. P. L. Luisi and R. Wolf, in *Solution Behavior of Surfactants: Theoretical and Applied Aspects*, Vol. 2, K. L. Mittal and E. J. Fendler, eds., Plenum, New York, 1982, p. 887.
122. A. V. Levashov, Y. L. Khmel'nitsky, N. L. Klyachko, V. Y. Chernyak, and K. Martinek, *J. Colloid Interface Sci. 88*: 444 (1982).
123. A. V. Levashov, Y. L. Khmel'nitsky, N. L. Klyachko, and K. Martinek, in *Surfactants in Solution*, K. L. Mittal and B. Lindman, eds., Plenum, New York, 1984, p. 1069.

124. K. Martinek, A. V. Levashov, Y. L. Khmel'nitsky, N. L. Klyachko, and I. V. Berezin, *Science 218*: 889 (1982).
125. K. Martinek, A. V. Levashov, N. L. Klyachko, and I. V. Berezin, *Dokl. Acad. Nauk SSSR 236*: 920 (1977).
126. A. V. Levashov, N. L. Klyachko, and K. Martinek, *Bioorg. Khim. 7*: 670 (1981).
127. K. Martinek, A. V. Levashov, N. L. Klyachko, V. I. Pantin, and I. V. Berezin, *Biochim. Biophys. Acta 657*: 277 (1981).
128. F. M. Menger and K. Yamada, *J. Am. Chem. Soc. 101*: 6731 (1979).
129. R. Wolf and P. L. Luisi, *Biochem. Biophys. Res. Commun. 89*: 209 (1979).
130. S. Barbaric and P. L. Luisi, *J. Am. Chem. Soc. 103*: 4239 (1981).
131. C. Grandi, R. E. Smith, and P. L. Luisi, *J. Biol. Chem. 256*: 837 (1981).
132. P. Douzou, E. Key, and C. Balny, *Proc. Natl. Acad. Sci. U.S.A. 76*: 681 (1979).
133. P. L. Luisi, P. Meier, E. Imre, and M. Fleschar, 185th American Chemical Society Meeting, Seattle, Washington, March 20–25, 1983, Abstract, Coll., 117.
134. P. Meier and P. L. Luisi, *J. Solid-Phase Biochem. 5*: 269 (1980).
135. L. M. Gierasch, J. E. Lacy, K. F. Thompson, and A. L. Rockwell, 185th American Chemical Society Meeting, Seattle, Washington, March 20–25, 1983, Abstract, Coll., 118.
136. K. Kon-no, M. Tosaka, and A. Kitahara, *J. Colloid Interface Sci. 79*: 581 (1981).
137. K. Kon-no, M. Tosaka, and A. Kitahara, *J. Colloid Interface Sci. 86*: 288 (1982).
138. K. Kon-no, M. Tasaka, Y. Saratani, and A. Kitahara, *Nippon Kagaku Kaishi* 543 (1982).
139. N. L. Klyachko, A. A. Baykov, A. V. Levashov, K. Martinek, and S. M. Nvaeva, *Bioorg. Khim. 6*: 1707 (1980).
140. C. Balny, G. Hui Bon Hoa, and P. Douzou, *Jerusalem Symp. Quantum Chem. Biochem. 12*: 37 (1979).
141. R. L. Misiorowski and M. A. Wells, *Biochemistry 13*: 4921 (1974).
142. K. Martinek, Y. L. Khmel'nitskii, A. V. Levashov, I. V. Berezin, *Dokl. Akad. Nauk SSSR 263*: 737 (1982); English transl., p. 81.
143. K. Martinek, Y. L. Khmel'nitskii, A. V. Levashov, N. L. Klyachko, A. N. Semenov, and I. V. Berezin, *Dokl. Akad. Nauk SSSR 256*: 1423 (1981); English transl., p. 143.
144. D. W. Armstrong, R. Seguin, C. J. McNeal, R. D. MacFarlane, and J. H. Fendler, *J. Am. Chem. Soc. 100*: 4605 (1978).

145. P. L. Luisi, F. J. Bonner, A. Pellegrini, P. Wiget, and K. Wolf, *Helv. Chim. Acta 62*: 740 (1979).
146. B. de Kruijff, P. R. Cullis, and A. J. Verkleijj, Trends in Biochemical Sciences 5: 79 (1980).
147. P. Douzou, *Cryobiochemistry*, Academic Press, New York, 1977.
148. A. L. Fink, *Acc. Chem. Res. 10*: 233 (1977).
149. K. Martinek, A. V. Levashov, N. Klyachko, Y. Khmelnitski, and I. V. Berezin, *Eur. J. Biochem. 155*: 453 (1986).

6

Water-in-Oil Emulsions

PAUL BECHER Paul Becher Associates Ltd., Wilmington, Delaware

I. INTRODUCTION

There is an interesting disparity in the literature of emulsions between the attention devoted to water-in-oil (W/O) emulsions and that devoted to oil-in-water (O/W) systems. Namely, the literature for O/W systems towers over that for W/O.* Two suggestive pieces of evidence may be presented in support of this.

First, the often-reproduced table of required HLBs (1) lists 72 entries for O/W emulsions and but 4 for W/O systems. Second, as a semantic matter, it is not uncommon to refer to W/O emulsions as *inverted* or *reversed* emulsions, although what makes them specifically inverted or reversed with respect to O/W emulsions is not clear.

*This is not the case for microemulsions, however.

It is interesting (and possibly useful) to speculate on the causes of the disparity of interest. It has been suggested that this is a reflection of the relative utility of the two types of emulsions. This cannot be the case. It is likely that in a practical sense, more O/W emulsions are employed than W/O ones, although the margin cannot be large.

It is possible to speculate that the difference is related to the intensity with which DLVO theory has been applied to disperse systems over the past 30 years or so. Once Albers and Overbeek [2] demonstrated that the double layer in W/O emulsions makes no significant contribution to their stability, these systems were no longer of investigative interest, since there existed no theoretical basis for their behavior, beyond the admittedly primitive concepts involving a mechanical barrier to coalescence, as described, for example, by Becher [3].

However, with the development of theories of stability which go beyond DLVO, i.e., theories of so-called *steric* or *thermodynamic* stabilization, it might have been expected that the question of the stability of W/O emulsions would be extensively reopened. This does not, in fact, appear to be the case. In the subsequent discussion (see Sec. II,B) we shall attempt, in some measure, to rectify this situation.

However, even in the absence of a theory, applications of W/O emulsions have been extensive, and some of these applications will also be cited briefly (see Sec. IV).

II. STABILITY OF WATER-IN-OIL EMULSIONS

A. Concept of the Mechanical Barrier

1. Background

More than 50 years ago, evidence was presented by Serrallach and Jones [4] for the existence of well-defined interfacial films between mineral oil and water, in the presence of various surface-active and polymeric materials (e.g., tragacanth, acacia, gelatin). These workers attempted to estimate the relative strengths of the films formed.

A few years later, Serrallach et al. [5] devised a technique for quantitative measurement of the strength of these interfacial films by what is essentially a modification of the du Nouy tensiometer. The effect of concentration and of time was evaluated for a number of vegetable and mineral oils. The authors concluded that, since in many cases the strength of the film increased on aging, the films were probably not significant in stabilizing the emulsion in its early lifetime, with the possible exception of tragacanth, which produced strong films initially and which do not vary significantly with time.

It is interesting to note that Serrallach et al. [5] observed what may probably be described as an ultralow interfacial tension between cod liver oil and sodium glycocholate.

Another early paper which repays study is that of Wellman and Tartar [6]. For example, for benzene emulsions stabilized by sodium stearate, if the water/benzene phase ratio is less than 0.5, water-in-oil emulsions result. It is an interesting historical footnote to mention that Wellman and Tartar were evidently the first to identify the existence of a phase-inversion temperature (PIT) [7].

2. Surface Rheology and the Mechanical Barrier

Without engaging in an elaborate historical review, it is apparent that the excessively qualitative concepts involved in the idea of a mechanical barrier to coalescence were unsatisfactory. With the recognition that an adsorbed monolayer could exhibit two-dimensional rheological behavior, however, the idea of the *film strength*, proposed, e.g., by Serrallach et al. [4,5], could be quantified. Indeed, the interfacial rheology of the adsorbed monolayer could now be called upon to explain the stability of water-in-oil as well as oil-in-water systems.

Becher [8] briefly summarized the possible contributions of interfacial viscosity to emulsion stability and described some methods of measurement.

A more extensive review of surface rheology is due to Joly [9]. Joly reviewed the theory and experimental techniques in considerable detail. Although his discussion of the effect of interfacial rheology on the stability of emulsions is rather brief, a number of important points are made, which may be summarized here. The Marangoni effect, related as it is to the apparent interfacial dilational viscosity, is important in stabilizing the emulsion droplets [10]. Similarly, high rigidity and high surface shear viscosity also exhibit stabilizing influences, although it is true that this is far from the whole story, since many cases of stable emulsions in the absence of high surface viscosity are known. In these cases, of course, the other influences of surface charge (not of importance here, as we have seen) and so-called thermodynamic stabilization (discussed below) may be controlling factors.

A typical example of this effect is shown in Fig. 1 [11]. In this study, a sample of SPAN 60 (sorbitan monostearate) was treated to remove components (unreacted polyol) believed to be the effective formers of the viscous interfacial film. Interfacial viscosity measurements were carried out, using the oscillating-bob technique [12]. Figure 1a shows the damping curves obtained with the untreated emulsifier, as a function of the age T of the interface (in seconds). The curve for T=0 is a straight line, indicating that no significant interfacial viscosity is being observed. At T=60, however, the curve

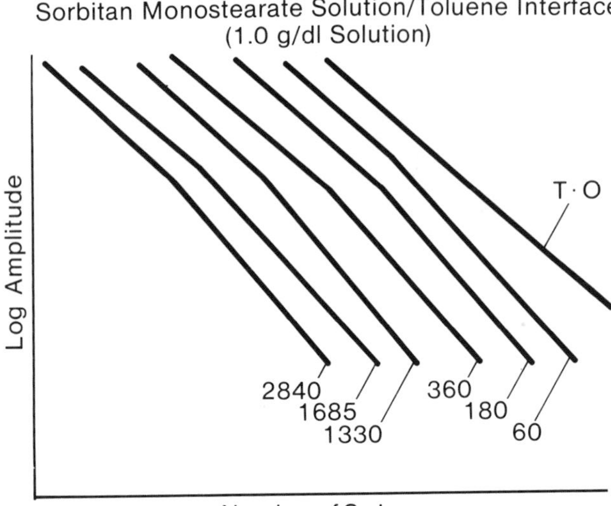

(a)

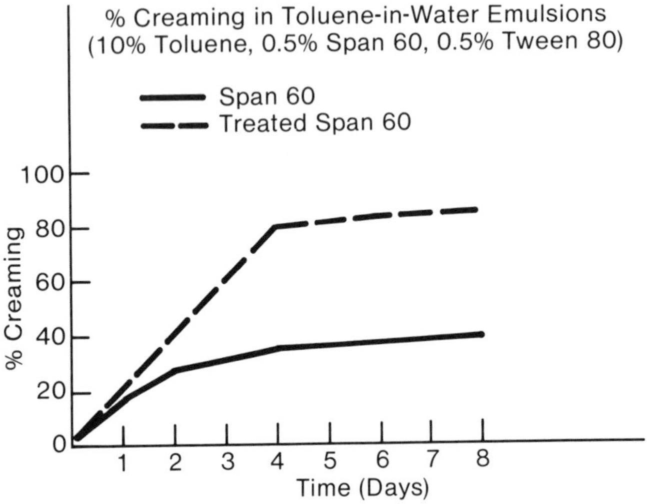

% Creaming in Toluene-in-Water Emulsions
(10% Toluene, 0.5% Span 60, 0.5% Tween 80)

Span 60
Treated Span 60

% Creaming
100
80
60
40
20
0
1 2 3 4 5 6 7 8
Time (Days)

(b)

FIG. 1 (a) Measurement of surface viscosity by oscillating-bob technique. Untreated SPAN 60 shows increasing interfacial viscosity with time (T in seconds). (b) Rate of creaming in emulsions made with and without treated SPAN 60. (From Ref. [11], courtesy of Allured Publishing Corp.)

is appreciably nonlinear, indicating an increase in surface viscosity with time. The treated emulsifier, when measured by the same technique, gave a dependence of the oscillation amplitude similar to the T=0 curve, and did not change with time.

Figure 1b shows the results of measurement of creaming in emulsions prepared with the treated and untreated emulsifiers. Clearly, the emulsion containing the treated (non-film-forming) emulsifier creamed at a significantly higher rate. Although this experiment was carried out on an O/W emulsion, application to the W/O emulsion case is obvious.

Surface viscosity measurements by Boyd and Sherman [13] show a similar increase in viscosity with time for solutions of SPAN 80 in mineral oil (Nujol). For a 1.0% solution, a small value for the modulus of elasticity was observed at the end of 24 hr. On the other hand, no increase in surface viscosity was observed for SPAN 20 in aqueous solution.

Sherman [14] presented an interesting discussion of the role of interfacial rheology in the coalescence of flocculated emulsion droplets. The net force acting on two approaching drops is taken to be

$$F = F_H + F_{DL} - F_{VW} + (F_{CB} - F_{CF}) + F_G \tag{1}$$

where F_H is the hydrodynamic force causing flow of continuous phase from between the droplets and distortion of their surface owing to the pressure developed between them, F_{DL} is the electrostatic double-layer repulsion (unimportant in the W/O case), F_{VW} is the van der Waals attractive force, $(F_{CB} - F_{CF})$ is the net deceleration produced by the continuous phase, and F_G is the force due to gravity. After flocculation, the term $(F_{CB} - F_{CF})$ should, of course, be equal to zero. However, the drops within a floc should still be subject to a net force made up of several components.

The magnitude of F will not be the same at each point on the surface of a flocculated drop where it is in contact with another drop, because the contribution of F_G will depend on the position of the drops relative to one another. The simplest case illustrating this point is that of a cubic close-packed array, where the flocculated drops are arranged so that each layer lies immediately over the other (Fig. 2). A drop will then experience a vertical force F_G [= (4/3)-$\pi r^3 \Delta\rho g$, where r is the drop radius, $\Delta\rho$ is the difference between the densities of the two liquids, and g is the gravitational constant] owing to the drop immediately above or below it, depending on whether we are considering a W/O or O/W emulsion. In the horizontal plane, the drops are located in the same layer on either side of the reference drop, but here F_G is zero. In contrast, at the other extreme is the case of closest packing in rhombohedral geometry

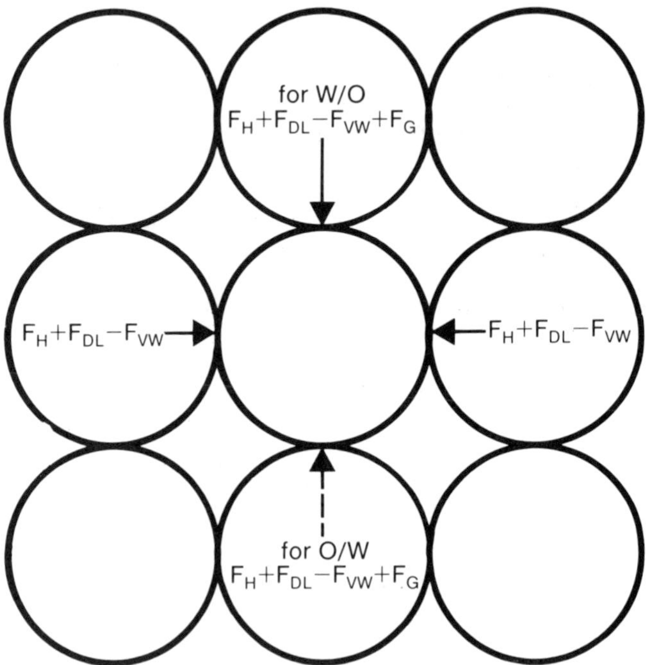

FIG. 2 Forces acting on a drop in an emulsion showing a cubic packing geometry. (From Ref. [14], courtesy of the *Journal of Colloid and Interface Science*.)

(Fig. 3). In this case, F_G is exerted in a nonvertical direction and at *two* points on the reference surface.

Sherman [14] showed that the compressive force acting on the reference droplet in either case is given by

$$q_0 = \frac{3F}{2\pi a_2^2} \tag{2}$$

where F is defined by Eq. (1) and a_2 is the radius of the surface of contact between the droplets, which may be calculated from a knowledge of F and of the Poisson ratios and elastic extension moduli of the two liquids.

If the interfacial film is viscoelastic, all that is necessary to induce film rupture is for the compressive force to exceed the instantaneous elastic compliance J_0. At this point, the adsorbed

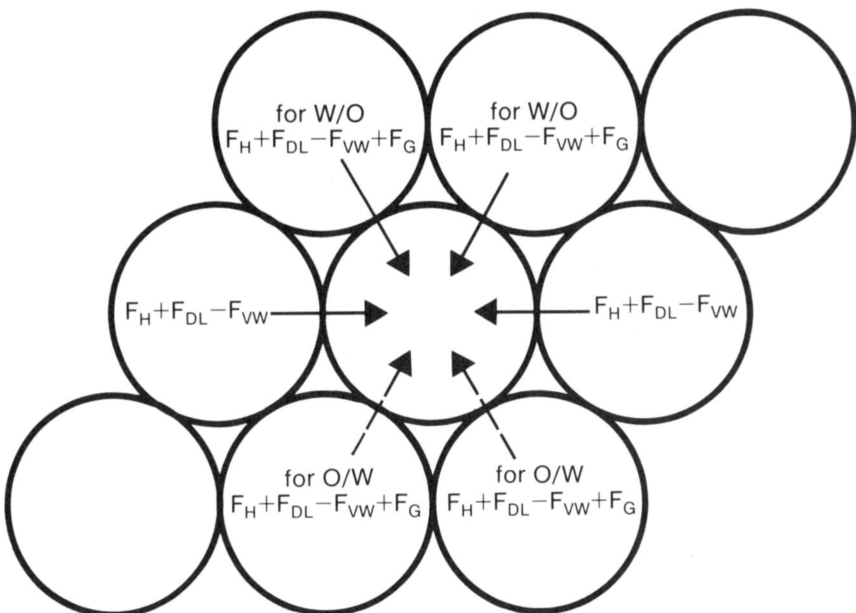

FIG. 3 Forces acting on a drop in an emulsion showing a rhombohedral packing geometry. (From Ref. [14], courtesy of the *Journal of Colloid and Interface Science*.)

emulsifier molecules are displaced, and coalescence may take place. Although quantitative applications of this theory are lacking, the results are suggestive.

Since Joly's review appeared, a number of significant papers on the thoery and measurement of surface rheological properties have appeared [15–18].

B. Thermodynamic Stabilization of Water-in-Oil Emulsions

The theory of thermodynamic stabilization has developed mainly in connection with studies carried out on dispersions of solids and, in particular, on dispersions of latices. However, the applicability of this approach to emulsions (both W/O and O/W) has been noted [19,20].

These concepts have been reviewed, with particular attention to nonaqueous media, by Vincent [21] and Parfitt and Peacock [22]. The entire area has been extensively reviewed by Napper [23].

A number of different approaches to the theory of thermodynamic stabilization have been offered [22]. Two such approaches, which will be seen to be complementary, will be discussed briefly.

1. Phase Separation

In the first approach, largely due to Napper [23], the conditions leading to flocculation are treated from the point of view of phase separation phenomena.

In binary systems, the phenomenon of phase separation arising from partial miscibility is well known, the systems phenol/water, aniline/hexane, and triethylamine/water being typical. In the first two cases, phenol and water are completely miscible above 70°C, while aniline and hexane are completely miscible above 60°C. This temperature is known as the upper critical solution temperature (UCST).

In the case of triethylamine/water, the system is completely miscible *below* 15°C, and this temperature is known as the lower critical solution temperature (LCST), since it appears in the lower area of the binary phase diagram.

Cases where both an LCST and a UCST are found, e.g., nicotine/water, are also known. In fact, any partially miscible system should exhibit both sorts of behavior, but one or the other of the limits may not be observable by reason of temperature limitations.

Similar phase separation phenomena can also be induced by a change of solvent. Thus, for the phenol/acetone system, phase separation occurs on the addition of about 20% of water.

The case is similar for polymers [24]. Advantage is taken of this phenomenon to enable the fractionation of these materials.

In the case of thermodynamic stabilization, the assumption is made that the dispersion medium is the solvent for the adsorbed or anchored polymeric stabilizer. Hence, under conditions which lead to phase separation (change of temperature or solvent), the polymer no longer confers stability on the system. This is, to be sure, a rather oversimplified description of the theory, but is essentially correct.

We should thus expect to find that on heating, for example, the dispersion should flocculate at some temperature above the UCST. Indeed, this is the case, although the so-called critical flocculation temperature is usually some 15–20°C above the UCST.

The quantitative basis for this approach is the thermodynamic relationship $\Delta G_M = \Delta H_M - T\,\Delta S_M$, where ΔG_M, ΔH_M, and ΔS_M are, respectively, the changes in free energy, enthalpy, and entropy of mixing of polymer and solvent when the layers of adsorbed polymer overlap during a particle collision or close approach. Since ΔG_M corresponds to the repulsive free energy, a positive value of the free energy corresponds to a stable system; i.e., the flocculation reaction is not favored.

The consequences of this approach may be summarized as follows:

1. ΔH_M positive, ΔS_M negative. In this case, ΔG_M is always positive, and the system is stable.
2. ΔH_M positive, ΔS_M positive. In this case, the enthalpic term contributes to stability, while the entropic term has a destabilizing effect. Such systems are destabilized by raising the temperature, to the point where the entropic-enthalpic ratio ($T\ \Delta S_M/\Delta H_M$) < 1. This corresponds to the theta temperature of the stabilizing polymer.
3. ΔH_M negative, ΔS_M positive. The repulsive free energy is always negative, and the system is unstable.
4. ΔH_M negative, ΔS_M negative. The entropic term promotes stability, but if the system is cooled, the entropic-enthalpic ratio will become <1, and the system will be unstable.

Clearly, then, a knowledge of the solution thermodynamics of the stabilizing polymer dissolved in the disperse phase would enable one to predict the stability regime. For O/W emulsions stabilized by nonionic ethoxylates, this would mean the system polyoxyethylene-water, which has been thoroughly investigated, e.g., by Napper and Natschey [25].

For W/O emulsions, on the other hand, data are less available. March and Napper [20] studied the stabilizing effect of block copolymers on toluene-in-water, *n*-pentanol-in-water, and cyclohexane-in-water emulsions, the last two of these being of special interest in connection with the present discussion.

March and Napper's results are summarized in Table 1, where the stabilizing polymers are described in terms of the anchor polymer and stabilizing moieties (the designations U and L in this table refer to upper and lower critical flocculation and theta temperatures). It will be noted that the W/O emulsions would be destabilized by *lowering* the temperature. The close agreement between the critical flocculation temperature and the theta temperature is striking evidence of the utility of this approach.

An interesting exercise would be to compare the information on the solution thermodynamics of, e.g., proteins with the experimental behavior of protein-stabilized W/O emulsions.

Napper summarized this approach in [23] and in a number of papers cited in that book.

2. Adsorption Models

The phase separation approach described in the preceding section involves no physical picture of the interfacial film of adsorbed polymer.

TABLE 1 Thermodynamic Limit to Stability of Sterically Stabilized Emulsions [21][a]

Anchor polymer	Stabilizing moieties	Dispersed phase	Dispersion medium	Emulsion type	U/L	CFT/K	θ/K	U/L
PVAc	POE	Toluene	0.39 M $MgSO_4$	Oil-in-water	U	318 ± 2	318 ± 2	θ_L
PVAc	PVA	Toluene	2.0 M NaCl	Oil-in-water	U	297 ± 3	299 ± 2	θ_L
PVAc	PAM	Toluene	2.1 M $(NH_4)_2SO_4$	Oil-in-water	L	297 ± 2	298 ± 3	θ_U
PVAc	PAA	Toluene	0.2 M HCl	Oil-in-water	L	298 ± 2	287 ± 5	θ_U
PAA	PLM	Water	*n*-Pentanol	Water-in-oil	L	301 ± 3	303 ± 5	θ_U
POE	PS	Water	Cyclohexane	Water-in-oil	L	308 ± 3	307 ± 1	θ_U

[a]Key to polymers: PVAc, poly(vinyl acetate); POE, poly(oxyethylene); PVA, poly(vinyl alcohol); PAM, poly(acrylamide); PAA, poly(acrylic acid); PLM, poly(lauryl methacrylate); PS, polystyrene.

The other principal theoretical treatments of the stabilization of emulsions and dispersions by adsorbed polymers, however, are based on the creation of a model of the interfacial film, to which the methods of statistical thermodynamics and polymer solution theory may be applied. Napper [26] has referred to these treatments as *ab initio* theories (although he argues that none of these theories is truly *ab initio*).

In general, as noted above, these theories initially require some model for the structure of the adsorbed layer. This involves assumptions about the way in which the polymer is anchored to the substrate (e.g., the water droplet in a W/O emulsion) and the conformation of the chains in the dispersion medium. In many cases, for the sake of simplicity, bond angles are assumed to be 90°, and the chains are assumed to have only extension, with zero dimensions in the directions parallel (more or less) to the droplet. Although such assumptions are unrealistic, they are suprisingly useful.

There is a subtle distinction in discussing the orientation and conformation of the adsorbed polymer between solid substrates (e.g., latices) and liquid substrates (e.g., emulsions). In the case of solids, the polymer is adsorbed (generally) on the surface of the particle, the exception being latices in which the stabilizing polymer is physically incorporated in a latex during its manufacture.

The conformation of a polymer adsorbed at the surface of a solid will depend on the structure of the polymer. For example, the polymer could be made up of hydrophilic (H) and lipophilic (L) monomer segments with a structure $(H)_n(L)_m$ or $(H)_n(L)_m(H)_{n'}$, or other variations of a like sort. In the case of a solid-liquid dispersion (e.g., a hydrophilic solid dispersed in an organic dispersion medium), the situation may be as envisioned in Fig. 4. The hydrophilic moieties are assumed to lie flat on the solid surface and are termed *trains*. The lipophilic moieties extend into the dispersion medium, and may be in the form of either *tails* (Fig. 4a,c) or *loops* (Fig. 4b,c).

In the case of emulsions, however, it is reasonable to assume that the polymer moiety corresponding to the disperse phase (water in the case of W/O emulsions) will be dissolved in the disperse phase. If the polymer is of the type $(H)_m(L)_n$, then the hydrophilic moiety will dissolve in the aqueous phase, while the lipophilic moiety will extend into the dispersion medium as a tail (see Fig. 5). With more complex polymers, e.g., proteins, the existence of loops as well as tails is possible, as in Fig. 5b,c.

In order to apply this model to estimate stability, it is necessary to calculate the average extension of the stabilizing moieties and then apply an appropriate theory of polymer solution.

We shall describe one such theory, that of Hesselink et al. [27, 28]. Although a number of possible changes in the conformation of

Lipophilic Dispersion
Medium

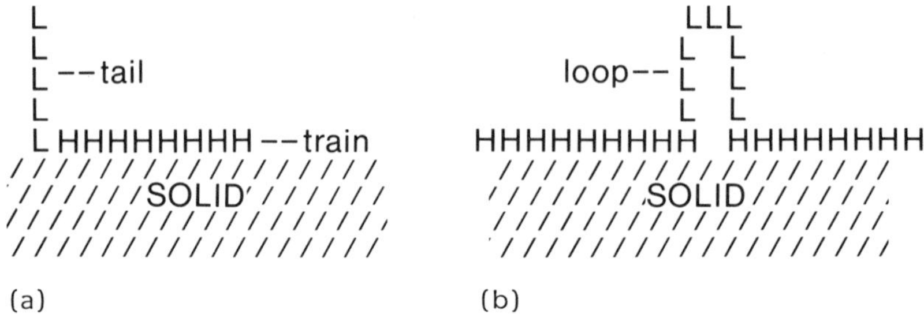

Lipophilic Dispersion
Medium

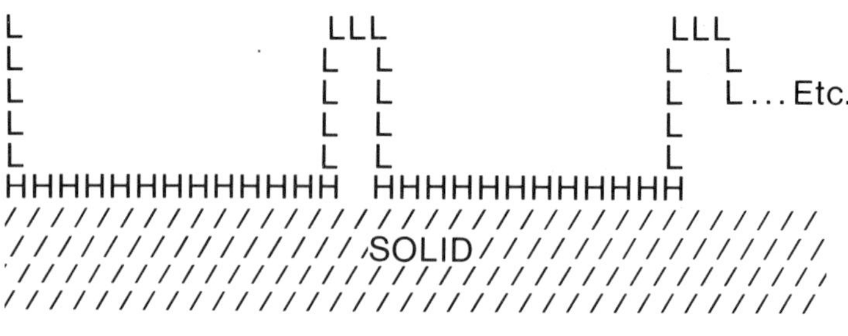

FIG. 4 Polymer conformation at interface between a lipophilic solid (e.g., latex) and a liquid. (a) Simple HL copolymer; only trains and tails are found. (b) HLH block copolymer; trains and loops are found. (c) LHLHL . . . block copolymer; loops, trains, and tails.

adsorbed polymer may be induced by the approach of two particles [29], Hesselink and co-workers considered two principle situations:

1. A polymer molecule adsorbed on a colloidal particle loses configurational entropy on the approach of a second particle. This is called the *volume restriction effect*.
2. When the layers of adsorbed polymers on the two particles interpenetrate, the higher polymer segment concentration thus de-

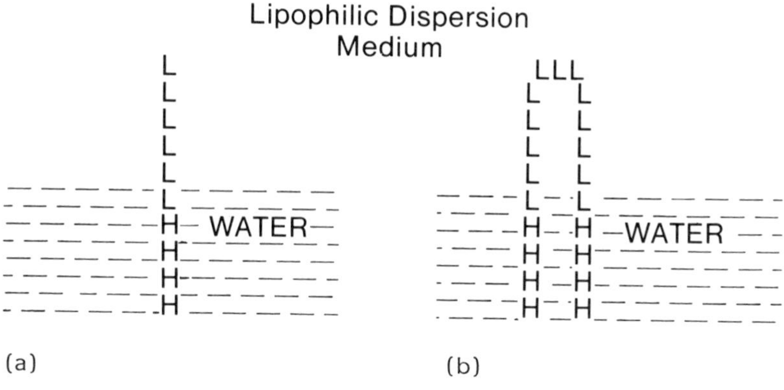

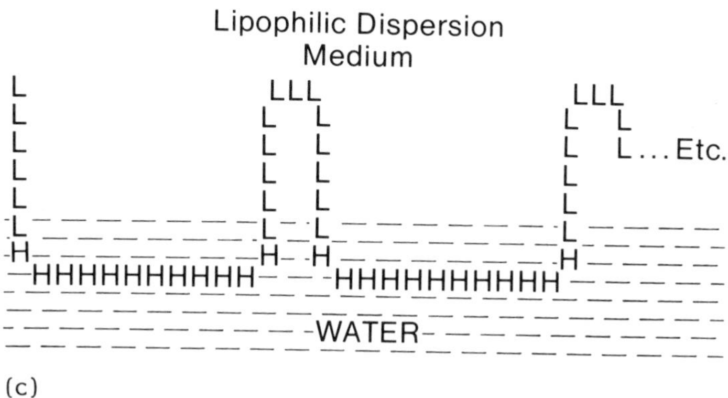

FIG. 5 Polymer conformation in the case of emulsions. Note that at the liquid-liquid interface trains do not, in effect, exist. Also, the spacing of loops or tails may be variable, depending to some extent on the flexibility of the polymer chains. This is in distinction to the situation shown in Fig. 4, where the spacing is dictated by the manner in which the train is adsorbed on the solid surface.

veloped will lead to a local *osmotic pressure*, which in most cases counteracts the approach.

In order to arrive at a numerical result, the assumption is made that the polymer chain can be expressed by using random-flight statistics. Thus the spatial dimensions are expressed in units

$(il^2)^{1/2}$, where i is the number of polymer segments of length l in the polymer chain. The authors equate $(il^2)^{1/2}$ with the experimental root-mean-square (rms) end-to-end distance $\langle r^2\rangle^{1/2}$ of the chain. Thus, $il^2 = \langle r^2\rangle$. The expansion α of the chain owing to long-range molecular interactions [30] is taken into account by setting $\langle r^2\rangle = \alpha\langle r^2\rangle_0{}^{1/2}$, where $\langle r^2\rangle_0{}^{1/2}$ is the unperturbed rms end-to-end distance determined in a theta solvent.

The free energy of repulsion arising from the volume restriction effect is then given by

$$\Delta F_{VR} = 2\sigma kTV(i,d) \tag{3}$$

where σ is the concentration of tails or loops at the particle surface (expressed as grams of adsorbed polymer per weight of disperse phase), i is the average length of a polymer segment, and d is the distance of separation of the particles. The quantity V(i,d) equals the free energy of repulsion per chain, and depends on the distribution of polymer chains and whether they are tails or loops.

The free energy of repulsion arising from the osmotic pressure term is given by

$$\Delta F_M = 2(2\pi/9)^{3/2}(\alpha^2 - 1)kT\sigma^2\langle r^2\rangle M(i,d) \tag{4}$$

where the quantity M(i,d) is the free energy of repulsion corresponding to the osmotic effect. It should be noted that the expansion factor α appears explicitly in Eq. (4) and that the surface concentration of polymer is present as the square.

Values for V(i,d) and M(i,d), obtained by numerical integration of the appropriate equations [28], are given in Table 2 in dimensionless distance units $d/(il^2)^{1/2}$.

From the classical picture of the stability of colloids [31], we recall that the *attractive* forces opposing repulsion are given by the London-van der Waals interaction. For parallel plates, if we ignore the effects of retardation and of the finite dimensions and thickness of the polymer layer, the free energy of attraction for two flat plates is given by

$$\Delta F_A = \frac{-A}{12\pi d^2} \tag{5}$$

where A is the so-called *Hamaker constant* and d is the distance of separation. Although the flat-plate model may seem unrealistic as a

TABLE 2 Values for the Dimensionless Volume Restricted V(i,d) and Osmotic Function M(i,d) [28]

	V(i,d)			M(i,d)		
$d/\sqrt{il^2}$	Equal tails	Equal loops	Copolymer	Equal tails	Equal loops	Copolymer
0.6	2.996	1.476	2.030	3.723	2.974	3.428
0.8	1.284	0.339	0.760	2.397	1.716	2.078
1.0	0.582	0.0561	0.307	1.585	0.837	1.280
1.2	0.262	0.00578	0.1282	1.043	0.314	0.801
1.4	0.1118	0.00035	0.0542	0.667	0.0940	0.480
1.6	0.0439		0.0231	0.406	0.0204	0.281
1.8	0.01582		0.0109	0.232	0.0034	0.150
2.0	0.00497		0.0041	0.127		0.081
2.5	0.00018			0.018		

model for emulsion droplets, it is actually quite satisfactory, as pointed out by Becher and Tahara [32].

Accordingly, the total interaction between droplets or latex particles is given by

$$\Delta F = \Delta F_{VR} + \Delta F_{M} + \Delta F_{A} \tag{6}$$

by combining Eqs. (3)-(5).

Thus, if we have some information about the interactions between the stabilizing moiety and the dispersion, the surface concentration of stabilizer, and the dimensions of the polymer chain, it is a comparatively simple matter to calculate the repulsive free energy due to the adsorbed polymer. For example, Hesselink et al. [28] evaluated Eq. (6) for a number of cases, using polystyrene as the adsorbed polymer. The molecular weight M is taken to be between 10^3 and 10^5. From light scattering, the rms end-to-end distance lies between 2.12 and 21.2 nm. The surface concentration σ is taken to be between 10^{-10} and 5×10^{-7} g cm^{-2} and the quality of the solvent between a = 0.9 (very poor solvent) and a = 1.6 (very good solvent).

Figure 6, for example, shows the result of applying the above equations to the case of adsorbed polystyrene chains, for both

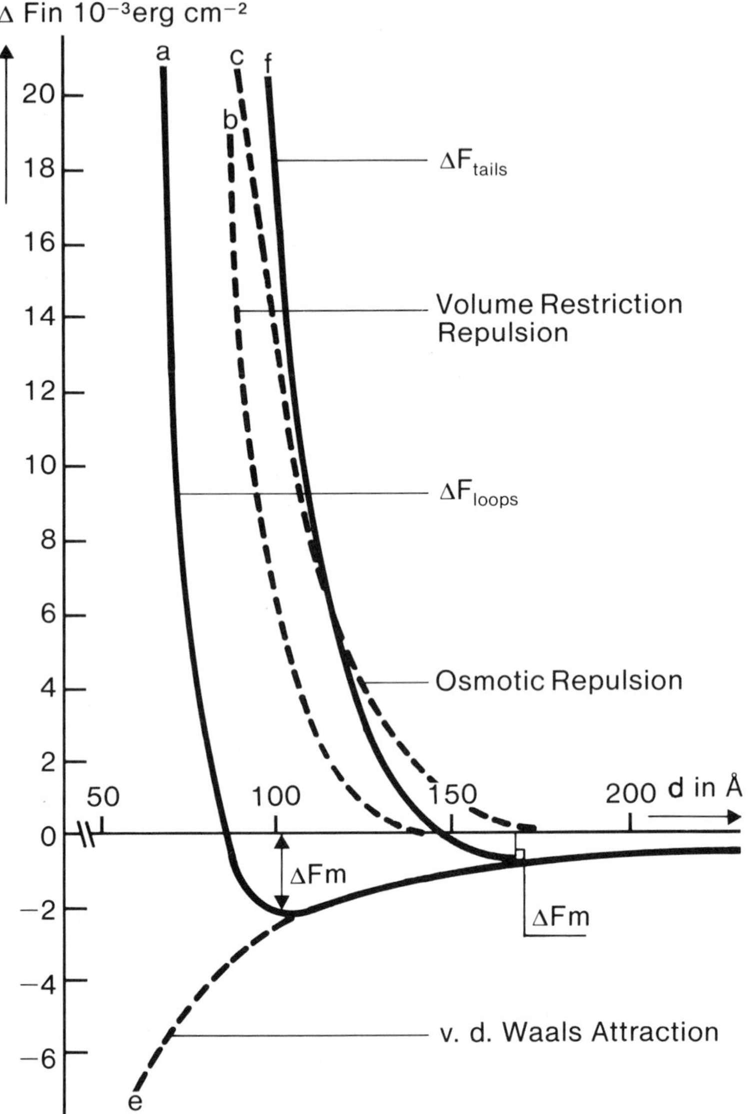

FIG. 6 Free energy of interaction versus distance between particles covered by equal tails (curve f) and equal loops (curve a). For particles covered by equal tails curve b gives the volume restriction contribution, and curve c the osmotic repulsion. Curve f is the effect of adding curves b and c and the van der Waals attraction curve e, i.e., using Eq. (6). (From Ref. [28], courtesy of the *Journal of Physical Chemistry*.)

loops and tails. It is assumed that the loops or tails are all equal, i.e., of the same length. The following parameters are assumed: $A = 10^{-13}$ erg, $\alpha = 1.2$, ω ($= \sigma M/N_A$, where M is molecular weight and N_A is Avogadro's number) $= 2 \times 10^{-8}$ g cm^{-2}, M = 6000 (hence $\langle r^2 \rangle_0^{1/2} = 5.2$ nm and the area per chain is 50 nm^2).

Curves b and c in Fig. 6 are, respectively, the results of applying Eqs. (3) and (4) for the case of tails; curve f is obtained by adding curves b, c, and e (the van der Waals attraction), according to Eq. (6). Curve a is the result of a similar calculation for equal loops. No primary minimum is observed (which, perhaps, is equivalent to saying that the primary maximum is off scale, even for very small distances of separation). In Fig. 6, the minimum is designated ΔF_m.

The depth of the minimum ΔF_m determines the stability of the colloidal system, since it is in this potential well that flocculation will occur. Note that curve c (tails) has a much shallower minimum than curve a (loops). From this, it may be concluded that, all other things being equal, adsorbed tails would contribute to a more stable system than loops.

We compare this minimal free energy per square centimeter, multiplied by the area of the flat plate h^2 (for a plate with edge length h), with the thermal energy of the particles. When $h^2 \Delta F_m < kT$, the particles will not coagulate and the system is stable. On the other hand, when $h^2 \Delta F_m > kT$, the particles will have a tendency to flocculate, which in the case of emulsions may lead to coalescence and total instability.

As a rule, if the depth of the minimum lies in the range $kT < h^2 \Delta F_m < 5kT$, the effect is minor, and even if some flocculation occurs the system may be redispersed by simple stirring or gentle shaking. However, when $h^2 \Delta F_m > 5kT$, the particles will remain flocculated, and this, as noted above, can lead to instability in an emulsion system. Put another way, in the system of Fig. 5, flat plates with an edge dimension of 0.1 μm will be stable when $\Delta F_m < 5kT/h^2 = 2 \times 10^{-3}$ erg cm^{-2}. On the other hand, if we double the edge length (i.e., 0.2 μm), a stable system is formed only if the depth of the minimum is $\leq 5 \times 10^{-4}$ erg cm^{-2}. This criterion cannot be met for the system shown in Fig. 6. This is clearly illustrated in Fig. 7, which shows the areas of stability in a plot of the depth of the minimum against the molecular weight of the stabilizing moiety, as a function of the particle size, the Hamaker constant A, and the expansion factor α. In this figure, $\sigma = 2 \times 10^{-8}$ g cm^{-2}. Note that an improvement in the quality of the solvent (α going from 1.2 to 1.6) reduces the required length of the stabilizing chain by a significant amount. An even more signfiicant reduction is obtained by a decrease in the attractive interaction, as measured by the Hamaker constant.

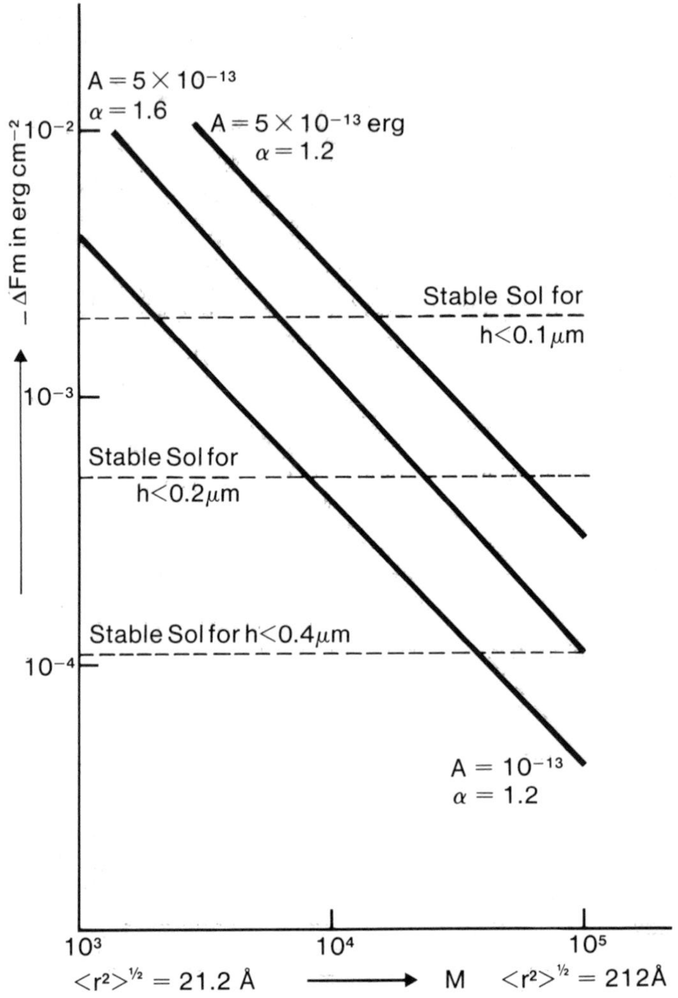

FIG. 7 Depth of the minimum $-\Delta F_m$ in the interaction curve as a function of the molecular weight of the adsorbed tails for $\omega = 2 \times 10^{-8}$ g cm^{-2}. The stability of the system is determined by the value of the quantity $h^2 \Delta F_m/kT$. The curves are calculated for M = 1000, $\langle r^2 \rangle_0^{1/2}$ = 21.2 Å and for M = 10^5, $\langle r^2 \rangle_0^{1/2}$ = 212 Å. (From Ref. [28], courtesy of the *Journal of Physical Chemistry*.)

It might be noted, by reference to Eq. (6), that when $\alpha = 1$ the osmotic effect vanishes, and that when $\alpha < 1$ the sign of the osmotic effect becomes positive, which is equivalent to an attraction. We see here an effect equivalent to that produced by a change of solvent in the phase separation approach.

Hesselink [33] has also computed the volume exclusion and osmotic (mixing) terms for spheres. For these results, it appears that the magnitude of these effects is smaller in the case of spheres than it is for flat plates. However, as pointed out above, for particles of the size of emulsion droplets, the difference between the two models may be minor.

A more important conclusion, drawn in the same paper, is related to the effect of a distribution of chain lengths in the stabilizing moiety. Recall that the calculations presented above include the assumption that the chain lengths are equal. In practice, this situation is rarely encountered. The assumption of a distribution of chain lengths results in a repulsion curve which is less steep than that for equal chains. The result is that the coagulation minimum is much shallower, and thus the region of stability is enhanced. This prediction is verified by the well-known experimental fact that highly purified surfactants are frequently poorer emulsifying agents than the random-chain-length products of commerce.

It is useful to attempt to relate these results to the behavior of water-in-oil emulsions. Examination of Figs. 6 and 7 reveals that, for the conditions assumed, only extremely small particles can be expected to be stabilized. The particle sizes considered by Hesselink et al. [28] are of the order of tenths of a micrometer. Although practical emulsions may well contain particles of this size, they usually consist of particles of the order of one to several micrometers.

Clearly, the effect of this larger droplet size could be counteracted (in comparison to the data presented in Figs. 6 and 7) by the use of longer stabilizing moieties. This condition could possibly be met by the use of polymeric emulsifiers. However, it is well known that water-in-oil emulsions may be stabilized by the use of conventional emulsifiers, for example, by the use of ethoxylated compounds based on C_{12}–C_{18} fatty chains.

If the emulsifier is positioned at the interface as envisioned in Fig. 5a, the stabilizing moiety is the lipophile, usually an alkyl or alkyl-aryl chain. For a C_{18} chain, the molecular weight is of the order of 253 and the chain length perhaps 3.0 nm. In a water-in-paraffin oil emulsion, even though α might be as high as 1.6, the Hamaker constant would be 5.97×10^{-13} erg cm^{-2} [34], slightly *larger* than that assumed in Fig. 6. It might be supposed, therefore, that steric stabilization could not possibly account for the stability of W/O emulsions.

Actually, the opposite is true. The concentration at the interface, for a conventional nonionic surfactant, may be determined (by the Gibbs adsorption equation) to be about 2×10^{-10} mole cm^{-2}. This translates, for a C_{18} chain, into $\omega = 5.1 \times 10^8$ g cm^{-2}, about the same as for polystyrene. However, since the surfactant lipophiles are so much shorter than the polystyrene chains, this mass concentration corresponds to a much *higher* concentration of tails. In fact, for the surfactant considered here the concentration of tails is equal to 1.2×10^{14} tails per square centimeter, as opposed to about 2×10^{12} tails per square centimeter for the polystyrene of molecular weight 6000. This is reflected in the area per molecule, which is about 85 Å^2 for the surfactants, as opposed to the 5000 Å^2 reported by Hesselink et al. [28].

This difference is enough to more than account for the stability of such emulsions. It might be noted further that for water-in-benzene emulsions, the Hamaker constant is nearly two orders of magnitude lower than for water in paraffin oil. One would expect that such emulsions would be even more stable.

It should, however, be noted that Albers and Overbeek [2] concluded that such short chains would be incapable of stabilizing water-in-oil emulsions, and that such emulsions would be readily redispersible. The contradiction between these and our own conclusions, stated above, arises from the fact that Albers and Overbeek did not include the effect of thermodynamic stabilization in their calculations.

C. Stabilization by Solids

Stabilization of W/O emulsions by finely divided solids is well known and proceeds by the same mechanism as for O/W systems. The older literature on this subject has been summarized by Becher [35] and, more recently, by Tadros and Vincent [36].

For the solid particles to stabilize an emulsion, it is necessary that they collect at the interface, forming, in a sense, a mechanical barrier to coalescence, but of a sort different from that discussed in Sec. II,A. For the interfacial concentration to occur, it is necessary that the particles be wettable by both liquid phases. The type of emulsion produced is dictated by the relative values of the contact angle which each liquid forms with the solid.

If we assume an interface with the particle situated at some point in the interface, it is clear that the sum of the two contact angles must be 180°. One contact angle is thus less than 90°, the other greater. (The case where the angles are the same, 90°, corresponds to a planar interface, and therefore an unstable emulsion.) The external phase is the one with which the solid has a contact angle less than 90°.

As Tadros and Vincent point out, the concentration of solid at the interface does not reduce the interfacial free energy, and the effect is purely mechanical.

Koretskii and co-workers [37,38] defined solid particle stabilization, not from the point of view of contact angle, but rather in terms of the heat of wetting. However, this amounts to the same thing, since the heat of wetting may be defined in terms of the contact angle. They used surfactants to modify the contact angle and showed that it is possible to predict the conditions for inversion from W/O- to O/W-type emulsions, as well as the condition corresponding to unstable systems.

III. PHYSICAL PROPERTIES OF WATER-IN-OIL EMULSIONS

In general, and as might be expected, the physical properties of water-in-oil emulsions differ little from those of oil-in-water systems. In general, however, such emulsions tend to be more viscous than O/W ones and, in the absence of a significant charge, are poorly conducting.

A. Rheology

In a series of papers, Sherman [39] showed that the rheology of W/O emulsions is affected by a number of factors, including the nature of the emulsifying agent, the disperse volume, and the particle size distribution. Sherman's reviews of emulsion rheology [40, 41] may be usefully referred to in this connection.

More recently, Suzuki et al. [42] studied the non-Newtonian flow of W/O emulsions. In the systems studied, i.e., benzene, cyclohexane, liquid paraffin, isopropyl myristate, and squalene in water, stabilized by Arlacel 83 or Arlacel 60, shear thinning was observed. They concluded that, in addition to the factors noted by Sherman, the nature of the disperse phase played a significant role in the rheological behavior of the emulsion.

B. Conductivity

The fact that W/O emulsions are nonconducting, or at least possessed of very low conductivity, has been used as a method of determining emulsion type [43]. However, it is possible to imagine systems in which the external phase may be a polar liquid (in which the internal phase would have to be insoluble). In such a case one would expect that droplet charge would be sufficient that appreciable conductivity could be experienced and, indeed, that electrostatic repulsion might make a contribution to stability.

It does not appear that such systems have been studied. However, a paper by de Rooy et al. [44] on dispersions of sols in nonaqueous media offers some food for thought. An extensive review of the dielectric properties of emulsions by Clausse [45] has appeared.

IV. APPLICATIONS

To discuss the many applications of W/O emulsions in detail would increase the length of this chapter inordinately. This section will be restricted to indicating some important fields and supplying references for further study.

One of the most important applications of W/O emulsions is in the food industry, e.g., for salad dressings, mayonnaise, and margarine. Food emulsions in general have been considered by Krog and Larsson [46] and by Jaynes [47]. Other works which may be consulted are those of Friberg [48] and Sherman [49].

Water-in-oil emulsions also find applications in cosmetics, where the use of the oil as the external phase provides esthetic advantages [50].

Other areas of application include medicinals [51], the petroleum industry [52], and paints.

REFERENCES

1. K. Shinoda and H. Kunieda, in *Encyclopedia of Emulsion Technology*, P. Becher, ed., Vol. 1, *Basic Theory*, Dekker, New York, 1983, pp. 360–361.
2. W. Albers and J. Th. G. Overbeek, *J. Colloid Sci. 14*: 501, 510 (1959); *15*: 489 (1960).
3. P. Becher, *Emulsions: Theory and Practice*, 2nd ed., Krieger, Melbourne, Fla., 1977, p. 111.
4. J. A. Serrallach and G. Jones, *Ind. Eng. Chem. 23*: 1016 (1981).
5. J. A. Serrallach, G. Jones, and R. J. Owen, *Ind. Chem. 25*: 816 (1933).
6. V. E. Wellman and H. V. Tartar, *J. Phys. Chem. 34*: 379 (1930).
7. K. Shinoda and H. Kunieda, in *Encyclopedia of Emulsion Technology*, P. Becher, ed., Vol. 1, *Basic Theory*, Dekker, New York, 1983, Chap. 5.
8. P. Becher, *Emulsions: Theory and Practice*, 2nd ed., Krieger, Melbourne, Fla., 1977, pp. 111–112.
9. M. Joly, in *Surface and Colloid Science*, E. Matijevic, ed., Vol. 5, Wiley-Interscience, New York, 1972, pp. 1–193.
10. P. A. Rehbinder and A. Taubmann, *Proc. Int. Congr. Surface Activity, 3rd, Cologne*, p. 209 (1960).
11. P. Becher, *Am. Perfumer 77*: 21 (1962).
12. M. Joly, in *Surface and Colloid Science*, E. Matijevic, ed., Vol. 5, Wiley-Interscience, New York, 1972, pp. 64–65.
13. J. Boyd and P. Sherman, *J. Colloid Interface Sci. 34*: 76 (1970).

14. P. Sherman, *J. Colloid Interface Sci. 45*: 427 (1973).
15. M. F. M. Osborne, *Kolloid Z. Z. Polym. 224*: 150 (1960).
16. F. C. Goodrich, *Proc. R. Soc. London Ser. A 130*: 359 (1969); F. C. Goodrich and A. K. Chaterjee, *J. Colloid Interface Sci. 34*: 36 (1970).
17. R. J. Mannheimer and R. S. Schechter, *J. Colloid Interface Sci. 32*: 195, 212, 225 (1970).
18. D. T. Wasan, L. Gupta, and M. K. Vora, *AIChE J. 17*: 1287 (1971).
19. P. Becher, S. E. Trifilletti, and Y. Machida, in *Theory and Practice of Emulsion Technology*, A. L. Smith, ed., Society of Chemical Industry, London, 1976, p. 271.
20. G. C. March and D. H. Napper, *J. Colloid Interface Sci. 61*: 383 (1977).
21. B. Vincent, *Adv. Colloid Interface Sci. 4*: 193 (1974).
22. G. D. Parfitt and J. Peacock, in *Surface and Colloid Science*, E. Matijevic, ed., Vol. 10, Plenum, New York, 1978, Chap. 4.
23. D. H. Napper, *Polymeric Stabilization of Colloidal Dispersions*, Academic Press, New York, 1983.
24. P. C. Hiemanz, *Polymer Chemistry: The Basic Concepts*, Dekker, New York, 1984, pp. 528–527.
25. D. H. Napper and A. Netschey, *J. Colloid Interface Sci. 37*: 518 (1971).
26. D. H. Napper, *Polymeric Stabilization of Colloidal Dispersions*, Academic Press, New York, 1983, Chap. 10.
27. F. Th. Hesselink, *J. Phys. Chem. 75*: 65 (1971).
28. F. Th. Hesselink, A. Vrij, and J. Th. G. Overbeek, *J. Phys. Chem. 75*: 2094 (1971).
29. D. H. Napper, *Polymeric Stabilization of Colloidal Dispersions*, Academic Press, New York, 1983, pp. 198–203.
30. P. J. Flory, *Principles of Polymer Chemistry*, Cornell Univ. Press, Ithaca, N.Y., 1953.
31. E. J. W. Verway and J. Th. G. Overbeek, *Theory of the Stability of Lyophobic Colloids*, Elsevier, Amsterdam, 1948.
32. P. Becher and S. Tahara, *Proc. Int. Congr. Surface Activity 6th, Zurich*, 1972, *2*(1): 519 (1973).
33. F. Th. Hesselink, *J. Polym. Sci. Polym. Symp. 51*: 439 (1977).
34. J. Visser, *Adv. Colloid Interface Sci. 3*: 331 (1972).
35. P. Becher, *Emulsions: Theory and Practice*, 2nd ed., Krieger, Melbourne, Fla., 1977, pp. 141–147.
36. Th. F. Tadros and B. Vincent, in *Encyclopedia of Emulsion Technology*, P. Becher, ed., Vol. 1, *Basic Theory*, Dekker, New York, 1983, pp. 272–276.
37. A. F. Koretskii and P. M. Kruglyakov, *Izv. Sib. Otd. Akad. Nauk SSSR Ser. Khim. Nauk 1971*: 139.
38. P. M. Kruglyakov, G. M. Glogoleva, and A. F. Koretskii, *Izv. Sib. Otd. Akad. Nauk SSSR Ser. Khim. Nauk 1972*: 9, 23.

39. P. Sherman, *J. Soc. Chem. Ind. 69*: S70 (1950); *Manuf. Chem. 26*: 306 (1955); *J. Colloid Sci. 10*: 63 (1955); *Proc. Int. Congr. Surface Activity, 3rd, Cologne, IIb*: 596 (1960).
40. P. Sherman, in *Emulsion Science*, P. Sherman, ed., Academic Press, New York, 1968, Chap. 4.
41. P. Sherman, in *Encyclopedia of Emulsion Technology*, P. Becher, ed., Vol. 1, *Basic Theory*, Dekker, New York, 1983, Chap. 7.
42. K. Suzuki, S. Matsumoto, T. Watanabe, and S. Oro, *Bull. Chem. Soc. Jpn. 47*: 2773 (1969).
43. P. Becher, *Emulsions: Theory and Practice*, 2nd ed., Krieger, Melbourne, Fla., 1977, p. 413.
44. N. de Rooy, P. L. de Bruyn, and J. Th. G. Overbeek, *J. Colloid Interface Sci. 75*: 542 (1980).
45. M. Clausse, in *Encyclopedia of Emulsion Technology*, P. Becher, ed., Vol. 1, *Basic Theory*, Dekker, New York, 1983, Chap. 9.
46. N. Krog and K. Larsson, in *Encyclopedia of Emulsion Technology*, P. Becher, ed., Vol. 2, *Applications*, Dekker, New York, 1985, Chap. 5.
47. E. N. Jaynes, in *Encyclopedia of Emulsion Technology*, P. Becher, ed., Vol. 2, *Applications*, Dekker, New York, 1985, Chap. 6.
48. S. Friberg, ed., *Food Emulsions*, Dekker, New York, 1976.
49. P. Sherman, ed., *Food Texture and Rheology*, Academic Press, New York, 1979.
50. M. Breuer, in *Encyclopedia of Emulsion Technology*, P. Becher, ed., Vol. 2, *Applications*, Dekker, New York, 1985, Chap. 7.
51. S. S. Davis, J. Hadgraft, and K. J. Palin, in *Encyclopedia of Emulsion Technology*, P. Becher, ed., Vol. 2, *Applications*, Dekker, New York, 1985, Chap. 3.
52. D. O. Shah, ed., *Surface Phenomena in Enhanced Oil Recovery*, Plenum, New York, 1981.

7

Adsorption of Polymers on Solids from Apolar Media

GORDON J. HOWARD* Visiting Scientist, 1983–85, Marshall Research and Development Laboratory, E. I. DuPont de Nemours & Company, Philadelphia, Pennsylvania

I. INTRODUCTION

The adsorption of polymers has been much studied over the past two decades and recent years have seen the introduction of novel experimental methods and more realistic theoretical analyses. The practical significance of a better understanding of polymer adsorption can hardly be overstated, as it is of direct importance to the stabilization

*Present address: 281 West Harvey Street, Philadelphia, Pennsylvania

(and destabilization) of colloidal dispersions [1] and underlies many chemical, technological, and biological phenomena [2–5]. Some applications of polymer adsorption, such as flocculation, are largely restricted to aqueous media and are beyond the terms of reference of this chapter; here we will relax the limitation to apolar systems only when discussion of results from polar systems illuminates the main theme.

Reviews of the adsorption of macromolecules have been given by Ash [6], who presents a well-balanced survey of work published in the early 1970s; by Vincent and Whittington [7], who pay particular attention to theoretical work and to problems of colloidal stability; by Takahashi and Kawaguchi [8], who include polyelectrolytes; and by Fleer and Lyklema [9], whose article is strongly directed to theory.

The basic objective of polymer adsorption studies is to specify the macromolecular conformation in the layer adsorbed at a phase interface—in particular, to describe how the adsorbed polymer conformation is related to its environment, especially to the nature of the surface, the solvent medium, the temperature, the composition of the adsorbate, the presence of competitors, and so on.

Excluded from this chapter are the vapor/solution interface (where most studies refer to the air/aqueous solution interface) and the liquid/liquid interface, despite the importance of polymer adsorption in emulsion stability and in emulsion polymerization; stability considerations are discussed by Vincent and Tadros [10]. Colloid stability, whether consequent on polymer adsorption or on depletion phenomena, is not reviewed here.

II. EXPERIMENTAL METHODS

Throughout the earlier stage of polymer adsorption research the primary experimental technique was to measure the amount of polymer taken up by the adsorbent from solution. Since this quantity is determined by comparison of solution concentrations before and after contact with the adsorbent, the measured adsorption is an apparent quantity and isotherms so derived are composite isotherms, in Kipling's nomenclature [11]. However, as polymer adsorption isotherms are usually restricted to dilute systems, the difference between absolute and individual adsorbances will be small. Mass adsorption may be quoted per unit mass of adsorbent or per unit surface area; for the latter, of course, a measured specific surface area is required, usually the BET area obtained from gas adsorption (N_2, Kr, Ar). As the reference adsorbate molecule is much smaller than the macromolecular adsorbate, conversion of experimental solution adsorption data to a surface area basis is justified only when the adsorbent is known to be nonporous or is otherwise fully accessible to adsorbing polymers; even with nonporous adsorbents there is the possibility that very long chains may bridge between adsorbent particles and thereby prevent entry to surface sites [12,13].

Valuable as the mass adsorption is for comparative studies, and its measurement can hardly be omitted, in itself it provides little insight into the state of the adsorbed polymer chains, particularly in the many cases in which the area of accessible surface is uncertain. Two other experimental approaches were used during earlier studies in attempts to learn more about adsorbed conformations. The first of these was the determination of the bound fraction (the fraction of chain repeat units of the adsorbed polymer that are in actual contact with the adsorbent surface) by the shift in infrared absorptions consequent on adsorbate-adsorbent interaction. This technique, introduced by Fontana and Thomas [14] and applied by other workers [15–18], is in practice a limited one, since for adequate transmission in the infrared, finely divided solids with minimal scattering of incident radiation are essential. Indeed, the great majority of reports of polymer-bound fractions refer to pyrogenic silica (Aerosil, Cab-o-Sil). Furthermore, the solvent employed must have suitable windows in their infrared spectra, thus ensuring another restriction. Most bound-fraction results have been measured by careful double-beam techniques, usually with an adsorbent-free supernatant solution in the reference path, but the precision of the data is not high. Modern infrared spectroscopy, with its facilities for digitalizing, storing, and manipulating spectra, has been surprisingly little used for polymer-bound fraction studies, although one example may be cited [19].

The second general method of characterizing adsorbed polymers that is well established is to measure the average thickness of the interfacial layer, and here two separate approaches were developed. In the first, the optical method of ellipsometry was adapted to polymer studies, initially by Stromberg and his colleagues [20]. By finding the changes in phase retardation and amplitude ratios when a monochromatic beam of light is reflected at a plane surface, the refractive index and thickness of an adsorbed layer may be ascertained. This thickness refers to a uniform layer, and it is usual to convert initial values to a root-mean-square (rms) distance by assuming a distribution function for the segments in the layer; a limitation of an experimental nature is that plane surfaces of high refractive index are necessary, chrome plates being most often used [21–23]. Hinkley [24] has used ellipsometry to study polystyrene adsorption on oxide-coated silicon wafers.

Another optical method for determining layer thickness was introduced by Peyser and Stromberg [25], who used attenuated total reflection (ATR) in the ultraviolet, but no further developments seem to have been made. Certainly, a recent review of the applications of ATR spectroscopic techniques for investigation of the solid-liquid interface [26] cites no other study of polymer adsorption.

Adsorbed layer thickness may also be estimated by transport measurements, usually viscometric, although sedimentation velocity in the ultracentrifuge has been used [27,28]. Again, there is uncertainty as to the kind of average thickness found hydrodynamically, although this question has been examined theoretically [29]. One of the earliest estimates of an adsorbed layer thickness was by Ohrn [30], who interpreted "anomalies" in the capillary viscometry of dilute polymer solutions as a flow restriction caused by adsorption on the capillary walls. Tuijnman and Hermans [31] also discussed the precision viscometry of dilute polymer solutions and considered adsorption effects. Ohrn reported very thick layers, of order 100 nm; other groups have made careful capillary viscosity measurements [32,33] and reported thick layers [33]. Flow through a porous bed, which has a much larger surface than a single capillary, was first investigated by Rowland and Eirich [34,35] and more latterly by Gramain [36]. An alternative approach, using colloidally stable dispersions of particulate adsorbents, was introduced by Doroszkowski and Lambourne [37]; here the viscosity/particle volume fraction relation for polymer-covered particles is compared to that for bare particles and the layer dimensions thus calculated.

All the sorts of layer thickness measurements seemed to be in general agreement with one another in that dimensions similar to the root-mean-square radius of gryation of free coils were reported. Taken together with the infrared estimates of bound fractions, typically around 0.25, a loop-tail conformation as exemplified by Eirich's well-known "pearl necklace" illustration [38] seemed to be reasonable. However, more recently applied experimental methods, together with new theoretical considerations, have forced modifications of this concept of macromolecules at interfaces. When resonance methods (electron spin resonance [39] and nuclear magnetic resonance [40–42]) were used, it was found that the fraction of polymer chain segments close to the surface, as evidenced by their restricted motion relative to "normal" segments, was extremely high, values of 0.80 or higher being commonly reported. Later studies tend to confirm the earlier observations [43,44]; there is virtue in referring to a "restricted" fraction of chain segments, confining the term bound fraction to the parameter derived from the infrared shift. In this way a direct segment-surface interaction may be distinguished from the restricted motion of segments close to, but not necessarily "on," the surface.

Clearly, if some 80% of the chain repeat units are within a few solvent lattice planes of the adsorbing surface, the picture of a "loopy" adsorbed conformation must be revised if the ellipsometric and hydrodynamic thickness values are valid. A powerful new optical procedure has also been used to measure layer thickness; in photon correlation spectroscopy [45] the diffusion constant of polymer-covered particles in dilute dispersion is found, compared to that from bare particles, and the diffusion constants are converted to radii by the Stokes-Einstein equations, permitting an estimate of the layer thickness. The adsorbent particles must be fairly monodisperse and close to spherical to ensure an exponential light-scattering correlation function. Ullmann and Phillies [46] have challenged the application of the Stokes-Einstein equations to the analysis of photon correlation spectroscopy; some adjustments of reported layer thicknesses may be necessary. Polymer latexes are the chosen model adsorbents for these studies, so it is not surprising that many aqueous systems are reported. The important result in the present context is that, whatever the dispersion medium, layer thickness estimations from photon correlation spectroscopy are in good accord with earlier hydrodynamic or ellipsometric values.

Other extended dimensions of adsorbed polymers are found from experiments designed to study the phenomenon of steric stabilization. The particle-particle distance at which strong repulsion sets in has been found by adaption of the Langmuir trough [47], by confining a stabilized dispersion within a membrane [48,49], by low-speed centrifugation [50], or by direct measurement of the forces between polymer-coated mica sheets [51–54] or quartz filaments [55].

Nonetheless, bound (or restricted) fractions coupled with average adsorbed chain dimensions still leave a gap in our understanding of interfacial conformations. What is needed is an experimentally determined segment density distribution normal to the surface. This is now possible through developments in neutron scattering [44], although there are good reasons why the method is likely to be applied to only a few systems. The intensity of neutrons scattered at a given angle depends on the distribution of scattering centers in the sample; further, nuclei differ in their ability to scatter neutrons and particular use is made of the very different coherent scattering lengths of ^{1}H and ^{2}H nuclei. For example, Barnett and co-workers [56] used a deuterated latex as the adsorbent and a protonic polymer as the dispersion stabilizer. By contrasting out the neutron scattering from the particle core by adjustment of the hydrogen/deuterium ratio in the solvent phase, the scattering of the adsorbed layer only is detected and, through appropriate transforms of the experimental data, the segment density distribution is obtained. Up to now, only aqueous dispersions have been studied, although a mean thickness, rather than the distribution, derived from the neutron scattering of a branched polymer chemically grafted to a deuterated poly(methyl methacrylate) latex in hexane has been measured [57]. The value of neutron scattering for investigation of adsorbed polymer conformations is so great that further results are awaited. Suitable isotropic contrast matching may enable copolymer conformations to be studied experimentally. The limitations of this new technique lie in the requirement for stabilized dispersions with suitably deuterated monodisperse particles (or deuterated adsorbates) and, above all else, access to neutron scattering facilities.

Before going on to review polymer adsorption from apolar media, a short digression on theoretical aspects is appropriate.

III. THEORETICAL ASPECTS

A. Survey

An early venture into polymer adsorption theory was the attempt by Simha, Frisch, and Eirich [58,59] to assess the influence of a reflecting wall on the conformation of a nearby isolated polymer chain. During the 1960s several groups further considered the adsorbed conformation of an isolated macromolecule by statistical mechanical [60–63] and lattice [64–69] methods or by Monte Carlo simulations [70–74]. Despite much theoretical dexterity, these and later efforts [75–80] have very limited utility since adsorbed chains will be isolated from their neighbors at the interface only at inaccessibly high dilutions.

Hoeve [81] extended earlier work [82] on the isolated chain to allow for interactions between adsorbed molecules. As the polymers were assumed to be so long that end effects could be neglected, Hoeve's model is a loop-and-train conformation, although it did incorporate excluded-volume effects [83–85]. Likewise, Silberberg [86] first considered isolated chains [64–67] and then introduced lateral interactions; again, the quasi-lattice treatment is for a loop-and-train model. Silberberg [86] defined the quantity $-\chi_s$ which, free of some qualifications (see [9]), may be described as the dimensionless free energy needed for the replacement of an adsorbed solvent molecule by a chain segment. Many later workers continued to use this convenient quantity, but it was Silberberg who showed that adsorption occurred only when χ_s exceeded a critical value, and he also showed that adsorption behavior at higher χ_s values was not very responsive to the magnitude of χ_s.

A fresh approach to polymer adsorption was the application of scaling theory [87] by de Gennes and his school [88,89]. In the dilute region the adsorbed polymer chains are isolated and the isotherm has a Henry's law form. In the plateau and semidilute regions, where osmotic effects are increasingly dominant, the isotherm equations, although explicit, are cumbersome: the average thickness of the relatively thin adsorbed layer is almost directly proportional to molecular weight in the plateau region, and to the three-fifths power in the semidilute range. Klein and Pincus [90] used a scaling approach in their calculation of interactions between polymer-covered surfaces in thermodynamically poor media, and they pointed out that although the supernatant solution is a single phase, in the adsorbed layer the conditions may favor a biphasic region.

Roe [91,92], instead of working with the isolated chain, calculated the number of ways in which solvent molecules and both free and adsorbed chain segments could be arranged in a region near the surface. The equilibrium segment density distribution is found by maximizing the derived partition function. The later studies of Scheutjens and Fleer [93–96] permit specification of the separate segment density distributions of trains, loop, and tails, whereas Roe's theory effectively neglects end effects, that is, the tails. Scheutjens and Fleer show the great importance of tails in determining the adsorbed concentration profile, especially in the outer part of the layer. In both these theories segment-solvent interactions in the volume beyond the immediate surface are accounted for by the Flory-Huggins χ parameter. Thus the adsorption behavior of a polymer of a given chain length is essentially specified by two free energy parameters, χ_s and χ; calculations show that, with both Roe's theory and that of Scheutjens and Fleer, which are lattice theories, the coordination number of the lattice has only a minor effect.

These theories do not provide analytical expressions for adsorption isotherms and other properties of the layer. In the Scheutjens and Fleer treatment a matrix formalixm of DiMarzio and Rubin [67,97] is used to obtain the equilibrium segment density profile by numerical iteration; consequently, limitations on computer time and memory restrict calculations to chains of moderate size, although some results are given for chains of 5000 segments [95]. A notable feature of these theories is that adsorption behavior over the whole concentration range is predicted in terms of individual and composite isotherms, as well as bound fractions, layer thickness, and segment density distributions.

The original papers are hard going for the nontheoretician, but two reviews are recommended [8,9] and a short summary by Vold and Vold [98] may be advised as prior reading. For the purposes of this chapter a summary of trends in polymer adsorption characteristics is presented below; here we follow the predictions of Scheutjens and Fleer, but include the features common to earlier treatments.

B. Predictions of Modern Theory

1. Isotherm

The adsorption isotherm is considered in terms of both the amount of polymer adsorbed from solution and the surface coverage of the adsorbent. Low molecular weight polymers are predicted to have "Langmuirian" isotherms, but as the chain length increases the isotherm takes on more and more the high-affinity shape. Adsorbance increases with decreasing solvent quality (the χ factor) and increases with molecular weight. In the plateau region the adsorbance is almost logarithmic with molecular weight for a theta (poor) solvent, but the slope (and the amount adsorbed) is smaller from athermal (good) solvents. At higher polymer concentrations the adsorption increases more as the square root of molecular weight, and at these higher concentrations the χ effect is less marked.

Changing $-\chi_s$ from 0.5 (weak segment-surface net free energy) to 5.0 (high interaction) increases the adsorbance from a theta solution twofold. The relations between adsorbance and chain length run essentially parallel as χ_s is changed; longer polymer molecules, when adsorbed, place more segments into loops and tails, where χ is the determining factor, while the fractional surface coverage stays essentially constant. As the volume fraction of solute increases toward unity the χ, χ_s influences become less noticeable. The strong preference for long chains over short which is shown in dilute and semidilute conditions is reversed at very high concentrations; the changeover is predicted at a volume fraction of about 0.7. The proportion of surface sites occupied by segments is practically independent of polymer concentration except for short chains; this fraction is greater with poorer solvents, as segments prefer their own company to that of the solvent.

Adsorption is finite only when $-\chi_s$ exceeds a critical value, which in the Scheutjens and Fleer theory is about 0.2 kT units.

2. Bound Fraction

In general, the bound fraction decreases somewhat with polymer concentration and with chain length. Compared at equal supernatant concentrations, the bound fraction increases as the solvent quality is improved, and the effect is more pronounced with longer chains. This superficially surprising result arises since the extension of the layer is less than that in poor solvents when compared at equal solution contents. At equal amounts of adsorbed polymer, however, the bound fraction increases with χ, that is, as the solvent quality worsens; this trend is more definite at lower χ_s values. To reach the same adsorbance requires higher solution concentrations in a good solvent than in a poor one.

The bound fraction increases with χ_s, and approximate estimates from the Scheutjens and Fleer theory are given in Table 1 for an equilibrium solution concentration of 0.1% and a chain length of 1000 segments. Thus in poor solvents the bound fraction at constant supernatant concentration, and under dilute conditions, is insensitive to χ_s, and in a good solvent a dependence on χ_s is found only at weak segment-surface interactions.

TABLE 1 Bound Fraction as a Function of χ and χ_s

$-\chi_s$	Theta ($\chi = 0.5$)	Athermal ($\chi = 0.0$)
1.0	0.32	0.43
2.0	0.34	0.62
3.0	0.35	0.65

3. Thickness and Segment Distribution

The mean layer thickness is an average of the segment contributions of loops and tails. Calculations give the root-mean-square thickness as being close to a square-root relation with molecular weight, actual values being somewhat smaller in better solvents when equal concentrations are compared; at a fixed amount of adsorbed polymer the layer is thicker because of the lower occupancy of the surface plane. The thickness is not very responsive to the value of χ_s, and when this parameter is high, solvent quality has a lesser effect. At supernatant concentrations around 1% the rms thickness is little more than half the radius of gyration of free chains. As the polymer concentration is raised, the layer thickness increases much more than the adsorbance, because of the greater contribution from the tails, and this is more pronounced in good solvency conditions.

In semidilute concentrations, and in a theta solvent, a chain of 1000 segments will have rather more than a third in trains, a little more than half in loops, and only some 6 to 7% in tails. However, although the tails represent the smallest proportion of the chain units, they dominate many of the properties of the adsorbed layer. For instance, in the example above, the exponential segment density distribution of the loops drops to a very low value beyond some 30 lattice planes and the layer becomes one of tails only. As a consequence, the hydrodynamic layer thickness is controlled by the tails fraction; since this, in turn, will be influenced by the values of χ and χ_s, these parameters maintain importance in situations such as the interaction of colloidal particles moderated by adsorbed polymers.

IV. EXPERIMENTAL RESULTS

A. Time Effects

Before reviewing the results on polymer adsorption from apolar media at equilibrium, it seems appropriate to discuss the rate of the adsorption process. Surprisingly little work has been reported on the kinetics of the process. General observations by many investigators indicate that the time to reach an apparent equilibrium adsorption depends markedly on the nature of the adsorbent; plane surfaces [99] or nonporous particles well dispersed in the liquid [100] may take up polymeric adsorbates in minutes or in a few hours, whereas porous substrates may be very much slower. According to Patel et al. [101], poly(methyl methacrylate) on a porous aluminum silicate adsorbed from benzene by a first-order kinetic scheme. When the adsorbent is nonrigid and is swellable by the medium, very lengthy adsorption periods may be encountered [102,103]; in the latter work, the data were treated by assuming a diffusion mechanism, but in neither study [102,103] was the system an apolar one.

In most cases polymeric adsorbates are polydisperse; homopolymers often have a broad molecular weight distribution, depending on the mode of polymerization, and copolymers may well carry an additional compositional distribution. As a result, equilibration of adsorption may involve displacement of first arrivals by subsequent adsorbing species that have a higher net adsorption energy. The consequences of polydispersity will be considered in some detail later, but for the present it is sufficient to say that apparent equilibrium in mass adsorption may hide slower exchange processes. Kolthoff and Kahn [104] measured the intrinsic viscosity of the supernatant and found that larger polymer chains were slowly displacing smaller ones long after the total amount of adsorbed polymer was sensibly constant; other workers also reported this effect [13,105].

Ellipsometry is a valuable technique for rate studies as the adsorption is onto a flat surface and the process is not complicated by considerations such as inter- or intraparticle porosity. Both the adsorbance, through the measured refractive index of the film, and the layer thickness may be found as a function of time of contact. Stromberg et al. [106] showed that, for polystyrene from cyclohexane on ferrochrome, the adsorbance and the thickness kept pace; the adsorbed layer was practically complete in 200–300 min when starting from an initial concentration of 0.05%. However, the rates of adsorption and layer expansion are not always found to stay in step, as in this example.

In some practical applications the time allowed for the adsorption process is very short and the conformation must be nonequilibrium; Wigsten and Stratton [107] applied a stopped-flow technique to investigate adsorption and associated flocculation in an aqueous system. A kinetic model for the irreversible adsorption of macromolecules has been proposed by Priel and co-workers [108] and is said to fit well with earlier experimental results [109]. A novel optical technique, based on electromagnetic surface waves, has been suggested for the study of the kinetics of polymer adsorption [110].

B. Isotherms

It is commonly observed that polymer adsorption isotherms for dilute systems have the high-affinity shape in which the initial branch is almost vertical but succeeded by a plateau, or semiplateau, at quite low equilibrium concentrations. Other things being equal, the higher the polymer molecular weight, the sharper the isotherm. However, not all polymer isotherms are high-affinity ones, and, indeed, many authors fit the Langmuir equation to experimental points, albeit only as a convenient form of representation [110,111,112]. Sometimes the fit to the Langmuir isotherm is regarded as showing "monomolecular" adsorption [113–115]. Rounded isotherms are associated with polydisperse polymers, especially when adsorbed from poorer solvents; Cohen-Stuart and co-workers [116] have explained the factors contributing to rounded isotherms. To anticipate a later section, there is much evidence that, at equilibrium in dilute systems, long chains are adsorbed in preference to short ones.

Thus, with a polydisperse adsorbate there will be a fractionation process, as shown experimentally by Felter et al. [117,118], higher molecular weight species being partitioned to the adsorbed phase, with lower species mostly in the supernatant. At low concentrations, where the surface is not too fully occupied, both low and high chains are able to become adsorbed with little fractionation, but, because of the smaller molecules, the amount adsorbed will be low. As the total amount of polymer is raised, but still keeping below the semidilute region, the dominance of the larger, higher-adsorbance molecules becomes more and more evident, smaller molecules are displaced, and the fractionation is enhanced. There is, of course, no opportunity for this kind of fractionation process with a monodisperse polymer, which therefore exhibits a much sharper isotherm.

Polydispersity has a further, rather unfortunate, effect on polymer isotherms. The level of adsorption is found, by careful measurement, to depend on the ratio of the mass (area) of adsorbent to the mass of adsorbing polymer. It is common practice to construct the isotherm by contacting a given mass of adsorbent with a fixed volume of solution, and repeating the pattern with a series of different polymer concentrations, or else by varying the amount of adsorbent and keeping the volume and concentrations fixed. A well-accepted criterion for the absence of surface-active impurities is that a common isotherm should result from either procedure [119]. With polydisperse polymers the isotherm may not be independent of the adsorbent/adsorbate ratio. When there is a fixed amount of polymer solution and a small quantity of adsorbent, only the longer chains will be adsorbed and the adsorbance will be high; however, with a larger ratio of adsorbent to polymer, more short chains will be able to get to the interface, but with a consequent reduction in the total level of adsorption. This phenomenon is entirely a result of the molecular weight distribution and does not reflect on thermodynamic reversibility, and there is no reason to suppose that segment-surface reversibility is any less than with the monomeric analog. On the other hand, this situation as interpreted by Cohen-Stuart is undoubtedly a contribution to the regrettably poor quality of many polymer adsorption isotherms.

Interestingly, some hypersharp polymer isotherms have been reported (e.g., [15,120]) for adsorbates which are clearly polydisperse in both molecular weight and composition; here the conditions must be such, by virtue of a high net adsorption free energy, that the adsorbance is practically independent of both chain length and copolymer composition. Turning to the relation between polymer isotherms and molecular size, it must be remarked that this is a topic that has been much studied, but often with polymers that were more or less broad in distribution. In hindsight, we may regard the trends as being established with polydisperse samples, with more exact results from later studies using narrowly distributed polymers. Perkel and Ullman [121], in one of the earliest systematic studies of polymer adsorption, showed that the plateau adsorbance of poly(dimethyl siloxane) or iron and on glass obeyed the equation

$$\Gamma = KM^{a} \tag{1}$$

and implied that adsorbed conformations of various forms would be mirrored in the value of the exponent a. Subsequently, many workers fitted their data into the style of Eq. (1), while recognizing that deviations might be found at very high and very low molecular weights.

Ash [6] listed some values for the exponent a reported during the period covered by his review; many values are low, between zero and 0.10, although some are much higher. For instance, Ashmead and Owen [122] found the exponent to be as high as 0.70 for poly(dimethyl siloxane) from benzene on glass; their polymers had blocked end groups and some commercial samples gave a lower slope, about 0.40, more in line with earlier results [121]. Ashmead and Owen reported an even higher exponent for adsorption of poly(dimethyl siloxane) from hexane, when the molecular weight was less than 10^5, but longer chains adsorbed independently of size. This apparent leveling off of adsorbance at molecular weights of around 1 million was also found by Hara and Imoto [123] for ethylene/vinyl acetate copolymers on glass from various solvents. Takahashi et al. [23] found from ellipsometry that polystyrene on chrome in cyclohexane (a theta solvent) increased in adsorbance almost linearly with molecular weight up to about 5×10^5 but thereafter became constant. From toluene and carbon tetrachloride, the same group [124] found polystyrene adsorption virtually indifferent to chain length. At least two authors [12,13] observed a complication with very high molecular weight adsorbates, in that localized bridging flocculation of otherwise nonporous adsorbent particles may create inaccessible surfaces during the adsorption process. This mechanism cannot apply to the ellipsometry result described above.

On a porous substrate, of course, an eventual decline in the amount of polymer adsorbed is anticipated because of inaccessibility, as shown by Furusawa et al. [125] for polystyrene in theta solution on Bioglass. Earlier, Howard and McConnell [126] found poly(ethylene oxide) to adsorb almost independently of molecular weight on a porous carbon, whereas on nonporous Graphon adsorbance increased with chain length. Indeed, Burns and Carpenter [127] correlated the adsorption of polystyrene from cyclohexane with the pore size distribution of the alumina adsorbent, and similar results have been reported by others [13,122].

Changes in the level of adsorption at low molecular weights are to be expected for polymers by virtue of end group effects, especially for polymers made by step growth mechanisms, which normally leave chemically dissimilar chain ends. Crowl and Malati [128] noted that the adsorption and colloidal stabilization of model alkyd resins on titanium dioxide and ferric oxide depended on the number of hydroxyl and carboxyl end groups. Along similar lines, Worwag and Hamann [129] found that low molecular weight polyesters with carboxyl ends absorbed much more strongly on titanium dioxide than those with esterified ends. Because the acid end groups are the principal anchors of the polyesters to the oxide surface, adsorbance actually decreased with chain length. Felter [130] discovered that the molecular weight fractionation on adsorption of polycaprolactone on calcium carbonate from chlorobenzene was biased by the preferential adsorption of hydroxyl-ended chains. The influence of end groups on the adsorption of oligomers has been documented in other papers [131–133].

Modern theories of polymer adsorption [91–95] suggest a more complex molecular weight dependence than the empirical Eq. (1). According to Scheutjens and Fleer [93–95], the adsorbance/molecular weight curve in the linear/log mode has a moderately sigmoidal shape, affected by the values of the χ and χ_s parameters. As χ approaches 0.5, the theta condition, the adsorbance is thought to increase without limit, although actual computations have not been made; this is in contrast to earlier theories. A comprehensive study of the adsorption of narrow molecular weight distribution polystyrenes over the range 6×10^2 to 2×10^6 was reported by Vander Linden and van Leemput [134]; the adsorbent was a pyrogenic silica and both a poor and a good solvent were used. Although a well-verified value for χ_s is not known, the data fit the Scheutjens and Fleer predictions, using correct χ values. On the other hand, the ellipsometry results of Takahashi et al. [23] are not in line. However, the theory, which requires numerical computation, has not been extended to the molecular weight range covered by the ellipsometric measurements (up to 10^7).

Experimentally, the Dutch school [135] investigated the molecular weight dependence of the adsorbance of poly(vinyl pyrrolidone) fractions. The adsorbance from apolar dioxane fit Eq. (1) poorly, tending to $a = 0.4$ below molecular weight 10^4 and $a = 0.03$ at higher molecular weights; the points fit the curve of the Scheutjens and Fleer theory better, and this is even more the case for the lower adsorbances found from water. Other sigmoidal versus log (molecular weight) plots are reported for polystyrene on Graphon from chloroform [132] and for the same polymer on silica from cyclohexane [136]. Further data from Dunn and Vold [137] for polystyrene/Graphon/toluene cover higher molecular weights and are also in general accord with the newer theories; their data points do not fit Eq. (1).

Even if the molecular weight dependence of adsorbance is not strong, when polydisperse macromolecules are presented to an adsorbing surface, considerable partitioning of species between adsorbed and unadsorbed phases will occur. Roe [92] points to the similarities between adsorption fractionation and the liquid-liquid phase separation of polymers as their dilute solutions exceed theta conditions. Reference has already been made to adsorption experiments demonstrating fractionation by changes in supernatant viscosity or by gel permeation chromatography [13,104,105,117,118], to which should be added more recent studies [130–140]. Competition between two polymers with narrow molecular weight distributions has been reported on several occasions. Howard and Woods [141] used a mixture of two anionically synthesized polystyrenes, one of which was tritium-labeled. On adsorption on silica from cyclohexane, almost complete displacement of the low molecular weight species was found as the total equilibrium polymer concentration neared the plateau region of the isotherm. Cohen-Stuart et al. [116,135] found the theoretically predicted isotherm for a mixture of two poly(vinyl pyrrolidones), actually from water onto silica. After the initial sharp rise a linear increase in the isotherm is inserted before the plateau, whereas the separate isotherms show only the high-affinity curve. A similar isotherm was found for polystyrene on silica from trichloroethylene [141]. Kawaguchi and co-workers [142] describe an extensive study of the preferential adsorption between monodisperse polystyrenes of different molecular weights on pyrogenic silica from carbon tetrachloride solution. Furusawa et al. [143–145] examined the competition in mixed polymer fractions by adsorption on a porous glass (polystyrene, cyclohexane); gel permeation chromatography of the supernatant was performed as a function of contact time. Preference for higher molecular weights was found, despite the complication of a porous substrate; the explanation advanced was, however, in terms of comparative rates of adsorption and desorption rather than an equilibrium behavior.

The difficulty of desorbing preadsorbed polymers is well known. For a monodisperse polymer, where the pseudoplateau of the isotherm is established at a low equilibrium concentration, dilution by more solvent to achieve appreciable desorption is almost impossible; calculations from the Scheutjens and Fleer theory suggest that dilution by several orders of magnitude would be necessary [95]. The apparent failure of a polymer to desorb on dilution is a consequence of the nature of the isotherm and is not indicative of irreversibility, nor should any partial desorption be taken as a measure of the strength of attachment to the surface. For heterogeneous polymers, which have rounded isotherms, the desorption isotherm is sharp; the adsorption isotherm is a composite of each separate species adsorbing with high affinity, adding up to a rounded curve, but the adsorbed molecules will only desorb when the external solution concentration becomes extremely dilute.

According to Pefferkorn et al. [146], exchange between labeled and unlabeled polymeric species of the same degree of polymerization is, apart from a small, loosely attached fraction, very slow and follows a second-order kinetic scheme.

This discussion of polymer adsorption isotherms has so far necessarily considered results from dilute solutions; there are very few reports of adsorption behavior at higher concentrations [147]. Adsorbance measurements become increasingly inaccurate as the adsorbate content is raised since the concentration difference caused by adsorption diminishes; with polymeric adsorbates the situation is exacerbated by high solution viscosities, slow adsorption, and bridging of particulate adsorbents. Lipatov and co-workers reported a number of polymer adsorption isotherms where the equilibrium concentration axis extends to about 10% (w/v); they explained the complex isotherms as due to the adsorption of macromolecular aggregates formed in solution at high concentrations. A summary of these studies is given in [148]. Certainly, considerable changes in the isotherm are expected to occur beyond the semidilute region as predicated by theory [95], in the context, however, of a homogeneous solution phase.

C. Bound and Restricted Fractions

Infrared spectroscopic determination of bound fractions relies on the shift in the characteristic absorption frequencies of the interacting groups of either the adsorbate or the adsorbent. Solvents must have suitable spectral windows, and the adsorbate-absorbent interaction must lead to an adequate frequency change. For adsorbates the methods of Fontana [14,149,150] are used, and the shift of the carbonyl peak on adsorption is an example. The hydroxyl absorption in pyrogenic silicas has often been used when the adsorbent interaction is followed. Fontana [14,149] showed the correspondence between bound fractions measured from adsorbate and adsorbent functional group shifts; in later studies this 1:1 relation was not always found [151,152]. A review of the measurement of adsorbed molecules by infrared spectroscopy, which includes consideration of polymeric adsorbates, has been published by Rochester [153].

The fraction of chain segments interacting with surface sites was found by Thies et al. [154] to decrease somewhat with surface coverage, and subsequent workers, e.g., Kawaguchi et al. [155], have mostly reported the same trend. The reduction in bound fraction with coverage is very small in some systems (e.g., [156]) but quite considerable in others. For instance, Killmann et al. [157] found that low molecular weight poly(ethylene oxide) on silica and adsorbed from carbon tetrachloride had a bound fraction of 0.50 at low coverage and 0.25 at the isotherm plateau. An even stronger decrease was observed for polybutadiene on silica [158]. Table 2 lists bound fractions for three polymers on silica adsorbed from several solvents. Agreement between different researchers is rather poor. The higher bound fractions of polymers adsorbed from poor solvents can be seen [poly(methyl methacrylate) in carbon tetrachloride and polystyrene in cyclohexane].

Dietz [151], following earlier studies by Joppien [18,159], investigated the adsorption of linear polyesters of modest molecular weight, as made by polycondensation, on silica. At the adsorption isotherm plateau, the bound fractions are as given in Table 3. There is a steady trend with composition of the polyester, but it is not certain whether the values reflect changes in the polymer-solvent interaction parameter, chain flexibility, or the geometric effects of placing the chain ester on the surface silanol.

Low molecular weight polymers may be expected to adopt a flatter profile at the surface with a consequent increase in the bound fraction. Vander Linden and van Leemput [134] obtained bound fractions for polystyrene on silica from cyclohexane of 0.30 above molecular weights of 2×10^4, but at 1×10^4 the bound fraction was 0.40, at 2×10^3 it was 0.87, and for the hexamer, of molecular weight 670, the bound fraction was unity. In the same paper they also reported bound fractions of the polystyrenes adsorbed from carbon tetrachloride; above 10^4 the value was about 0.17, but at 2×10^3 it increased to 0.33. Little change in the bound fraction of polybutadiene with molecular weight was found when these polymers were adsorbed on silica from cyclohexane [158]. However, the same group [136], working, like Vander Linden and van Leemput [134], with nearly monodisperse polystyrene in cyclohexane at the theta temperature, found a steady decline in bound fraction over the entire molecular

TABLE 2 Bound Fractions at the Isotherm Plateau[a]

Polymer	Solvent	Bound fraction	Reference
Poly(methyl methacrylate)	Chloroform	0.30	[16]
		0.28	[154]
		0.30	[160]
	Trichloroethylene	0.23	[16]
		0.32	[161]
	Carbon tetrachloride	0.55	[16]
		0.4-0.5	[156]
Polystyrene	Trichloroethylene	0.07	[16]
		0.13	[15]
	Carbon tetrachloride	0.34	[18]
		0.18	[134]
	Decalin	0.13	[154]
	Cyclohexane	0.30	[134]
Poly(vinyl pyrrolidone)	Dioxane	0.25	[162]
		0.35	[19]

[a]Adsorbent is pyrogenic silica in all cases.

TABLE 3 Bound Fraction of Polyesters on Silica from Chloroform [151]

Polyester	Bound fraction
2 G 4[a]	0.35
2 G P	0.35
6 G 8	0.48
10 G 12	0.64

[a]The first number represents the carbon atoms in the backbone of the diol component and the final number the chain carbons in the diacid; P signifies the isophthalate residue.

weight range, from 600 to 2×10^6, with very high bound fractions from the short chains. The actual values were less than those reported in [134], except for oligomers. Among other reports of infrared bound fractions, a definite dependence of bound fraction on molecular size was found with poly(dimethyl siloxane) on silica from carbon tetrachloride, particularly at low concentrations [163]. According to Kalnin'sh et al. [156], oligomers of poly(methyl methacrylate) have bound fractions of 0.70–0.85, depending on the solvent, but above molecular weights of a few thousand the values drop to around 0.4.

Killmann and Bergmann [164] showed that the infrared absorption spectrum of adsorbed poly(vinyl pyrrolidone) exhibits a splitting of the carbonyl band; the data were interpreted as indicating an additional interaction with the surface of the polymer segments neighboring the specifically bound groups.

Dietz [151] demonstrated the effect of surface modification of the adsorbent of the bound fraction; poly(ethylene succinate), 2 G 4, was adsorbed from chloroform to give the bound fractions shown in Table 4. A similar exercise was carried out earlier by Thies [161], using poly(methyl methacrylate) as the probe adsorbate and trichloroethylene as the solvent; his results are summarized in Table 5. Although Thies showed that the bound fraction of poly(methyl methacrylate) on the precipitated silica was high, the proportion of the surface silanols occupied was less than that with the pyrogenic silica; this may be related to the porosity of the precipitated grade. On alumina the bound fractions are low, even though this adsorbent has a high density of surface hydroxyls. Both Thies and Dietz observed that silylation of the silica surface reduces both adsorption and bound fraction; Thies considered that this surface modification reduces the total number of sites and also impairs access to the unreacted silanols.

TABLE 4 Bound Fractions at the Isotherm Plateau for 2 G 4 on Silica from Chloroform [151]

Adsorbent	Bound fraction	Adsorbance (mg/g)
Pyrogenic silica; heat treatment 200°C	0.35	91
Pyrogenic silica; 400°C	0.37	74
Pyrogenic silica; 800°C	0.39	64
Methyl silyl derivative	0.10	8
Phenoxy derivative	0.33	20

TABLE 5 Bound Fractions of Poly(methyl methacrylate) from Chloroform [161]

Adsorbent	Bound fraction	Adsorbance (mg/g)
Pyrogenic silica; heat treatment 110°C	0.32	48
Pyrogenic silica; 310°C	0.32	45
Precipitated silica; 110°C	0.47	57
Precipitated silica; 310°C	0.43	59
Methyl silyl derivative	0.12	—
γ-Alumina; 110°C	0.31	25
γ-Alumina; 310°C	0.27	24

The effect of heat treatment is more complex. With nonporous silica heating will reduce the number of surface hydroxyls but it may also affect the porosity of the alumina and the precipitated silica.

Comyn and co-workers [165] indicated that inelastic electron tunneling spectroscopy of poly(methyl methacrylate) and poly(vinyl acetate) adsorbed on aluminum oxide shows the probable mode of interaction to be an ester cleavage at the oxide surface, although the more commonly assumed hydrogen bonding of ester to surface hydroxyl is not discounted.

Microcalorimetry has been used [166] in polymer adsorption studies, and bound fractions may be extracted from the data [167]. The net heat of adsorption of a polymer from solution is compared to that of an appropriate monomeric analog. The experimental problem is that the heat change on adsorption is not only small but also slow; indeed, Cohen-Stuart et al. [162] doubt whether any reported heats of polymer adsorption include the contribution from conformational adjustments at long times. Nonetheless, calorimetric bound fractions are in reasonable agreement with those from infrared spectroscopy.

Robb and co-workers [168] first showed that the conformation of adsorbed polymers could be studied by electron spin resonance (ESR) techniques. It is necessary that spin labels be randomly bonded along the polymer chain and that the labeling groups do not disturb the adsorption process. Robb and co-workers started with *N*-vinyl pyrrolidone copolymerized with 3 mole % allylamine, which acted as the sites for the nitroxide labels. On adsorption of the polymer the ESR signal was resolved into contributions from low-mobility labels, having correlation times around 10^{-8} sec, while labels in loops and tails a short distance out from the surface had correlation times around 8×10^{-10} sec. Poly(vinyl pyrrolidone) adsorbed from chloroform onto silica was found to have a restricted fraction of approximately 0.6 at the isotherm plateau, whereas poly-(methyl methacrylate) gave values around 0.9 at all coverages. These workers concluded that, provided the solvent does not markedly compete with polymer for surface sites, change of solvent from good to poor tends to flatten the adsorbed layer, as judged by the increase in the ESR restricted fractions [169]. Tests were made to establish the extent to which the spin labels might influence adsorption behavior; it was concluded that the ESR method may be invalid when nitroxide-labeled polymers are adsorbed from a nonpolar solvent onto a polar surface.

Further work on the adsorption of spin-labeled poly(vinyl acetate) on silica surfaces has been described [170]. The ESR spectral behavior is consistent with a relatively flat adsorbed polymer conformation from poor (toluene) and indifferent (chloroform) solvents on a silica fairly high in surface silanol content, but a loopy conformation is found for adsorption from a good solvent (ethyl acetate). With a less hydroxylated silica, loopy adsorbed structures are indicated for polymers adsorbed from chloroform as well as from ethyl acetate.

An extensive survey of ESR for polymer adsorption was reported by Liang et al. [171]; in their work, poly(vinyl acetate) was lightly labeled, with 1 to 10 nitroxides per polymer chain. They found that with a chloroform solution, the ESR behavior varied a great deal with the adsorbent. At plateau coverage on alumina the restricted fraction was close to unity, while on silica most of the spin labels were in loops or tails, with titanium dioxide and glass occupying intermediate positions. On alumina the restricted fractions were very high from all the solvents studied, but on silica the fraction decreased with increasing solvency. In general, flatter conformations were found at lower coverages. A labeled polystyrene had only a small slow component in the signal after adsorption from chloroform on both silica and glass.

Sakai et al. [172] interpret the ESR spectrum of labeled poly-(methyl methacrylate) adsorbed on silica in terms of three components, corresponding to long loops (tails were not explicitly considered), short loops, and trains. From three solvents—benzene, chloroform, and carbon tetrachloride—adsorbed segments were predominantly in trains and only a few in long loops; the relative amounts of trains, short loops, and long loops remained almost constant with surface coverage, until at high coverage the fraction in trains dropped and the loops, especially long loops, increased. Restricted fractions from benzene and chloroform were identical, but from the poorer solvent, carbon tetrachloride, the adsorbed layer was compressed and had many short loops. The same authors compared the ESR restricted fractions with infrared bound fractions; the latter values are somewhat less dependent on surface coverage and were little more than half the ESR fractions.

In another ESR study, Hommel et al. [173] described another way to probe surface conformations. Poly(ethylene oxide) molecules were grafted by one end to the surface of spherical silica particles and the other end group was spin-labeled. The resulting ESR spectra were interpreted as showing that, up to the tetramer, the grafted molecules adopted a "brush-bristle" conformation, but longer chains behaved more like random-walk molecules.

Observations of a high restricted fraction from the NMR spectrum were first reported by Roe et al. [40]. Soon afterward Miyamoto and Cantow [41] showed that, because of excessive line broadening, there was no ^1H resonance signal from poly(methyl methacrylate) adsorbed from $CDCl_3$ on silica at high adsorbent/polymer ratios. Diaz-Barrios and Howard [42] analyzed the NMR spectra of silica dispersed in polymer solutions in a manner similar to infrared bound fraction measurements. A polystyrene of molecular weight 4000 had a restricted fraction of 0.78, and for samples of longer chain length the fractions were only a little lower. Similar values were found with styrene/vinyl ferrocene copolymers. Lipatov et al. [174] investigated the molecular mobility of polystyrene and a polycarbonate adsorbed on silica from concentrated solutions; the decrease in the mobile NMR signal was attributed to the effect of the adsorbent particles on the structure formation of concentrated polymers and its consequences for the adsorption process.

More sophisticated analyses of polymer restricted fractions by pulsed NMR methods have been published. Barnett and Cosgrove [43] describe variants of pulsed NMR techniques applicable to adsorbed polymers in which signals from species with strong dipolar coupling ("solid-like") and those which are weakly coupled ("liquid-like") are separated, and at the same time the large solvent signal is eliminated. Polystyrene adsorbed on graphitized carbon from carbon tetrachloride has a restricted fraction of 0.30 [175], reflecting the weak phenyl-carbon interaction. Terminally grafted polystyrene (molecular weight 10^4) on a carbon surface, which has an all-tails conformation, showed an NMR signal with a negligible contribution from restricted chain segments. In general, and including for this purpose studies in polar media, NMR restricted fractions are similar to the ESR values and are much higher than the bound fractions found from infrared spectroscopy.

D. Layer Thickness

When Stromberg et al. [20,106] first applied ellipsometry to polymer adsorption, they discovered that the adsorbed layers were relatively thick, values of several nanometers being found. As previously mentioned, calculation of an average layer thickness requires the assumption of a segment density distribution; usually an rms average is quoted from an assumed exponential distribution. For polystyrene in cyclohexane at the theta temperature, the average thickness on chrome was proportional to the square root of the molecular weight; incidentally, Peyser and Stromberg [25] obtained similar thicknesses by ATR measurements on quartz. Gebhard and Killmann [22] found a square-root dependence for polystyrene on metal surfaces when adsorbed from theta solvents, a dependence also recorded by Takahashi et al. [23]. Under theta conditions the radius of gyration of an isolated polymer coil in solution (that is, in the unperturbed state) is proportional to the square root of molecular weight. There is therefore a temptation to assume that the ellipsometry data point to an adsorbed conformation of isolated Gaussian coils, but this conclusion is not warranted. As emphasized by Fleer and Lyklema [9], the rms thickness predicted from the Scheutjens and Fleer theory for the semidilute region is proportional to $M^{1/2}$ for athermal as well as theta solutions; in both cases this dependence arises from the tails fraction of the segment density distribution. Takahashi and Kawaguchi [8], however, consider that the proportionality factor of the Scheutjens and Fleer theory does not fit their ellipsometric thickness results, preferring the fit of Silberberg's theory [86]. Of course, the fit will depend on the value adopted for the χ_s parameter.

Although polystyrene layer thickness is proportional to the square root of the molecular weight, Gebhard and Killmann [22] found that this was not so with poly(methyl methacrylate) under theta conditions, and that the layer was more compressed. When polystyrene was measured at temperatures a little over the theta value, Kawaguchi and Takahashi [176] found that while the amount adsorbed decreased as the solvency improved, the thickness increased. Although the molecular weight dependence of the mean thickness followed the power law predicted by Hoeve's loop-train model [84,85], the numerical accord was poor, probably because of the neglect of tails in the model. Further measurements in good solvent conditions have been reported [177]. For cyclohexane at 40 and 45° and for carbon tetrachloride at 35° the thickness of polystyrene on chrome depends on the molecular weight to the power 0.40, distinctly lower than the 0.50 power at theta conditions.

The average thickness of a polymer adsorbed at an interface is a point of distinction between the various theories of polymer adsorption. Silberberg [86], Scheutjens and Fleer [93,94], and Hoeve [84,85] predict thinner films than in the theta condition, while others [63,178,179] suggest the opposite. The present results clearly favor the former view but a detailed agreement is absent; in particular, the Scheutjens and Fleer prediction that the square-root dependence on molecular weight is independent of solvent is not verified experimentally. Kawaguchi and Takahashi [177] present an argument based on de Gennes scaling theory [88,89] to show that the exponent should be 0.40 in good solvents and also show that 0.50 is to be expected in theta solvents.

At about the time when Stromberg was pioneering ellipsometry for polymer adsorption studies, Rowland and Eirich [34,35] developed a flow method for layer thickness. Polymer preadsorbed on the faces of a sintered glass filter disk narrows the capillary and hence restricts liquid flow. In principle this technique can be very sensitive, since the flow rate through a capillary of circular cross section is proportional to the fourth power of the radius. Rowland and Eirich estimated the layer thickness of polystyrene on glass at theta conditions as being approximately equal to the radius of gyration in solution. They also found the thickness to be proportional to the square root of the molecular weight; however, in a good solvent (benzene at 30 and 50°) the hydrodynamic layer thickness did not show a simple power-law dependence, although the layer was noticeably thinner than in theta conditions, especially at the higher molecular weights. With other good solvents the layer thickness was directly proportional to the solution intrinsic viscosity, but had values less than the radii of gyration.

Using a cellulose ester membrane filter, Gramain [36] showed that the equilibrium thickness of poly(vinyl acetate) adsorbed from chloroform varied linearly with the intrinsic viscosity. With this system equilibration was slow and the thickness was not independent of the pore dimensions of the adsorbent. The measured layer thickness decreased when the pore size was less than the hydrodynamic volume of the polymers in solution. The layer thickness of poly(vinyl acetate) reported by Gramain was about half that found by Rowland and Eirich on glass, and this may reflect the higher polarity of the cellulose ester substrate. Later work by Gramain and Myard [180] focused on the effect of shear rate on the flow restriction caused by polymers adsorbed within a porous bed; for this study polystyrene in toluene and polyacrylamide in water were chosen. A reversible dilation of adsorbed dimensions with shear rate was observed. Somewhat surprisingly, the thickness increased with shear rate, which was attributed to circulation of solvent within the partially draining layer leading to swelling in the axial direction. Clearly, the nature of the "hydrodynamic" thickness measured by permeability experiments is not self-evident and even for laterally homogeneous adsorbed films the averaging process is uncertain.

Lee and Fuller [181,182] described how the thickness, measured by ellipsometry, of adsorbed polymer layers changes with the flow rate of an external fluid.

Varoqui and Dejardin [29] calculated the average hydrodynamic thickness for a polymer layer having an exponential segment distribution. They found that the thickness corresponds to the outermost region from the surface, where the local segment density is very low. Since this calculation essentially neglected the special contribution of tails, which, according to Scheutjens and Fleer, will dominate this outer region of the adsorbed film, it seems clear that hydrodynamically measured thicknesses refer to very high averages of the segment density distribution; in effect, these thicknesses are measures of the most extended part of the adsorbed conformation. This conclusion was verified in an examination of measured and predicted hydrodynamic thicknesses of adsorbed layers [183].

Doroszkowski and Lambourne [37] deduced the hydrodynamic thickness from the viscosity of dispersed particles whose surfaces were coated with polymer. Well-dispersed and colloidally stable particles are necessary for success. In the original paper [37] the effect of partial flocculation was corrected by extrapolation of the viscosity data to infinite shear rate; in view of Gramain and Myard's later observations [180], the validity of this procedure might be questioned. However, for sterically well-stabilized particles the shear rate dependence is found to be slight. Doroszkowski and Lambourne found relatively thick layers from low molecular weight polymers, some of which were branched copolymers. For instance, a linear lauryl methacrylate/methyl methacrylate copolymer of molecular weight 9000 had a 21-nm layer on titanium dioxide when adsorbed from an aliphatic hydrocarbon medium. These authors did not report any systematic study of layer thickness with molecular weight or with solvent, but their results were helpful in the design of polymeric steric stabilizers. Barron and Howard [184] applied the method of [37] and found the thickness of poly(methyl methacrylate) on silica from carbon tetrachloride to be almost independent of molecular weight between 1.4×10^4 and 2.0×10^5, with an average value about 12 nm. Howard and McGrath [185], using the same technique, reported that in trichloroethylene, a better solvent, adsorbed poly(methyl methacrylate) was about half as thick as in carbon tetrachloride.

Sedimentation velocities may be analyzed similarly; Fontana and Thomas [14] found relatively thin layers (2.0–4.0 nm) for poly(lauryl methacrylate) on carbon, whereas a vinyl pyrrolidone/stearyl methacrylate copolymer was much thicker (21.0 nm). Some later aqueous measurements in the ultracentrifuge [27] gave a square-root molecular weight relation for the layer thickness of poly(vinyl alcohol).

A novel method for adsorbed layer thickness applicable to anisotropic magnetic particles has been proposed by Scholten [186]; here the decay of the magnetic birefringence on cessation of flow enables calculation of the rotational diffusion coefficient and, hence, the hydrodynamic layer dimension. Photon correlation spectroscopy also measures the diffusion coefficient (translational), from which the effective particle size is deduced; subtraction of the bare particle dimensions gives the layer thickness. So far this technique has been limited to studies in aqueous systems.

E. Solvent Effects and Displacement

In the preceding sections some reference was made to adsorption behavior under differing solvency conditions. Here we make a more comprehensive survey of the influence of solvents on polymer adsorption.

Two exchange free energies control the way in which polymer adsorption varies with solvent: χ, the polymer-solvent interaction parameter, and χ_S, the segment-surface net interaction parameter. Thus the solvent effect is a balance between solvency and the relative strengths of attachment of segments and solvent molecules at the surface. If it were not for the χ_S parameter, a change to a thermodynamically better solvent, where polymer-solvent contacts are favored at the expense of polymer-polymer and solvent-solvent contacts, would decrease the level of adsorption. A solvent change will also induce conformational changes in chain both in solution and in the adsorbed layer, so the net solvent dependence is not readily predictable a priori. It might be supposed that, at low surface coverage, the χ_S parameter will largely determine adsorption behavior, whereas as the isotherm is ascended and lateral interactions operate more and more, the χ parameter will become increasingly dominant [163]. However, as calculations by Scheutjens and Fleer show [95], the former condition is virtually inaccessible to experimental measurement.

A number of earlier studies [187,188] pointed to a connection between adsorbance and solvent quality. A close relationship between adsorption and the polymer-solvent interaction parameter was reported by Mizuhara et al. [189]. For poly(vinyl acetate) on a glass substrate, the plateau adsorbance varied smoothly with the intrinsic viscosity and with the difference in solubility parameters between polymer and solvent.

Koral et al. [188] found proportionality between the plateau adsorption of poly(vinyl acetate) on tin and iron powders and the reciprocal of the intrinsic viscosity in all but one of the solvents used in their study; the exception was the most polar solvent, acetonitrile, from which the polymer was unadsorbed, demonstrating blanketing of the surface sites. When high molecular weight polystyrene adsorbed from six apolar solvents onto graphitized carbon, Schick and Harvey [190] observed a smooth variation of adsorbance with χ. Hara and Imoto [123] considered that their results on the adsorption of an ethylene/vinyl acetate copolymer on glass powder were consistent with a model of adsorbed conformations which retain the Gaussian coil of chains in solution. In a second paper [191], concerning the adsorption of ethyl cellulose on glass, the same analysis was applied. As this analysis is based on fitting the hydrodynamic volume of solution polymer coils onto the surface layer, it should also predict the molecular weight dependence of adsorption, but this was not tried. Indeed, the adsorption of ethyl cellulose on glass was reported to be independent of molecular weight.

Howard and Ma [112] also found a linear plot for the solubility parameter difference against adsorbance of a commercial poly(methyl methacrylate) onto a titanium dioxide bearing a basic surface coating. In this case an acid-base interaction between surface and polymer controls χ_s, which may be relatively indifferent to the solvents used. This may also be applicable to some recent results for a polyester on iron oxide [143]. A more extensive survey shows clearly that the solvent dependence of polymer adsorption must refer to specific molecular interactions. Ellerstein and Ullman [192] measured the adsorption of poly(methyl methacrylate) on iron powder from 11 solvents, 9 of them halogenated compounds, of which ethyl bromide was the most polar. Although the general trend was for the adsorbance to be greater from the more apolar solvents, and less when the solubility parameters of polymer and solvent were closer together, other factors contributed substantially. Little polymer adsorbed from chloroform, and the authors supposed this to be due to acidic hydrogen of the solvent; a slightly hydrolyzed poly(methyl methacrylate) was strongly adsorbed from benzene because of its acidity.

Howard and McConnell also surveyed several solvents in their study of poly(ethylene oxide) adsorption onto silica [193] and carbon [126]. For six solvents the ranking, based on intrinsic viscosities, runs from chloroform to methanol; the affinities for the silica surface, determined by vapor adsorption, are largely in the opposite direction, polar to apolar. As the solvency effects tend to nullify the solvent affinity, the net influence of the solvent on poly(ethylene oxide) adsorption is a nice balance between both factors. On a carbon surface [126] the adsorption correlates better with the afifnity series than with solvency, but again specific factors are invoked for some of the solvents.

Negative adsorption of polymer is observed from some solvents. For instance, Hamori et al. [196] found no adsorption of poly(methyl methacrylate) on silica from the polar solvents acetone and acetonitrile; these basic solvents compete successfully with the polymer segments for the acidic silica surface. Similarly, Ashmead and Owen [122] failed to detect finite adsorption of poly(dimethyl siloxane) from 2-butanone onto glass, even though this is a theta solvent. However, butanone interacts strongly with a glass surface. For instance, a 1% (v/v) addition to benzene substantially reduced poly(dimethyl siloxane) adsorption; further, butanone vapor shifted the hydroxyl stretching band in the infrared. Tetrahydrofuran and ethyl acetate also both reduced the polymer adsorption from benzene and caused shifts in the infrared hydroxyl absorption [122].

A study of adsorption from mixed solvents, as indicated above, has been used by several workers to make a tentative discrimination between solvent competition and solvency. Addition of 10% acetonitrile prevents poly(dimethyl siloxane) from adsorbing to iron from benzene

[121], while a larger quantity of this polar displacer (approximately 60%) is needed to eliminate the adsorption of poly(methyl methacrylate) on iron [192]. Another polar competitor, dimethyl formamide, stops the adsorption of poly(ethylene oxide) from benzene onto carbon when added to about 10% [126].

Thies [120] looked at the effect of several polar species on the adsorption of poly(vinyl acetate) and poly(co-ethylene-vinyl acetate) on pyrogenic silica. Although the adsorption of poly(vinyl acetate) was somewhat decreased by acetonitrile, methanol, cyclohexanol, anthracene triol, and dioxane, it was still quite substantial at additions of displacer up to 10%. Only at the higher displacer concentrations was the bound fraction of acetate groups reduced. The copolymer (29% vinyl acetate) dissolved in trichlorethylene was completely displaced from the silica surface by less than 0.5 M methanol and cyclohexanol, although with adsorption from cyclohexane solution these levels of addition were insufficient to remove all the copolymer from the surface. Thies [195] also found that phenol had only a slight effect on the adsorption of the two polymers and was a much poorer displacer than the two alcohols; however, infrared spectroscopy showed that phenol interacted strongly with both polymers in solution.

Chan et al. [196] took mixed solvent adsorption further and determined the "point of zero adsorption," that is, the mole fraction of displacer at which the polymer is just not adsorbed. As an example, they found the point of zero adsorption for poly(vinyl acetate) from benzene onto fibrous cellulose to be 0.52 (mole fraction) for acetone, 0.43 for dioxane, 0.20 for 2-ethyl hexanol, and 0.10 for methanol. According to Chan et al., the point of zero adsorption is independent of polymer molecular weight and of the pore accessibility of the substrate and hence should be a valuable measure of the net interaction of adsorbate groups with the absorbent in a given solvent environment. In the example cited above, the points of zero adsorption for the displacers are in the order of their hydrogen-bonding ability. Patel et al. [197] also detected the zero adsorption composition for benzene/methanol and benzene/dioxane mixtures; the polymer was poly(methyl methacrylate) and the adsorbents aluminum and calcium silicates. Methanol was more efficient than dioxane in displacing polymer, the actual amounts depending on the adsorbent. Nefedov and Zhmakina [198] reported that an increase in the chloroform content of polystyrene solutions in carbon tetrachloride weakens the interaction energy of adsorption on macroporous glass beads; these workers were interested in the adsorption process in column chromatography.

Dietz and Hamann [199] found that polar competitors for the 2 G 4 polyester reduced its adsorption on silica from chloroform solution. In this system, acetonitrile proved to be a mild displacer, but *n*-butanol and pyridine reduced the polyester adsorption to zero at volume fractions of about 0.09 and 0.07, respectively; an ethylene oxide oligomer was also a good displacer. The bound fraction of ester groups decreased progressively as the adsorption was reduced by the polar competitor. The desorption of physically adsorbed polymer by addition of low molecular weight species is useful in the study of chemical grafting onto particle surfaces. Laible and Hamann [200], for instance, showed that physically adsorbed polystyrene was readily removed from pyrogenic silica, whereas grafted polymer was unaffected by washing with displacer liquids.

Theoretical insight into displacer action has been given by Cohen-Stuart et al. [201–203]. These authors refer to the point of zero adsorption as the critical diplacer concentration but, unlike Chan et al. [196] use volume fractions. In the three-component system, polymer, displacer, and solvent, three χ_s parameters may be defined, one for each pair, although these are not independent since

$$\chi_s^{po} = \chi_s^{pd} + \chi_s^{do}$$

where the superscripts d, o, p refer to displacer, solvent, and polymer respectively. As an efficient displacer will adsorb more strongly than a polymer segment, χ_s^{pd} will be negative, and the greater its numerical value, the more efficient it is. By adding the separate contributions to the free energy when a polymer adsorbs onto a surface, displacing solvent molecules, and including the analogous addition for exchange of solvent and displacer species at the surface, an expression is deduced which allows calculation of χ_s^{po} from the measured ϕ_{cr}, the critical displacer concentration. This expression is complex, even when some approximations are introduced; as well as ϕ_{cr} the equation involves χ_s^{do} and three Flory-Huggins parameters, χ^{po}, χ^{do}, and χ^{pd}. Cohen-Stuart et al. show how, by locating four ϕ_{cr} values from the combination of two solvents and two displacers, the solvency terms may be simplified, leaving only χ_s^{do} to be measured by separate adsorption of displacer from the solvent. A lattice term is also included, but this, in principle, may be calculated from the lattice specifications. Experimental displacement isotherms have been reported for poly(vinyl pyrrolidone) adsorbed on silica from the polar solvents dioxane and water. Values for the net segmental adsorption exchange free energy χ_s^{po} are about −4.0 (kT units), which is indicative of a strong adsorption; bound fraction data and calorimetric enthalpies are consistent with a high χ_s^{po} [162].

Not only is this the first experimentally reported value of χ_s^{po}, but a further result of the analysis shows that the lattice critical adsorption energy previously taken to be a simple lattice parameter must be modified by inclusion of rotational degrees of freedom. Clearly, the work by Cohen-Stuart et al. [201–203] is an important development. With accurate numerical values for the χ and χ_s parameters, the Scheutjens and Fleer theory may be used to describe characteristic features of the adsorption behavior of practical polymer/solvent/adsorbent systems.

Competition between adsorbate species leads to the next major topic, that between polymer segments differing in chemical nature.

F. Polymer Mixtures and Copolymers

There are two ways of arranging segment-segment competition in a polymer adsorption system. One way is to study the adsorption behavior of mixed polymers, and the other is to see how the composition of copolymers influences adsorption.

Thies [15] systematically studied the adsorption of polystyrene/poly(methyl methacrylate) mixtures from trichloroethylene onto Cab-o-Sil. Polystyrene in competition with poly(methyl methacrylate) is not adsorbed unless the latter polymer is unable to saturate the surface. If polystyrene is preadsorbed on the silica and then excess poly(methyl methacrylate) is introduced, rapid displacement occurs and all the polystyrene is replaced at the interface by poly(methyl methacrylate). Bound fractions (by infrared spectroscopy) of both the carbonyl and phenyl groups are relatively unaffected by coadsorption of either polymer. It is well understood that dissimilar polymers in a common solvent usually phase separate above a critical, and low, concentration. In Thies's experiments the initial concentrations of the mixed polymers were below the critical level and the solutions were homogeneous. On the other hand, in mixed adsorption layers the local concentration is likely to exceed the phase separation condition. Displacement of polystyrene by poly(methyl methacrylate) was rapid despite the mutual incompatibility of the polymers.

Schick and Harvey [204] also compared polystyrene (labeled with ^{14}C) and poly(methyl methacrylate) adsorption, but on the nonpolar Graphon surface, and reported both consecutive and concurrent adsorptions. Polystyrene is the polymer preferentially adsorbed on Graphon (see [13]), but its adsorption from 2-butanone was reduced by about one-third by poly(methyl methacrylate) and by poly(butyl methacrylate). Further measurements were made with poly(vinyl acetate) as the competitor for polystyrene, but it was found unable to displace preadsorbed polystyrene from either a good (toluene) or a poor (2-butanone) solvent. When poly(vinyl acetate) was preadsorbed, polystyrene displaced it successfully in the presence of toluene but only partially from 2-butanone.

Botham and Thies [205] also reported on other polymer mixtures and their adsorption onto silica. An ethylene/vinyl acetate copolymer rapidly replaces both polystyrene and ethyl cellulose; among the copolymers, those higher in vinyl acetate content displace those lower. When mixed polymers were used at concentrations around the compatibility limit it seemed that adsorption of the preferred polymer was enhanced by the presence of the unadsorbed polymer. The bound fractions were low, suggesting a voluminous layer. In a later paper [206] the same authors described the adsorption from perchloroethylene on silica of polystyrene, polybutadiene, and some styrene/butadiene copolymers. Polystyrene was preferred to polybutadiene and to di- and tri-block (S/B/S) copolymers. Polybutadiene was displaced from silica by the tri-block polymer, which itself was displaced by polystyrene.

The adsorption of a methacrylate polymer carrying long alkyl side groups was measured by Steinberg [100], who also measured competitive adsorption with a methacrylate/vinyl pyridine copolymer. When poly(lauryl methacrylate) was adsorbed on iron powder at saturation coverage and the copolymer then introduced, it adsorbed without much desorption of the homopolymer. In this system a mixed adsorbed layer is formed with the loopy copolymer overlying the polymethacrylate. If the vinyl pyridine copolymer was adsorbed first, no subsequent adsorption of the homopolymer was found.

Lipatov et al. also measured adsorbances and bound fractions in polymer mixtures: poly(butyl methacrylate/poly(co-butadiene-acrylonitrile)/silica/chloroform [207] and poly(butyl methacrylate)/polystyrene/silica/carbon tetrachloride [208]. Adsorption was studied at relatively high concentrations, above the compatibility limit, and both the adsorbance and bound fraction dependence on concentration were very complex. The authors discuss their observations in terms of molecular aggregation in solution; it is interesting to note that, with a single polymeric adsorbate, the fact that the adsorption isotherm is sensitive to the ratio of adsorbent to polymer solution is ascribed by Lipatov et al. to structure formation in the solution phase, whereas Cohen-Stuart et al. [116] would consider the effect as due to polydispersity.

Another question posed by competition between two polymeric adsorbates is whether adsorption is controlled by the chemical nature of segment functional groups or whether stereochemical placement also has a role. To address this question, three sets of workers have examined the adsorption of chemically identical polymers differing in tacticity. Botham and Thies [209] compared isotactic, syndiotactic, and atactic samples of poly(isopropyl acrylate) as adsorbates on silica; the adsorptions from trichloroethylene were equal within experimental error, although the infrared bound fraction was slightly higher for the isotactic polymer. However, this isotactic sample was unable to displaced preadsorbed atactic polymer. Conventional poly(methyl methacrylate) is partly syndiotactic; Hamori et al. [194] compared its adsorption behavior to that of an isotactic version. Both adsorbed to about the same extent on silica from toluene and from 1,2-dichloroethane; from a polar solvent, acetonitrile, the isotactic polymer adsorbed, whereas the conventional polymer did not. Miyamoto et al.

[210] also used isotactic and syndiotactic poly(methyl methacrylate). The separate isotherms are quite similar, with plateau adsorptions of 120 mg/g for the isotactic sample and 105 mg/g for the syndiotactic polymer, measured from chloroform onto silica. Nonetheless, good separation was achieved from a 1:1 mixture added to adsorbent just sufficient to be saturated by adsorption of the isotactic polymer.

Competition between different chemical structures in adsorbates may also be investigated with copolymers. Indeed, copolymers provide an interesting example of the influence of structure on adsorption, since a progressive variation of composition and sequential arrangements may be obtained fairly easily. That changes in polymer composition would show up in adsorption was noticed early in polymer studies. Kolthoff et al. [211] measured the adsorption from benzene solution of several styrene/butadiene copolymers on Graphon. The adsorption of polystyrene was low, but incorporation of a small amount of butadiene in the copolymer increased the value to a level which was also shown by copolymers of higher butadiene content. Polybutadiene was slightly less well adsorbed than the copolymers, although the difference was barely beyond experimental error. In other work, Kolthoff and Gutmacher [212] showed that the adsorption of styrene/butadiene copolymers (commercial GR-S rubbers) on carbon blacks was almost independent of copolymer composition within the range 5 to 75% butadiene. Some other synthetic rubbers were studied by Binsford and Gessler [213]; these were butyl rubbers, copolymers of 2-butene with a little isoprene. When adsorbed onto channel black, a carbon with an oxygenated surface, the pickup was independent of isoprene content.

Another early study, already cited, was that by Koral et al. [188], who showed that the adsorption of poly(vinyl acetate) onto iron powder could be increased threefold by hydrolysis of 13% of the repeat units to the vinyl alcohol form; a further increase in the hydroxyl content had little effect. The literature contains several other references to the alteration of adsorption behavior by incorporation of polar comonomers. Ellerstein and Ullman [192], using the same iron powder as Koral et al., examined how slight hydrolysis of poly(methyl methacrylate) modified its adsorption from benzene solution. When 0.90% of the chain units were hydrolyzed to methacrylic acid groups the adsorbance was 2.0 mg/g, increasing to only 2.1 mg/g at 1.5% hydrolysis, whereas the original homopolymer was adsorbed to only 0.8 mg/g. Erman et al. [214] reported that a carboxyl-containing copolymer was the best adsorbed, onto rutile and barite, of a series, and confirmed that the stronger bonding was associated with the acid group by infrared spectroscopy.

Howard and Ma [112] found that some commercial poly(methyl methacrylates) adsorbed well onto titanium dioxide from solvents such as 2-butanone, but that laboratory-made homopolymers were almost unadsorbed. This effect was traced to a small acidic component in the nominally uncopolymerized product, and confirmatory experiments showed that a few copolymerized carboxylic acid groups made an otherwise indifferent polymer adsorbable. Fontana [149] observed that a copolymer of lauryl methacrylate with vinyl pyrrolidone adsorbed more than the methacrylate homopolymer, and demonstrated the participation of the pyrrolidone groups in surface interactions. Fontana and Thomas [14] also found that ethylene oxide side chains grafted onto a methacrylate backbone led to high adsorption, with the ether oxygens able to displace some of the carbonyl groups from the silica surface.

Perhaps the most systematic studies of copolymer adsorption have been those of Howard and his co-workers. In this work, series of copolymers covering the full composition range were synthesized and variation of comonomer sequential arrangement was achieved by block or graft copolymerization. When the relative adsorption of two polymers is assessed, the net segment-solvent free energy of adsorption and the polymer-solvent interaction differ. By suitable selection, some examples can be presented in which one or the other effect is predominant. The first copolymer series reported [126,193,215] was a set of ABA block copolymers of ethylene oxide (A) and propylene oxide (B); these polymers are rather low in molecular weight, and the hydroxyl end groups will have some influence on the adsorption level [126]. Adsorption of these copolymers on carbon from benzene increased smoothly with ethylene oxide content; poly(ethylene oxide) is less soluble than poly(propylene oxide) in this solvent. A random copolymer (70% ethylene oxide) fitted on the adsorbance/composition curve. This example is one where the primary surface interaction is between ether groups and oxygenated sites, and will be little altered by composition. If the trend of adsorption with composition is a solubility effect, then changing to a solvent in which the homopolymer relative solubilities are reversed should also reverse the slope of the adsorbance/composition curve. This was the case. Again, the random copolymer fits in with the block polymers. The observations made with the carbon surface were also found [215] when these copolymers were adsorbed onto a nylon powder.

Another type of copolymer is one in which the comonomers are expected to have very different affinities for the surface sites. Insertion of a more polar comonomer may change the solubility characteristics, but it is possible to choose components in which solubility complications are minimized. Copolymers of styrene and methyl methacrylate were used in trichloroethylene as this solvent has nearly equal solubilities for the two parent homopolymers. Copolymers with 10 to 92% styrene had the same adsorbance on silica, but a little more than the adsorbance of poly(methyl methacrylate) [16]. As expected from Thies's polymer competition experiments [15], styrene homopolymer was adsorbed to a much smaller extent. Thus, very few methacrylate units are necessary to obtain strong adsorption, and incorporation of more than the critical amount (about 5%) seems superfluous as far as the mass adsorbance is concerned. The same general effect was shown with other solvents (benzene and carbon tetrachloride). Further, the adsorbance/composition relation holds, within experimental error, for both ABA block and random copolymers. This research group [16] was able to show that differences in adsorption properties of block and random polymers, as indicated by the bound fractions, were also minimal. The adsorption of the copolymer series on a porous carbon was similar to that on silica; both are high-energy surfaces. However, when the copolymers were adsorbed on a lower-energy surface (Graphon), several changes were reported. The polystyrene is now the favored homopolymer, and poly(methyl methacrylate) is unadsorbed, an observation in line with the competitive polymer adsorption study by Schick and Harvey [204]. Block copolymers with more than 50% styrene adsorbed a little better than polystyrene, but at lower styrene contents adsorbance gradually dropped to the zero value of poly(methyl methacrylate).

Random copolymers of 50% or more styrene adsorbed a little less than the block polymers, but without much change with composition. However, at low styrene contents the adsorbance dropped severely, so that at 10% styrene the random copolymer was not adsorbed, whereas the corresponding block polymer was adsorbed to about the same extent as homopolystyrene. This may represent a situation where, because the overall surface interaction is low, a cooperative effect of a sequence of higher-affinity groups, as in the block copolymers, exerts an influence.

Barron and Howard [184] also examined graft copolymers in which short polystyrene chains were attached to the poly(methyl methacrylate) backbone. As well as investigating the adsorption on precipitated silica, they studied the action of all three classes of styrene/methyl methacrylate copolymers in the steric stabilization of the silica. From the poor solvent, carbon tetrachloride, where adsorption is high, all sequential arrangements provided an adequate anchor of the polymer to the surface. In trichloroethylene, graft and block copolymers stabilized over almost the entire composition range, and with little to distinguish one series from the other, but the stabilizing power of the random copolymers declined beyond 40% styrene, and was lost completely at 60%. The adsorption of block and random copolymers of styrene and methyl methacrylate under progressively worsening solvency has been described by Guthrie and Howard [216] as part of a study of solvent effects in steric stabilization.

Howard and McGrath [185] worked with styrene/2-vinyl pyridine copolymers; here the solubility parameters of the parent homopolymers are similar but the different polarities of the comonomers might be expected to be evident in adsorption behavior. Adsorption on silica remained high and almost constant with composition until the styrene content approached 100%, when the polymer was not adsorbed. AB block copolymers performed much as the random copolymers.

Laible and Hamann [200] compared 2-vinyl pyridine and methyl methacrylate as comonomers in polystyrenes and, like Barron and Howard [184] and Howard and McGrath [185], measured steric stabilization as well as adsorption on silica. Block and random copolymers were synthesized at the 2 mole% level, and plateau adsorbance values are summarized in Table 6. As found by Howard and McGrath [185], 2-vinyl pyridine was a stronger anchor than methyl methacrylate on the silica surface. The higher adsorption of 98:2 styrene/methyl methacrylate block from toluene, compared to the random copolymer, is similar to the results of Hopkins and Howard [13]. Laible and Hamann also included *n*-butanol as a displacer to illustrate the superiority of the pyridine segment; the tabulated results also point to the greater resistance to displacement shown by the methacrylate blocks (Table 6).

Another example of the influence of polar comonomeric units involves vinyl acetate/ethylene copolymers. Copolymers of 15 and 30% vinyl acetate adsorbed on glass microspheres [123] to the same extent from carbon disulfide and benzene, but with adsorption from chloroform a solubility factor entered. This class of copolymer was investigated subsequently [155], and infrared bound fractions of the acetate groups were found to increase slightly from approximately 0.20 at 0.10 mole fraction to 0.26 at 0.60 mole fraction of acetate; the total fraction of bound segments went from 0.04 to 0.21 over this composition range. Diaz-Barrios and Paredes [217] compared the adsorption of styrene/butadiene copolymers from carbon tetrachloride on silica with that of the parent homopolymers. On four grades of silica, there was a steady increase of adsorbance as the styrene content went up. Earlier, Sadakne and White [139] reported that a styrene/butadiene copolymer adsorbed less on carbon from toluene than the homopolymers; of course, polymers from diene monomers take

TABLE 6 Plateau Adsorption (mg/g) of Copolymers (2% Minor Component) from Toluene on Silica [200]

	Styrene/methyl methacrylate		Styrene/2-vinyl pyridine	
Solvent	Block	Random	Block	Random
Toluene	56	23	99	101
Toluene, 0.4 M in *n*-butanol	29	2	62	62

up different configurations on polymerization, and the effect on adsorption has not been considered. Nonetheless, there seems to be some discrepancy with other reports [211,212]. Kawaguchi et al. [218] adsorbed random copolymers of styrene and butadiene from cyclohexane at 35° onto silica. The adsorbance increased with the styrene content; note that the solvent is a theta solvent for polystyrene. Bound fractions for both comonomers were measured; at a low (10%) mole fraction of styrene the phenyl bound fraction was 0.24, decreasing to about 0.15 at higher (70%) contents, compared to a value of 0.18 for polystyrene. Very few butadiene segments were bound, 0.05 to 0.08 over the composition range studied, and always less than the bound fraction (0.19) of polybutadiene. The overall bound fraction, whatever the segment, is rather low, 0.09 to 0.13, as the styrene content is raised, suggesting a somewhat extended layer. Adsorption of metallocene-containing copolymers has been reported by Diaz-Barrios and co-workers [42,219]; however, these vinyl ferrocene copolymers show no unusual features in their characteristic adsorption behavior.

Two sets of AB block copolymers were used by Dawkins et al. [220] and the adsorption studied on silica from trichloroethylene. The plateau adsorbance of the blocks lay between those of the homopolymers, and it was suggested that both blocks, polystyrene with poly(dimethyl siloxane) or poly(ethylene oxide), contributed to the adsorbed layer and both formed loops. Styrene/ethylene oxide copolymers high in styrene have greater adsorptions, and a conformation with the ether segments at the surface and the styrene segments excluded was proposed.

General conclusions from these several copolymer studies are that smooth changes in adsorption behavior result from the concomitant solubility and affinity factors; compositional and sequential variations will normally produce strong effects only when critical conditions are approached—either a solubility limit or a low net adsorption energy. As a final example of the combination of both polymer-solvent and segment-surface influences, the adsorption of three random copolymers of acrylonitrile and butadiene may be examined [221]. Table 7 shows the mass adsorbance from trichloroethylene onto silica, the Flory-Huggins parameter, and the difference between the solubility parameters of copolymer and solvent. The enthalpy of adsorption of the copolymers was not known; the heat of adsorption into the first layer was deduced by vapor adsorption of segment analogs. The solvent value, 35 kJ/mole, fell between that of ethyl cyanide, as the analog of an acrylonitrile segment (44 kJ/mole), and 2-butene, the butadiene analog (29 kJ/mole). As the acrylonitrile fraction in the copolymers was reduced, the solvent quality improved; at the same time the net segment-surface interaction was weakened, so the adsorption characteristics could be attributed to either or both influences.

TABLE 7 Adsorption of Acrylonitrile/Butadiene Copolymers from Trichloroethylene on Silica [221]

Acrylonitrile (%)	Adsorbance (mg/g)	Nitrile bound fraction	χ	$\Delta\delta$
39.9	496	0.77	0.51	1.04
32.9	309	0.70	0.42	0.68
23.1	264	0.48	0.25	0.13

V. CONCLUSIONS

This survey of polymer adsorption from predominantly apolar solvents shows that some of the theoretical predictions have been fulfilled. Nonetheless, subtle molecular effects are found in the more complex systems, such as copolymeric adsorbates, mixed solvents, and higher concentrations, and these topics are, to date, not adequately covered by the more physical approaches of adsorption theory. However, recent theoretical developments are encouraging, especially in their ability to handle the full adsorbate concentration range. It is hoped that future research will consider copolymers and nonhomogeneous adsorbents and provide a more thorough treatment of polydispersity and adsorbate competition. The Flory-Huggins lattice theory of solutions is deficient in some respects, and theoreticians will need to consider whether an improved treatment of the solution phase, such as the corresponding states approach [222] or the equation of state theory [223], should be incorporated.

On the experimental side, there is still a lack of information on the kinetic aspects of polymer adsorption. Further, to obtain the maximum knowledge about adsorbed conformations, several new advanced experimental techniques should be applied to some well-defined systems; too often in the past it was necessary to extrapolate from one system to another. Many experimental problems are found in the study of adsorption at high polymer concentrations, which has considerable industrial interest. To establish that a theoretical description of adsorption in dilute solution is consistent with fact presupposes an accurate knowledge of the χ and χ_s factors; but theory, established at low concentrations, may then be used with some confidence in experimentally inaccessible regions.

There is also a need for a semiempirical but predictive theory [224]. In this connection the acid-base approach of Fowkes and Mostafa [225] may be the key. The role of acid-base interactions in polymer adsorption has been considered by several authors [226,

227], most recently by Cremer [228]. Fowkes, however, adopts a quantitative approach in which the two-parameter characterization of Drago et al. [229] for the acid-base nature of solvents is extended to polymers (segments) and, with more difficulty, to surfaces [230, 231]. Developments along these or similar lines seem to have considerable promise.

REFERENCES

1. Th. F. Tadros, in *The Effect of Polymers on Dispersion Properties*, Th. F. Tadros, ed., Academic Press, London, 1982, p. 1.
2. B. J. Fontana, in *Chemistry of Biosurfaces*, Vol. I, M. L. Hair, ed., Dekker, New York, 1971, p. 83.
3. K. J. Ives, *The Scientific Basis of Flocculation*, Sijthoff and Noordhoff, Alphenaan den Rijn, Netherlands, 1978.
4. R. L. Patrick, *Treatise on Adhesion and Adhesives*, Vol. I, Dekker, New York, 1967.
5. G. Kraus, *Reinforcement of Elastomers*, Wiley (Interscience), New York, 1965.
6. S. G. Ash, in *Colloid Science*, Vol. I, D. H. Everett, ed., The Chemical Society, London, 1973, p. 103.
7. B. Vincent and S. Whittington, in *Surface and Colloid Science*, Vol. 12, E. Matijevic, ed., Plenum, New York, 1981, p. 1.
8. A. Takahashi and M. Kawaguchi, *Adv. Polym. Sci. 46*: 1 (1982).
9. G. J. Fleer and J. Lyklema, in *Adsorption from Solution at the Solid/Liquid Interface*, G. D. Parfitt and C. H. Rochester, eds., Academic Press, London, 1983, p. 153.
10. B. Vincent and Th. F. Tadros, in *Encyclopedia of Emulsion Technology*, Vol. I, P. Becher, ed., Dekker, New York, 1983, p. 220.
11. J. J. Kipling, *Adsorption from Solutions of Non-Electrolytes*, Academic Press, London, 1965.
12. G. Kraus and J. T. Gruver, *Rubber Chem. Technol. 41*: 1256 (1968).
13. A. Hopkins and G. J. Howard, *J. Polym. Sci. Part A-2 9*: 841 (1971).
14. B. J. Fontana and J. R. Thomas, *J. Phys. Chem. 65*: 480 (1961).
15. C. Thies, *J. Phys. Chem. 70*: 3783 (1966).
16. J. M. Herd, A. Hopkins, and G. J. Howard, *J. Polym. Sci. Part C 34*: 211 (1971).
17. E. Killmann, *Polymer 17*: 864 (1976).
18. G. R. Joppien, *Makromol. Chem. 175*: 1931 (1974).
19. J. C. Day and I. D. Robb, *Polymer 21*: 408 (1980).
20. R. R. Stromberg, E. Passaglia, and D. J. Tutas, *J. Res. Natl. Bur. Stand. Sect. A 67*: 431 (1963).

21. E. Killmann and H. G. Wiegand, *Makromol. Chem. 132*: 239 (1970).
22. H. Gebhard and E. Killmann, *Angew. Makromol. Chem. 53*: 171 (1976).
23. A. Takahashi, M. Kawaguchi, H. Hirota, and T. Kato, *Macromolecules 13*: 884 (1980).
24. J. A. Hinkley, *Polym. Prepr. Am. Chem. Soc. Div. Polym. Chem. 25*(1): 178 (1984).
25. P. Peyser and R. R. Stromberg, *J. Phys. Chem. 71*: 2066 (1967).
26. J. W. Strojek, J. Mielczarski, and P. Nowak, *Adv. Colloid Interface Sci. 19*: 309 (1983).
27. R. H. Ottewill and T. Walker, *Kolloid Z. Z. Polym. 227*: 108 (1968).
28. M. J. Garvey, Th. F. Tadros, and B. Vincent, *J. Colloid Interface Sci. 49*: 57 (1974).
29. R. Varoqui and P. Dejardin, *J. Chem. Phys. 66*: 4395 (1977).
30. O. E. Ohrn, *J. Polym. Sci. 19*: 199 (1956).
31. C. A. F. Tuijnman and J. J. Hermans, *J. Polym. Sci. 25*: 385 (1957).
32. Z. Priel and A. Silberberg, *J. Polym. Sci. Polym. Phys. Ed. 16*: 1917 (1978).
33. Y. Cohen and A. B. Metzner, *Macromolecules 15*: 1425 (1982).
34. F. W. Rowland and F. R. Eirich, *J. Polym. Sci. Part A-1 4*: 2033 (1966).
35. F. W. Rowland and F. R. Eirich, *J. Polym. Sci. Part A-1 4*: 2401 (1966).
36. P. Gramain, *Makromol. Chem. 176*: 1875 (1975).
37. A. Doroszkowski and R. Lambourne, *J. Colloid Interface Sci. 26*: 214 (1968).
38. F. W. Rowland, R. Bulas, E. Rothstein, and F. R. Eirich, *Ind. Eng. Chem.* 57(9): 46 (1965).
39. I. D. Robb and R. Smith, *Eur. Polym. J. 10*: 1005 (1974).
40. R. J. Roe, D. D. Davies, and T. K. Kwei, *Polym. Prepr. Am. Chem. Soc. Div. Polym. Chem. 11*(2): 1263 (1970).
41. T. Miyamoto and H. J. Cantow, *Makromol. Chem. 162*: 43 (1972).
42. A. Diaz-Barrios and G. J. Howard, *Makromol. Chem. 182*: 1081 (1981).
43. K. G. Barnett and T. Cosgrove, *J. Magn. Reson. 43*: 15 (1981).
44. K. G. Barnett, T. Cosgrove, B. Vincent, A. N. Burgess, T. L. Crowley, T. A. King, J. D. Turner, and Th. F. Tadros, *Polymer 22*: 283 (1981).
45. Th. van den Boomgaard, T. A. King, Th. F. Tadros, H. Tang, and B. Vincent, *J. Colloid Interface Sci. 61*: 68 (1978).
46. G. Ullmann and G. D. J. Phillies, *Macromolecules 16*: 1947 (1983).
47. A. Doroszkowski and R. Lambourne, *J. Polym. Sci. Part C 34*: 253 (1971).

48. L. Barclay, A. Harrington, and R. H. Ottewill, *Kolloid Z. Z. Polym. 250*: 655 (1972).
49. A. Homola and A. A. Robertson, *J. Colloid Interface Sci. 54*: 286 (1976).
50. M. J. Garvey, Th. F. Tadros, and B. Vincent, *J. Colloid Interface Sci. 55*: 440 (1976).
51. J. N. Israelachivili, R. K. Tandon, and L. R. White, *J. Colloid Interface Sci. 78*: 43 (1980).
52. J. Klein, *Nature 288*: 248 (1980).
53. J. N. Israelachivili, M. Tirrell, J. Klein, and Y. Almog, *Macromolecules 17*: 204 (1984).
54. Y. Almog and J. Klein, *J. Colloid Interface Sci. 106*: 33 (1985).
55. L. Knapschinsky, W. Katz, B. Ehmke, and H. Sonntag, *Colloid Polym. Sci. 260*: 1153 (1982).
56. K. G. Barnett, T. Cosgrove, T. L. Crowley, Th. F. Tadros, and B. Vincent, in *The Effect of Polymers on Dispersion Properties*, Th. F. Tadros, ed., Academic Press, London, 1982, p. 183.
57. D. J. Cebula, J. W. Goodwin, R. H. Ottewill, G. Jenkin, and J. Tabony, *Colloid Polym. Sci. 261*: 555 (1983).
58. R. Simha, H. L. Frisch, and F. R. Eirich, *J. Phys. Chem. 57*: 584 (1953).
59. H. L. Frisch, R. Simha, and F. R. Eirich, *J. Chem. Phys. 21*: 365 (1953).
60. R. J. Roe, *J. Chem. Phys. 43*: 1591 (1965).
61. R. J. Roe, *J. Chem. Phys. 44*: 4264 (1966).
62. K. Motomura and R. Matuura, *J. Chem. Phys. 50*: 1281 (1969).
63. W. C. Forsman and R. E. Hughes, *J. Chem. Phys. 38*: 2130 (1963).
64. A. Silberberg, *J. Phys. Chem. 66*: 1872 (1962).
65. A. Silberberg, *J. Phys. Chem. 66*: 1883 (1962).
66. A. Silberberg, *J. Chem. Phys. 46*: 1105 (1967).
67. R. J. Rubin, *J. Chem. Phys. 43*: 2392 (1965).
68. R. J. Rubin, *J. Res. Natl. Bur. Stand. Sec. B 69*: 301 (1965).
69. R. J. Rubin, *J. Res. Natl. Bur. Stand. Sec. B 70*: 237 (1966).
70. E. A. DiMarzio and F. L. McCrackin, *J. Chem. Phys. 43*: 539 (1965).
71. F. L. McCrackin, *J. Chem. Phys. 47*: 1980 (1967).
72. E. J. Clayfield and E. C. Lumb, *J. Colloid Interface Sci. 22*: 269 (1966).
73. E. J. Clayfield and E. C. Lumb, *J. Colloid Interface Sci. 22*: 285 (1966).
74. S. Bluestone and C. L. Cronan, *J. Phys. Chem. 70*: 306 (1966).
75. D. Chan, I. D. Mitchell, and B. W. Ninham, *J. Chem. Soc. Faraday Trans. 2 71*: 235 (1975).
76. M. Lax, *Macromolecules 7*: 660 (1974).
77. M. Lax and J. Gillis, *Macromolecules 10*: 334 (1977).

78. M. Lal, M. A. Turpin, K. A. Richardson, and D. Spencer, in *Adsorption at Interfaces*, K. L. Mittal, ed., ACS Symp. Ser. 8, American Chemical Society, Washington, D.C., 1975, p. 16.
79. E. Eisenrieglar, K. Kremer, and K. Binder, *J. Chem. Phys. 77*: 6296 (1982).
80. T. Tanaka, *Macromolecules 10*: 51 (1977).
81. C. A. J. Hoeve, *J. Chem. Phys. 44*: 1505 (1966).
82. C. A. J. Hoeve, E. A. DiMarzio, and P. Peyser, *J. Chem. Phys. 42*: 2558 (1965).
83. C. A. J. Hoeve, *J. Chem. Phys. 43*: 3007 (1965).
84. C. A. J. Hoeve, *J. Polym. Sci. Part C 30*: 361 (1970).
85. C. A. J. Hoeve, *J. Polym. Sci. Part C 34*: 1 (1971).
86. A. Silberberg, *J. Chem. Phys. 48*: 2835 (1968).
87. P. G. de Gennes, *Scaling Concepts in Polymer Physics*, Cornell Univ. Press, Ithica, N.Y., 1979.
88. P. G. de Gennes, *Macromolecules 14*: 1637 (1981).
89. P. G. de Gennes, *Macromolecules 15*: 492 (1982).
90. J. Klein and P. Pincus, *Macromolecules 15*: 1129 (1982).
91. R. J. Roe, *J. Chem. Phys. 60*: 4192 (1974).
92. R. J. Roe, *Polym. Sci. Technol. 12B*: 629 (1980).
93. J. M. H. M. Scheutjens and G. J. Fleer, *J. Phys. Chem. 83*: 1619 (1979).
94. J. M. H. M. Scheutjens and G. J. Fleer, *J. Phys. Chem. 84*: 178 (1980).
95. J. M. H. M. Scheutjens and G. J. Fleer, in *The Effect of Polymers on Dispersion Properties*, Th. F. Tadros, ed., Academic Press, London, 1982, p. 145.
96. G. J. Fleer and J. M. H. M. Scheutjens, *Adv. Colloid Interface Sci. 16*: 341 (1982).
97. E. A. DiMarzio and R. J. Rubin, *J. Chem. Phys. 55*: 4318 (1971).
98. R. D. Vold and M. J. Vold, *Colloid and Interface Chemistry*, Addison-Wesley, Reading, Mass., 1983.
99. C. Petersen and T. K. Kwei, *J. Phys. Chem. 65*: 1330 (1961).
100. G. Steinberg, *J. Phys. Chem. 71*: 293 (1967).
101. V. M. Patel, C. K. Patel, and R. D. Patel, *Angew. Makromol. Chem. 13*: 195 (1970).
102. D. Cole and G. J. Howard, *J. Polym. Sci. Part A-2 10*: 993 (1972).
103. W. A. Kindler and J. W. Swanson, *J. Polym. Sci. Part A-2 9*: 853 (1971).
104. I. M. Kolthoff and A. Kahn, *J. Phys. Chem. 54*: 251 (1950).
105. T. M. Polonski, I. I. Maleyev, M. N. Soltys, and M. D. Opainich, *Polym. Sci. USSR 8*: 2099 (1966).
106. R. R. Stromberg, D. J. Tutas, and E. Passaglia, *J. Phys. Chem. 69*: 3955 (1965).

107. A. L. Wigsten and R. A. Stratton, in *Polymer Adsorption and Dispersion Stability*, E. D. Goddard and B. Vincent, eds., ACS Symp. Ser. 240, American Chemical Society, Washington, D.C., 1984, p. 429.
108. B. Aizenbud, V. Volterra, and Z. Priel, *J. Colloid Interface Sci. 103*: 133 (1985).
109. G. Penner, Z. Priel, and A. Silberberg, *J. Colloid Interface Sci. 80*: 437 (1981).
110. J. C. Loulergue, Y. Levy, and C. Allain, *Macromolecules 18*: 306 (1985).
111. T. Sato, *J. Appl. Polym. Sci. 15*: 1053 (1971).
112. G. J. Howard and C. C. Ma, *J. Coat. Technol. 51*: 47 (1979).
113. K. Sumiya, T. Taii, K. Nakamae, and T. Matsumoto, *Kobunshi Ronbunshu 38*(3): 139 (1981).
114. K. Nakamae, K. Sumiya, and T. Matsumoto, *Prog. Org. Coat. 12*: 143 (1984).
115. K. Nakamae, K. Sumiya, T. Taii, and T. Matsumoto, *J. Polym. Sci. Polym. Symp. 71*: 109 (1984).
116. M. A. Cohen-Stuart, J. M. H. M. Scheutjens, and G. J. Fleer, *J. Polym. Sci. Polym. Phys. Ed. 18*: 559 (1980).
117. R. E. Felter, E. S. Moyer, and L. N. Ray, *J. Polym. Sci. Part B 7*: 529 (1969).
118. R. E. Felter, *J. Polym. Sci. Part C 34*: 227 (1971).
119. R. Ullman, in *Encyclopedia of Polymer Science and Technology*, Vol. 1, H. F. Mark, N. G. Gaylord, and N. M. Bikales, eds., Wiley (Interscience), New York, 1964, p. 551.
120. C. Thies, *Macromolecules 1*: 335 (1968).
121. R. Perkel and R. Ullman, *J. Polym. Sci. 54*: 127 (1961).
122. B. V. Ashmead and M. J. Owen, *J. Polym. Sci. Part A-2 9*: 331 (1971).
123. K. Hara and T. Imoto, *Kolloid Z. 237*: 297 (1970).
124. A. Takahashi, M. Kawaguchi, and T. Kato, *Kenkyu Hokoku—Asahi Garasu Kogyu Gijutsu Shoreikai 40*: 149 (1982).
125. K. Furusawa, H. Nakanishi, and A. Kotera, *Proc. Int. Conf. Colloid Surf. Sci., Budapest* (1975), p. 227.
126. G. J. Howard and P. McConnell, *J. Phys. Chem. 71*: 2981 (1967).
127. H. Burns and D. K. Carpenter, *Macromolecules 1*: 384 (1968).
128. V. T. Crowl and M. A. Malati, *Discuss. Faraday Soc. 42*: 301 (1966).
129. R. Worwag and K. Hamann, *Ber. Bunsenges. Phys. Chem. 71*: 291 (1967).
130. R. E. Felter, *J. Polym. Sci. Part B 12*: 583 (1974).
131. J. Chudoba, B. Hrncir, and E. J. Remmelzwaal, *Acta Hydrochim. Hydrobiol. 6*: 153 (1978).
132. G. J. Howard, C. C. Ma, and C. W. Yip, *Polym. Commun. 24*: 182 (1983).

133. N. N. Filatova, D. Yu. Rossina, V. V. Evreinov, and S. G. Entelis, *Vysokomol. Soedin. Ser. A 25*: 1221 (1983).
134. C. Vander Linden and R. van Leemput, *J. Colloid Interface Sci. 67*: 48 (1978).
135. M. A. Cohen-Stuart, G. J. Fleer, and B. H. Bijsterbosch, *J. Colloid Interface Sci. 90*: 310 (1982).
136. M. Kawaguchi, K. Hagakawa, and A. Takahashi, *Polym. J. 12*: 265 (1980).
137. V. K. Dunn and R. D. Vold, in *Adsorption at Interfaces*, K. L. Mittal, ed., ACS Symp. Ser. 8, American Chemical Society, Washington, D.C., 1975, p. 96.
138. R. E. Felter and L. N. Ray, *J. Colloid Interface Sci. 32*: 349 (1970).
139. G. S. Sadakne and J. L. White, *J. Appl. Polym. Sci. 17*: 453 (1973).
140. C. Vander Linden and R. van Leemput, *J. Colloid Interface Sci. 67*: 63 (1978).
141. G. J. Howard and S. J. Woods, *J. Polym. Sci. Part A-2 10*: 1023 (1972).
142. M. Kawaguchi, K. Maeda, T. Kato, and A. Takahashi, *Macromolecules 17*: 1666 (1984).
143. K. Furusawa, K. Yamashita, and K. Konno, *J. Colloid Interface Sci. 86*: 35 (1982).
144. K. Furusawa and K. Yamamoto, *Bull. Chem. Soc. Jpn. 56*: 1958 (1983).
145. K. Furusawa and K. Yamamoto, *J. Colloid Interface Sci. 96*: 268 (1983).
146. E. Pefferkorn, A. Carroy, and R. Varoqui, *J. Polym. Sci. Polym. Phys. Ed. 23*: 1997 (1985).
147. L. Duloq, *Makromol. Chem. Suppl. 12*: 265 (1985).
148. Yu. S. Lipatov, *Prog. Colloid Polym. Sci. 61*: 12 (1976).
149. B. J. Fontana, *J. Phys. Chem. 67*: 2360 (1963).
150. B. J. Fontana, *J. Phys. Chem. 70*: 1801 (1966).
151. E. Dietz, *Makromol. Chem. 177*: 2113 (1976).
152. E. Killmann, M. Korn, and M. Bergmann, in *Adsorption from Solution*, R. H. Ottewill, C. H. Rochester, and A. L. Smith, eds., Academic Press, London, 1983, p. 259.
153. C. H. Rochester, *Adv. Colloid Interface Sci. 12*: 43 (1980).
154. C. Thies, P. Peyser, and R. Ullman, *Proc. 4th Int. Congr. Surface Active Substances*, Vol. II, Gordon & Breach, New York, 1967, p. 1041.
155. M. Kawaguchi, A. Inoue, and A. Takahashi, *Polym. J. 15*: 537 (1983).
156. K. K. Kalnin'sh, A. N. Krasovskii, B. G. Belen'kii, and G. A. Andreyeva, *Polym. Sci. USSR Ser. A 18*: 2636 (1976).
157. E. Killmann, J. Eisenlauer, and M. Korn, *J. Polym. Sci. Polym. Symp. 61*: 413 (1977).

158. M. Kawaguchi, T. Sano, and A. Takahashi, *Polym. J. 13*: 1019 (1981).
159. G. R. Joppien, *Makromol. Chem. 176*: 1129 (1975).
160. A. V. Kiselev, V. I. Lygin, I. N. Solomanova, D. O. Usmanova, and Yu. A. Eltekov, *Kolloid. Zh. 30*: 386 (1968).
161. C. Thies, *J. Polym. Sci. Part C 34*: 201 (1971).
162. M. A. Cohen-Stuart, G. J. Fleer, and B. H. Bijsterbosch, *J. Colloid Interface Sci. 90*: 321 (1982).
163. K. I. Brebner, G. R. Brown, R. S. Chahal, and L. E. St. Pierre, *Polymer 22*: 56 (1981).
164. E. Killmann and M. Bergmann, *Colloid Polym. Sci. 262*: 372 (1985).
165. R. R. Mallik, R. G. Pritchard, C. C. Horley, and J. Comyn, *Polymer 26*: 551 (1985).
166. E. Killmann and R. Eckart, *Makromol. Chem. 144*: 45 (1971).
167. E. Killmann and M. Bergmann, *Colloid Polym. Sci. 262*: 381 (1985).
168. K. K. Fox, I. D. Robb, and R. Smith, *J. Chem. Soc. Faraday Trans. 1 70*: 1186 (1974).
169. A. T. Clarke, I. D. Robb, and R. Smith, *J. Chem. Soc. Faraday Trans. 1 72*: 1489 (1976).
170. A. T. Bullock, G. G. Cameron, I. More, and I. D. Robb, *Eur. Polym. J. 20*: 951 (1984).
171. T. M. Liang, P. N. Dickenson, and W. G. Miller, in *Polymer Characterization by e.s.r. and n.m.r.*, A. E. Woodward and F. A. Bovey, eds., ACS Symp. Ser. 142, American Chemical Society, Washington, D.C., 1980, p. 1.
172. H. Sakai, T. Fujimori, and Y. Imamura, *Bull. Chem. Soc. Jpn. 53*: 1749, 3457 (1980).
173. H. Hommel, A. P. Legrand, H. Balard, and E. Paperir, *Polymer 24*: 959 (1983).
174. Yu. S. Lipatov, T. S. Khramova, T. T. Todosiichuk, and L. M. Sergeeva, *Colloid J. USSR 39*: 148 (1977).
175. K. G. Barnett, T. Cosgrove, B. Vincent, D. S. Sissons, and M. A. Cohen-Stuart, *Macromolecules 14*: 1018 (1981).
176. M. Kawaguchi and A. Takahashi, *J. Polym. Sci. Polym. Phys. Ed. 18*: 2069 (1980).
177. M. Kawaguchi and A. Takahashi, *Macromolecules 16*: 1465 (1983).
178. M. Lal and R. F. T. Stepto, *J. Polym. Sci. Polym. Symp. 61*: 401 (1977).
179. I. S. Jones and P. Richmond, *J. Chem. Soc. Faraday Trans. 2 73*: 1062 (1977).
180. P. Gramain and P. Myard, *Macromolecules 14*: 180 (1981).
181. J. J. Lee and G. G. Fuller, *Macromolecules 17*: 374 (1984).
182. J. J. Lee and G. G. Fuller, *J. Colloid Interface Sci. 103*: 569 (1985).

183. M. A. Cohen-Stuart, F. H. W. H. Waajen, T. Cosgrove, B. Vincent, and T. L. Crowley, *Macromolecules 17*: 1825 (1984).
184. M. J. Barron and G. J. Howard, *J. Polym. Sci. Polym. Chem. Ed. 12*: 1269 (1974).
185. G. J. Howard and M. J. McGrath, *J. Polym. Sci. Polym. Chem. Ed. 15*: 1705 (1977).
186. P. C. Scholten, *Faraday Discuss. Chem. Soc. 65*: 242 (1978).
187. H. H. G. Jellinek and H. L. Northey, *J. Polym. Sci. 14*: 583 (1951).
188. J. Koral, R. Ullman, and F. R. Eirich, *J. Phys. Chem. 62*: 541 (1958).
189. K. Mizuhara, K. Hara, and T. Imoto, *Kolloid Z. Z. Polym. 229*: 17 (1969).
190. M. J. Schick and E. N. Harvey, *Adv. Chem. Ser. 87*: 63 (1968).
191. K. Hara and T. Imoto, *Kolloid Z. Z. Polym. 237*: 438 (1970).
192. S. Ellerstein and R. Ullman, *J. Polym. Sci. 55*: 123 (1961).
193. G. J. Howard and P. McConnell, *J. Phys. Chem. 71*: 2974 (1967).
194. E. Hamori, W. C. Forsman, and R. E. Hughes, *Macromolecules 4*: 193 (1971).
195. C. Thies, *J. Colloid Interface Sci. 27*: 734 (1968).
196. F. S. Chan, P. S. Minhas, and A. A. Robertson, *J. Colloid Interface Sci. 33*: 568 (1970).
197. V. M. Patel, K. C. Patel, and R. D. Patel, *Angew. Makromol. Chem. 62*: 177 (1977).
198. P. P. Nefedov and T. P. Zhmakina, *Polym. Sci. USSR 23*: 304 (1981).
199. E. Dietz and K. Hamann, *Angew. Makromol. Chem. 51*: 53 (1976).
200. R. Laible and K. Hamann, *Adv. Colloid Interface Sci. 13*: 65 (1980).
201. M. A. Cohen-Stuart, G. J. Fleer, and J. M. H. M. Scheutjens, *J. Colloid Interface Sci. 97*: 515 (1984).
202. M. A. Cohen-Stuart, G. J. Fleer, and J. M. H. M. Scheutjens, *J. Colloid Interface Sci. 97*: 526 (1984).
203. M. A. Cohen-Stuart, J. M. H. M. Scheutjens, and G. J. Fleer, in *Polymer Adsorption and Dispersion Stability*, E. D. Goddard and B. Vincent, eds., ACS Symp. Ser. 240, American Chemical Society, Washington, D.C., 1984, p. 53.
204. M. J. Schick and E. N. Harvey, *J. Polym. Sci. Part B 7*: 495 (1969).
205. R. A. Botham and C. Thies, *J. Polym. Sci. Part C 30*: 369 (1970).
206. R. A. Botham and C. Thies, *J. Colloid Interface Sci. 45*: 512 (1973).

207. Yu. S. Lipatov, G. M. Semenovich, L. M. Sergeeva, and L. V. Dubrovina, *J. Colloid Interface Sci. 86*: 432 (1982).
208. Yu. S. Lipatov, G. M. Semenovich, L. M. Sergeeva, and L. V. Dubrovina, *J. Colloid Interface Sci. 86*: 437 (1982).
209. R. A. Botham and C. Thies, *J. Colloid Interface Sci. 31*: 1 (1969).
210. T. Miyamoto, S. Tomoshige, and H. Inagaki, *Polym. J. 6*: 564 (1974).
211. I. M. Kolthoff, R. G. Gutmacher, and A. Kahn, *J. Phys. Chem. 55*: 1240 (1951).
212. I. M. Kolthoff and R. G. Gutmacher, *J. Phys. Chem. 56*: 740 (1952).
213. J. S. Binsford and A. M. Gessler, *J. Phys. Chem. 63*: 1376 (1959).
214. V. Yu. Erman, A. V. Uvarov, S. N. Tolstaya, and N. A. Alexandrova, *Makromol. Granitse Razdela Faz.* 100 (1971).
215. G. J. Howard and P. McConnell, *J. Phys. Chem. 71*: 2991 (1967).
216. I. F. Guthrie and G. J. Howard, in *Polymer Adsorption and Dispersion Stability*, E. D. Goddard and B. Vincent, eds., ACS Symp. Ser. 240, American Chemical Society, Washington, D.C., 1984, p. 297.
217. A. Diaz-Barrios and E. Paredes, *J. Appl. Polym. Sci. 27*: 4387 (1982).
218. M. Kawaguchi, M. Aoki, and A. Takahashi, *Macromolecules 16*: 635 (1983).
219. A. Diaz-Barrios and A. Rengel, *J. Polym. Sci. Polym. Chem. Ed. 22*: 519 (1984).
220. J. V. Dawkins, M. J. Guest, and G. Taylor, in *The Effect of Polymers on Dispersion Properties*, Th. F. Tadros, ed., Academic Press, London, 1982, p. 39.
221. P. Clarke and G. J. Howard, *J. Polym. Sci. Polym. Chem. Ed. 11*: 2305 (1973).
222. I. Prigogine, A. Bellemans, and V. Mathot, *The Molecular Theory of Solutions*, North-Holland, Amsterdam, 1957.
223. P. J. Flory, R. A. Orwell, and A. Vrij, *J. Am. Chem. Soc. 86*, 3507, 3515 (1964).
224. J. Schrober, *Prog. Org. Coat. 12*: 339 (1984).
225. F. M. Fowkes and M. A. Mostafa, *Ind. Eng. Chem. Prod. Res. Dev. 17*: 3 (1978).
226. P. Sorensen, *J. Paint Technol. 47*:(602): 31 (1975).
227. H. P. Schreiber, *J. Paint Technol. 46*:(598): 35 (1974).
228. M. Cremer, , *Proc. XVII FATIPEC Congress, Lugano* (1984), p. 259.
229. R. S. Drago, G. C. Vogel, and T. E. Needham, *J. Am. Chem. Soc. 93*: 6014 (1971).

230. F. M. Fowkes, in *Microscopic Aspects of Adhesion and Lubrication*, J. M. George, ed., Elsevier, Amsterdam, 1982, p. 119.
231. F. M. Fowkes, D. O. Tischler, J. A. Wolfe, L. A. Lannigan, C. M. Adamu-John, and M. J. Halliwell, *J. Polym. Sci. Polym. Chem. Ed. 22*: 547 (1984).

8

Lubrication

B. BRISCOE Department of Chemical Engineering and Chemical Technology, Imperial College, London, England

D. TABOR Department of Physics, Cavendish Laboratory, University of Cambridge, Cambridge, England

I. INTRODUCTION

The term lubrication covers a very wide field of concepts and applications. It describes processes which reduce the friction and wear generated at the interfaces between contacting bodies in relative motion. The principle is simple but the practice is extremely complex.

II. SIMPLE MODELS OF LUBRICATION

A lubricant is a mechanically weak material which is interposed at the interface between two stronger bodies. The purpose of this layer is to prevent extensive solid-solid contact and to form a weak interface layer in which all the relative motion is accommodated. Supressing direct solid-solid contact reduces the damage to the solid surfaces and hence the rate at which they are worn away. The weak interfacial layer also reduces the frictional work. Thus the frictional energy dissipation is contained within a thin interfacial layer which has the capacity to accommodate large rates of strain without suffering significant mechanical or molecular breakdown. In practice, virtually all the frictional work appears as heat [1]. The contributions from other forms of work such as light and sound emission and entropic and structural changes are quite small.

The means by which these requirements are achieved will be described briefly, but the first point to be recognized is that the surfaces of most solid bodies are contaminated with weak layers; if this were not the case, few of the relative movements which we see between solid bodies would occur. For example, if two iron bars are carefully cleaned and maintained in a high vacuum (better than 10^{-7} torr) so that contaminating surface films are completely removed, it is almost impossible to slide them over one another [2]. The ratio of the frictional force to the normal load, that is, the coefficient of friction, may easily exceed 10. Furthermore, examination shows that there is actual welding at the regions of contact even if the experiment is carried out at room temperature. By contrast, if the experiment is performed in air, sliding may be carried out with relative ease, the coefficient of friction being about one-half. However, some minute welds will have formed at many of the contacting asperities but the natural contamination (mainly oxide films) will prevent appreciable growth in the number and size of the adhesive welds [3]. The practice of lubrication seeks to improve upon the action of the natural surface films. Thus a thin grease film may reduce the coefficient of friction μ to less than 0.1 and the surface damage may be largely eliminated, and under appropriate conditions it may be possible to separate the surfaces completely with a fluid lubricant so that μ may be less than 0.001 and the surfaces may make no contact at all with one another. While there are very many applications of lubrication, including bearings and material forming operations, all share a common general grouping of lubrication mechanisms. These groups have become known as regimes of lubrication. As we shall see, some of the regimes are ill defined; furthermore, the distinctions between them are not precise.

III. REGIMES OF LUBRICATION

There are three basic forms of lubrication. The recent history of the subject has been chronicled by Dowson [4]. The type of lubrication which maintains a continuous fluid film between rigid and relatively widely spaced bodies is termed hydrodynamic (liquid) or aerodynamic (gas) lubrication.

A. Hydrodynamic Lubrication

Hydrodynamic lubrication was first studied analytically by Osborne Reynolds almost 100 years ago [5]. There have been many improvements in the analysis since then but the general principles have remained unchanged [6]. The geometry of the surfaces must be such that the contact region forms a convergent wedge into which the lubricating fluid can be drawn. As a result of the finite viscosity of the fluid a hydrodynamic pressure is developed in the film which is sufficient to keep the surfaces apart at some equilibrium thickness. The shear stresses which have to be applied to achieve this are primarily involved in overcoming the viscous resistance of the film, and they determine the magnitude of the friction of the system. With a liquid medium the fluid is rather incompressible and only shallow angles of convergence (less than 1°) are sufficient to provide effective hydrodynamic lubrication. With a gas very much smaller angles of convergence are required. The important parameters are the entry conditions for the fluid, the geometry of the contact, the relative speed V of the bodies, the imposed normal load W, and the viscosity η of the fluid. Hydrodynamic lubrication may be considered to be fully effective if it produces a separation of the two bodies in excess of typical asperity dimensions. Under these conditions the coefficient of friction is very small (0.001) and in principle there is no wear. Figure 1a illustrates the results for a journal bearing. The geometry of the clearance is exaggerated to show the hydrodynamic wedge.

A typical experimental pressure distribution is sketched [7]. It is interesting to note that this pressure distribution was first measured by Beauchamp Towers [8] using a series of manometers; it was Towers' work which stimulated Reynolds to study this problem analytically. Figure 1b shows the minimum film thickness as a function of the dimensionless parameter $\eta N/P$, where N is the angular velocity or rotational speed of the journal and P is the nominal pressure, i.e., the normal load divided by the apparent bearing area of the journal. The parameter $\eta N/P$ is an important feature in the presentation of hydrodynamic lubrication data [4]; in the hydrodynamic region the friction is proportional to $\eta N/P$ and the minimum film thickness is a unique function of this parameter. The hydrodynamic regime operates at larger values of $\eta N/P$, where the film thickness is greater than the surface roughness. Figure 1b, which is a plot of the friction coefficient μ against $\eta N/P$, is sometimes termed the Stribeck–Hersey curve. The other regimes of lubrication are indicated in Fig. 1b. They come into existence at smaller values of surface separation.

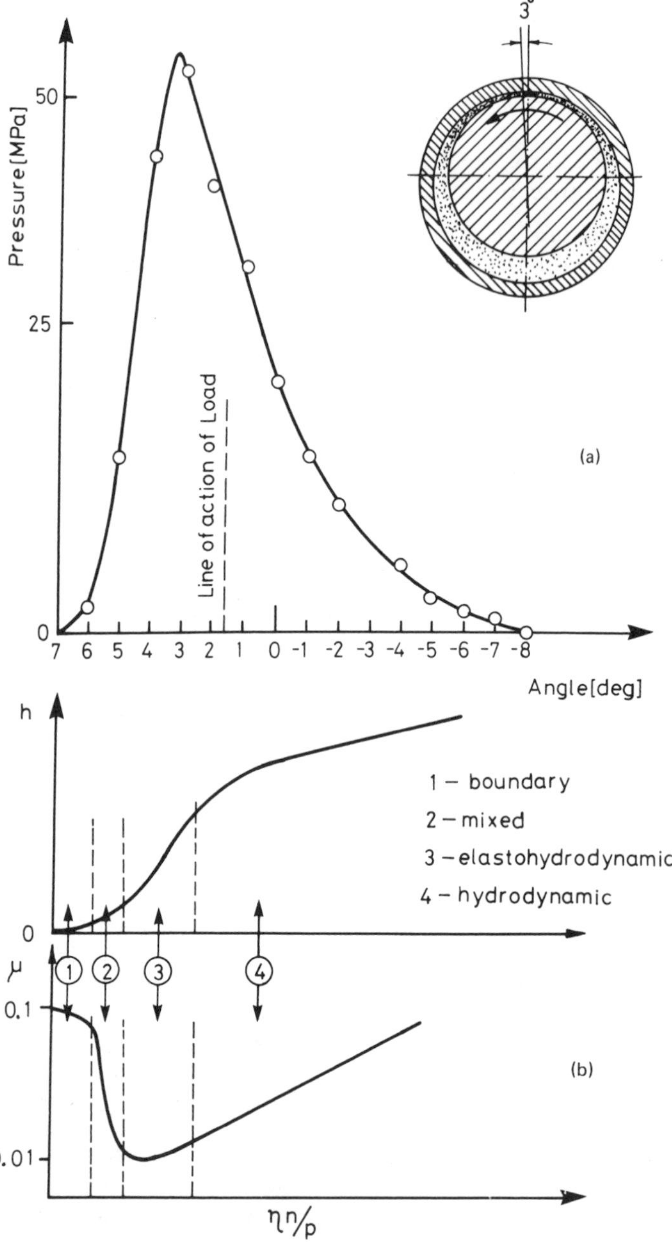

FIG. 1 (a) Hydrodynamic lubricating action produced by a journal bearing. The clearance shown is exaggerated to illustrate the converging wedge of lubricating fluid. The load is acting downward on the bearing housing. The maximum pressure generated is about

As mentioned above, the only material parameter in hydrodynamic lubrication is the viscosity η of the fluid, and we now consider three important factors that can affect it. First, the rates of shear are large and this may produce appreciable adiabatic heating of the fluid; as a result the viscosity will fall. Similarly, if the fluid is non-Newtonian it may shear thin. Third, the viscosity of most fluids increases under pressure, though in hydrodynamic lubrication the pressures are generally too small to have an appreciable effect. This may be contrasted with elastohydrodynamic lubrication (see below), where the effect of pressure on η plays a crucial part. Broadly speaking, the effects of adiabatic heating, shear thinning, and pressure can be incorporated satisfactorily into a theoretical analysis of hydrodynamic lubrication [6].

A more contentious issue concerns the conditions at the boundary between the solid surface and the fluid. In classical hydrodynamic theory, the surface is assumed to be molecular smooth and there is assumed to be a zero slip between the surface and the adjacent lubricant molecules. If there is slip this will reduce the shear gradient in the lubricant film and this in turn will reduce the equilibrium film thickness [9]. In spite of many investigations on this point, wall slip has not been established as a general phenomenon in hydrodynamic lubrication even where there is poor affinity between fluid and solid. It may exist in other sliding systems, and we shall return to this point later. There are, however, cases where indifferent fluid-solid interactions may adversely influence hydrocynamic behavior but these exist in the lubricant entry regions. Poor wetting may reduce the amount of fluid entering the converging wedge and produce a condition of "starved" lubrication.

Having made these general comments, we may now deal briefly with the microscopic origins of viscosity. Viscous work is performed at a microscopic level by the processes of moving molecular entities with respect to each other. For simple low molecular weight fluids it may be the whole molecule. With higher molecular weight species such as those which are in commercial oils, parts or segments of the molecular chain may be involved. These relative motions will do work against intermolecular forces, which in most cases will be van der Waals forces. In order to calculate the viscous energy dissipated during shear we must prescribe the size of the molecular units involved, their relative pathways, and the increment of distance which

---

55 MPa at a position of 3° from the vertical. Significant pressures are only generated in small arc of approximately 15°. (b) Variation of minimum film thickness h and coefficient of friction μ as a function of the parameter $\eta N/P$. The four regimes of lubrication are indicated approximately.

is achieved in each relaxation process. We must also show how the energy close to the shear planes is converted into heat. The first part of the problem is usefully described by the Eyring theory, which we shall consider briefly after we have reviewed the two other regimes of lubrication.

B. Elastohydrodynamic Lubrication

A new factor in hydrodynamic lubrication emerged about 30 years ago [4,10a,10b]. It was recognized then that with real solids, as distinct from the ideal rigid solids assumed in classical hydrodynamic lubrication, appreciable elastic deformations of the surfaces or of the surface asperities could occur in the contact region. With materials of low modulus the pressures may be relatively small but the elastic deformations produced may be sufficient to change the geometry of the convergent wedge in a significant way. The rheological equation must be made consistent with the equations of elastic deformation. This is the field of elastohydrodynamic lubrication. The effect is well illustrated with a soft rubber sphere sliding over a glass plate in the presence of a lubricating fluid [11]. Figure 2 is an example of the characteristic shape of the interfacial region. The "nip" at the rear of the contact is a common feature of elastohydrodynamic lubrication. In this region the possibility of solid-solid contact is greatest; in addition, the surface and fluid temperatures are at their maximum [12]. It is the engineers' challenge to compute the value of this minimum film thickness and to ensure that it exceeds the mean value of the composite surface asperity heights, σ^*. Such calculations are now commonplace; in practice, the minimum calculated film thickness is not very sensitive to surface roughness provided the film is reasonably thick (see later). The analysis is naturally more complex than that for hydrodynamic lubrication as the compliance of the bodies must be included.

With rubberlike materials the pressures are too small to have an appreciable effect on the viscosity. The position, however, is very different for metals, where local contact pressures up to several gigapascals may be reached (see below). Further, the equilibrium film thickness is usually very small (h of order 1000 Å) so that extremely high shear rates, up to 10^8 sec^{-1}, may easily be reached. These in turn may lead to appreciable adiabatic heating. In a later section we shall consider interfacial temperatures in the context of lubricant failure. Finally, we must also note that these conditions are produced in a transient manner where the contact times for the fluid may be of very short duration. The contact time is given by the contact length divided by the surface velocity, and typically the value may be a few milliseconds. It is evident that the rheological conditions are very complex.

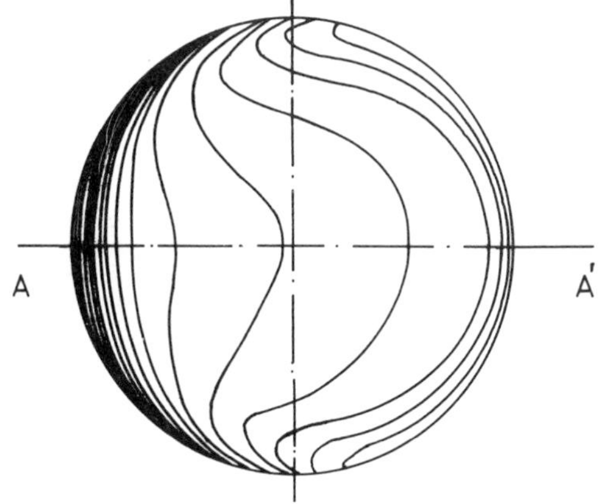

(a)

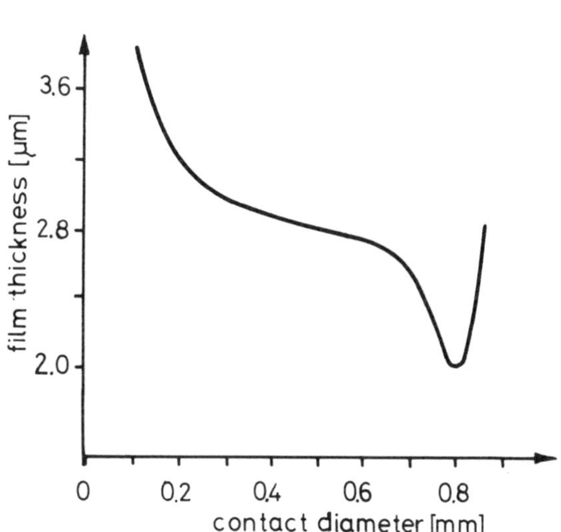

(b)

FIG. 2 Example of an elastohydrodynamic contact formed between a spherical rubber surface and a smooth glass plate lubricated with a silicone fluid. (a) Form of the optical interference fringes generated with monochromatic light; sliding direction A' to A. (b) Film thickness profile along AA' showing the classical "nip" at the rear of the contact. Further inspection of (a) shows that the minimum film thickness region is in the shape of a horseshoe.

The first variable to be addressed by the elastohydrodynamicist was pressure. The viscosity η of most fluids increases rapidly with pressure and an equation due to Barus is often accurate [13]. This equation has the form

$$\eta = \eta_0 \exp(\alpha P) \tag{1}$$

where for a typical hydrocarbon fluid α is of the order of 10^{-8} Pa^{-1}. Thus a hydrostatic stress of 1.5 MPa may produce a millionfold increase in viscosity. Consequently, the higher the normal load, the more difficult it becomes to extrude "fluid" from the contact. This is an important feature of elastohydrodynamic lubrication, and it poses an interesting question as to how the fluid film is ever ruptured.

Most of our information on fluid rheology *in* lubrication comes from what are called disk machines, which simulate the rolling contact produced by ball or roller bearing assemblies and gears [14,15]. The contact configuration is shown in Fig. 3; the contacting surfaces are very smooth and accurately aligned. If the disks have exactly the same surface velocity, the shear rate in the contact zone is virtually zero. The shear rate may then be varied by adjusting the

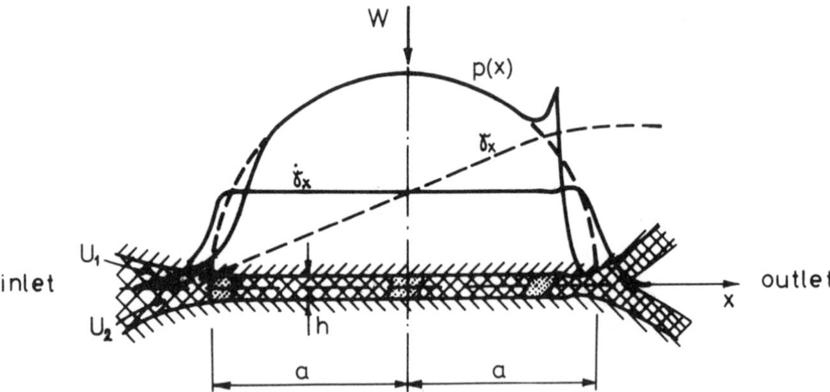

FIG. 3 Contact produced by a disk machine where two disks rotate with velocities U_1 and U_2 when the liquid film of thickness h is sheared with a strain rate $\dot{\gamma}_x = |(U_1 - U_2)/h|$. The accumulation of strain, γ_x, is shown along the contact length 2a. The pressure profile in the fluid, p(x), is shown and may be compared with the profile in Fig. 1. The most marked difference is the pronounced pressure "spike" just before the "nip" in the film thickness.

relative values of these velocities. Suitable instrumentation allows the measurement of the film thickness and the rotating torque. The shear area is calculated assuming elastic deformation of the disks. Hence a curve of shear stress τ against shear rate $\dot{\gamma}$ may be plotted. Figure 4 is a typical example [15]; the quantity μ_T is the traction coefficient, defined as shear stress/normal pressure. Rather than plotting the ordinate as $\dot{\gamma}$, it is given as the ratio of the relative surface velocity divided by the mean surface velocity; for a constant film thickness, which is a reasonable approximation, this quantity is equal to $\dot{\gamma}$.

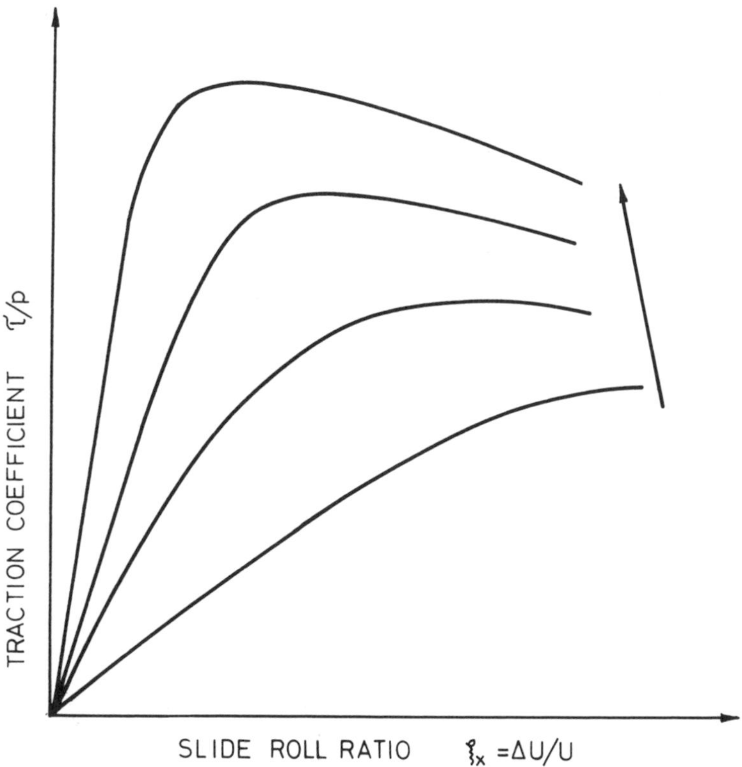

FIG. 4 Schematic set of traction coefficients against the slide roll ratio ξ_x for a typical lubricant as obtained in a disk machine. The quantity $\xi_x = \Delta U/U$, where $\Delta U = |U_1 - U_2|$ and $U = (U_1 + U_2)/2$ and is a measure of γ_x or $\dot{\gamma}_x$ (Fig. 3); τ is the friction per unit area and p is the contact pressure. The direction of the arrow indicates an increase in pressure and decreases in temperature and rolling speed U.

The main features of Fig. 4 are as follows: (1) a linear portion at low values of shear rate $\dot{\gamma}$ where the shear stress τ is proportional to $\dot{\gamma}$, (2) a marked deviation from linear behavior, (3) a trend toward a plateau where τ has a constant maximum value (it may decrease slightly at the highest values of $\dot{\gamma}$).

The first portion suggests the shear of a Newtonian liquid possessing a viscosity appropriate to the operating pressure according to Eq. (1). The departure from linear behavior could be ascribed to viscous heating since the viscosity decreases rapidly with increasing temperature [16]. Another possible explanation involves the concept of viscoelastic retardation in compression: although the viscosity increases rapidly with pressure, a finite time is needed to achieve the equilibrium value [17,18]. This is the time the molecules need to accommodate the volume contraction produced by the high pressure. A similar type of relaxation time may also be involved in shear. Although these effects are genuine, the simplest approach is to recognize that, under high pressures and high rates of shear, the liquid behaves in a non-Newtonian manner.

Johnson [19] has elegantly reviewed these various causes for the effects shown in Fig. 4. Its behavior may indeed be described by the Eyring approach to viscosity. Although this is strongly criticized by most physicists working in the field of liquids, it provides a general equation which "mimics nature" over an extraordinarily wide range of parameters, in particular, temperature, pressure, and shear rate. The basic ideas are simple. The liquid is envisaged as an assembly of molecules which generally have a high degree of mobility; they can swap places as a result of thermal fluctuations (see Fig. 5a). For small molecules the potential barrier ε that must be overcome to enable place swapping to occur is about one-third of the heat of vaporization. This may be considered as the energy required to open up a hole large enough for the molecule to move into.

For convenience we will restrict our discussion to single molecules; for long-chain molecules the moving species may correspond to a part or segment of the molecular chain. The main chain may often be considered to be oriented in the shear direction, and then side groups or relatively free portions of the main chain are the mobile units. If a shear stress τ is applied to the liquid, a certain amount of mechanical work can be done on a molecule in the course of its movement to a neighboring site. If a is the cross-sectional area of molecule and the average distance it moves is λ, the work done is $\tau a \lambda$ (Fig. 5b), where λ is the distance to the peak of the potential energy curve, i.e., about one-half of the molecular separation. The quantity $a\lambda$ has the dimensions of volume and for a spherical molecule is, in this model, approximately equal to one-half of the volume of the molecule itself. This quantity is referred to as the stress activation volume Ω. Clearly, this mechanical work favors place swapping in one direction and hinders it in the opposite direction. The movement of the

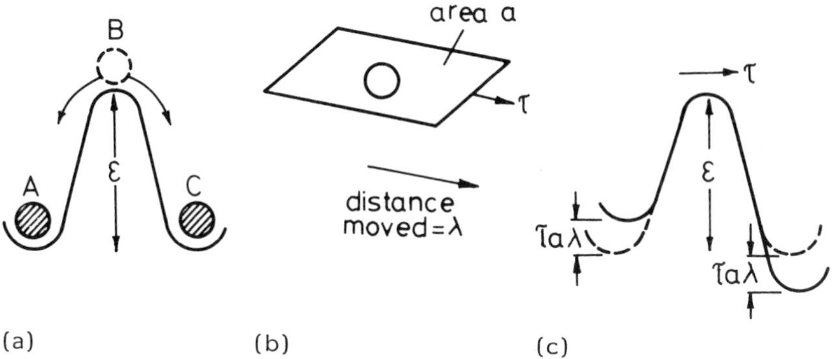

FIG. 5 Eyring approach to stress-aided thermally active processes for material flow. In (a) a species A may move position to a vacent site at C by passing through a metastability denoted by B. An activation energy ε is required, but since sites A and C have the same energy no net flow occurs. The application of a shear stress τ to the system does work $\tau a\lambda$ on A and hence the potential energy section in (a) is distorted (c). Net flow of species now occurs and a macroscopic strain is accommodated.

molecule is thus a stress-aided thermally activated process biased in the direction of the applied shear stress (Fig. 5c). If a pressure p is now applied, the space available for molecular movement is reduced and the thermal energy needed to enable place swapping to occur is increased by $p\phi$, where ϕ is referred to as the pressure activation volume. This leads to the final result that the shear rate $\dot{\gamma}$ in the liquid is of the form

$$\dot{\gamma} = A \left[\exp\left(-\frac{\varepsilon + p\phi}{kT}\right)\right] \sinh\left(\frac{\tau\Omega}{kT}\right) \tag{2}$$

For small values of τ, $\sinh(\tau\Omega/kT) \approx 1/2\tau\Omega/kT$, so that at fixed temperature and pressure $\dot{\gamma}$ is proportional to τ, i.e., the liquid shows Newtonian viscosity. If, however, the mechanical work $\tau\Omega$ is very large compared with the thermal energy kT, the sinh function approximates an exponential function and Eq. (2) becomes

$$\dot{\gamma} = B \exp\left(-\frac{\varepsilon + p\phi - \tau\Omega}{kT}\right) \tag{3}$$

On taking logarithms of both sides we see that the shear stress τ is proportional to $\ln \dot{\gamma}$; that is, it shows very little variation with shear rate. The Eyring equation then reduces to

$$\tau = \tau_0 + \alpha P \tag{4}$$

where $\alpha = \phi/\Omega$. This agrees well with the experimental results and corresponds to the plateau region of Fig. 4. Thus the shear behavior over the whole range may be described in terms of a non-Newtonian fluid by using the Eyring relation, modified perhaps to allow for certain structural relaxations.

An alternative approach is to regard the film as a solid with elastoplastic properties. The reason for this view is that pressure raises the glass transition temperature of the lubricant [20]. If the pressure is sufficiently high the lubricant at the temperature of the experiment may be below its glass transition temperature. It may then be regarded as a solid. Of course, the lubricant in the liquid state has no long-range order and in the solid state is also amorphous or glassy. Thus the transition is a second-order one. Nevertheless, there is convincing evidence for elastic behavior at low strains and plastic behavior at high strains. For example, in the disk machine the total shear strain produced in the contact zone is a linear function of the contact length, the rate of strain, and the flow of velocity. Thus Fig. 4 may be regarded as a plot of shear stress against shear strain, so that the initial linear part provides a measure of the shear modulus of the film.

A typical value of the shear modulus for a contact pressure of 10^9 is 10^8 Pa, which is comparable with equilibrium values measured by Winer and co-workers [21] but a factor of 5 to 10 times smaller. This may be because the relaxation times at these high pressures are very large compared with the contact time. There are also experimental uncertainties due to the elastic yielding of the disks themselves [22]. At high strains the shear stress is almost independent of shear rate and this is the behavior to be expected of a waxlike solid. However, the distinction between a waxlike solid and a non-Newtonian Eyring liquid in the high shear rate regime is largely a matter of semantics [19]. Indeed, the best model to describe the behavior of these films is an elastic spring in series with a dashpot filled with an Eyring fluid. As the pressure is increased the modulus of the spring increases linearly with pressure while the viscosity of the dashpot increases exponentially until it enters the nonlinear regime. At this stage the lubricant film provides a shearable solidlike protective barrier which prevents solid-solid contact of the substrates. As we noted before, it is hard to see how the lubrication of such a contact can fail.

So far we have assumed that the solids are smooth. However, at the film thicknesses generated in this regime of lubrication the peak-to-valley heights of the surface roughness may be comparable with the film thickness. This type of lubrication is termed *mixed* or *partial* elastohydrodynamic lubrication as it involves contributions from elastohydrodynamic effects and boundary lubrication (see later). This realization does not help us in our understanding of lubrication failure as each asperity contact can be considered as a micro elastohydrodynamic contact [23]. It does, however, greatly complicate the modeling of the elastohydrodynamic lubrication of real surfaces [10,24,25]. The presence of surface roughness does not appear to significantly influence the entry conditions for the fluid. The mean film thickness is also largely unchanged provided that there are relatively few asperity contacts and also no leakage channels generated along furrows which are aligned in the direction of fluid flow. The asperity interactions do have a pronounced effect upon the wear of the contact. The original correlation, which was made by Dawson, showed that the extent of surface damage may be related to a parameter, h_0/σ^* (the Dawson number), where h_0 is the film thickness calculated for a smooth contact and σ^* is a measure of the combined roughness of the two surfaces. The study and analysis of mixed lubrication has received considerable attention in recent years [25].

We will now describe an alternative means of producing weak solid layers on the surfaces of contacting bodies, the process of boundary lubrication.

C. Boundary Lubrication

There are ambiguities in the definition of the regime of boundary lubrication. Sir William Hardy [26] coined the phrase to describe what he considered to be lubrication by monomolecular layers. The materials he used were neat amphipathic molecules. During this period the close-packed monomolecular film on aqueous substrates was receiving attention following the work of Agnes Pockel and Lord Rayleigh [27]. Irvin Langmuir and Kathleen Blodgett [28] had begun their work on the deposition of these films on solid substrates to produce the Langmuir-Blodgett layers, which are currently exciting the interest of electronic engineers. Langmuir's interest in these films, in later years at least, was as models for lubrication studies. The idea grew that any close-packed molecular layer with a terminal group with a high affinity for the substrate exposing low-energy terminal groups at the other end of the molecule would provide efficient lubrication if slid against similar low-energy terminal groups on the other solid surface. These species could be dry films of the Langmuir-Blodgett type or films formed by adsorption from a suitable fluid medium. The important point was that the film should be "solid." It was believed that if a condensed (solid) film could be formed under pressure on the Langmuir trough a similar film could be formed on a solid substrate with similar mechanical properties. However, the simple picture of a well-oriented monomolecular film sliding over a similar monolayer soon proved to be unacceptable. Consider, for example, stearic acid regarded as the classical Langmuir-type boundary lubricant. When it is dispersed in neutral oils it provides an efficient interfacial barrier to prevent gross asperity welding. However, in most metallic contacts the acid produces thick soap layers rather than a bifilm, and we have a soapy "mush" [29,30]. This is perfectly effective as a lubricant. It is possible that less reactive functional groups may produce monomolecular films but in practice these films are less effective; aliphatic alcohols are in this category.

If we begin to include in our definition of boundary lubrication thick layers of organic species such as soaps formed in situ, we should also have to include a variety of organic chalcogenides and halides [31–34]. These species are designed to react chemically with the substrates under intense levels of energy dissipation; in these cases it is often found that traces of oxygen and water have important synergistic effects [35,36]. Such materials are termed extreme pressure lubricants. This is a misnomer; it is extreme temperatures which cause these materials to react to produce complex surface residues. These residues contain high fractions of halogens or chalcogenides [37] and even polymeric species [38]. Some additives are organometallic and as a result of chemical reaction may form thin metallic films. These chemically formed layers have features in common with dry films deposited directly from organic polymers [39] or transition metal dichalcogenides such as MoS_2 [40]. We also should include these materials in the generic class.

Boundary lubrication is thus provided by weak solid layers whose thickness is perhaps 100 nm or less. The structure and chemistry of these layers is often entirely governed by physical and chemical interactions between the lubricant and substrate. It is interesting to inquire whether the rheology of these layers is very different from that of the highly compressed fluids discussed in a previous section. As we shall see, the shear responses of both types of film are often remarkably similar.

The experimental investigation of boundary lubrication has two parts: rheological studies similar to those described for elastohydrodynamic lubrication and the examination of the criteria for film rupture and hence lubrication failure. For the moment we will deal only with the rheological part. The experimental methods aim to confine and shear the film in a model contact. The important rheological parameter is the quantity τ, the shear strength of the film introduced in the section dealing with elastohydrodynamic lubrication. The central problem is the estimation of the real area of contact undergoing shear; rarely can this quantity be measured directly with sufficient precision [41]. The experimental approach is to use very smooth substrates, usually in the configuration of a spherical surface against a plane surface, and to calculate the contact area by

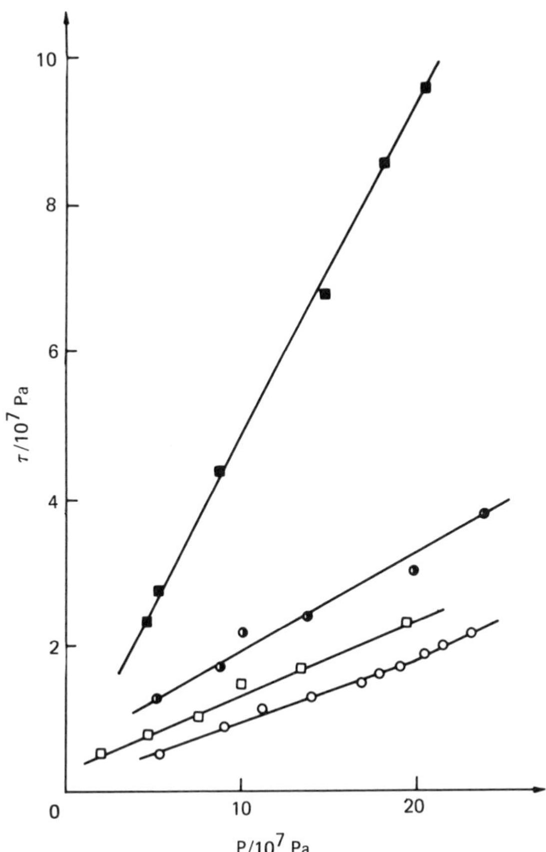

FIG. 6 Interfacial shear stress τ as a function of contact pressure P for a range of organic polymeric films at 20°C. (□) High-density polythene; (◑) low-density polythene; (■) polystyrene; (○) PTFE. Very similar data are obtained for many organic materials [41].

using elastic contact mechanics. The same method is used in elastohydrodynamic studies. However, instead of polished metals, inorganic glasses and cleaved mica are preferred. The surfaces are much smoother and the materials have a higher elastic limit [42,43]. These surfaces are covered with a thin film (2.5–100 nm in thickness) of the boundary lubricant: it is sheared by sliding one body over the other under a constant load. Much of this type of work is done in the absence of a superincubent fluid, as even relatively small sliding velocities produce some hydrodynamic or elastohydrodynamic lift and hence a load attenuation. Typical data for a variety of solid organic films are shown in Fig. 6 plotted as τ against the mean calculated contact pressure. The tractions produced by highly compressed hydrocarbon-based liquids are very similar to those of thin solid films of boundary lubricants [44,45].

IV. RHEOLOGY OF INTERFACIAL FILMS

A good deal is known about the interfacial shear characteristics of very thin solid films when they are sheared between solid substrates [46,47]. Equation (2) is general for solids. The other general functional relationships, as determined empirically for a range of solids, are

$$\tau \propto \exp\left(\frac{Q}{RT}\right)$$

$$\tau \propto \ln\left(\frac{V}{V_0}\right) \qquad (5)$$

$$\tau \propto \exp\left(\frac{t_c}{t_0}\right)$$

where T, V, and t_c are, respectively, the temperature, sliding velocity, and contact time; Q, V_0, and t_0 are material constants which may be functions of other variables; i.e., V_0 is a function of T. If these organic layers are heavily swollen with a solvent the relationships of Eq. (5) remain, as far as is known, applicable, but the relationship of Eq. (1) is a more accurate description of the pressure dependence [48]. It is seen that the shear properties of these solid films closely resemble the shear behavior deduced from the Eyring equation in the high-pressure regime, where the Eyring liquid behaves like a waxy solid. Furthermore, detailed application of these equations yields activation energies and activation volumes which are all physically acceptable [43,48,49]. This is a remarkable result.

We now make a number of comparisons between thin-film shear and bulk shear. This involves certain problems, since the conditions which exist within contacts cannot be reproduced during bulk deformation. Bulk deformation is usually carried out at low rates of strain, less than 1 sec^{-1}, while strain rates in the contact experiment often exceed 10^6 sec^{-1}. On the other hand, the influence of pressure (contact or true hydrostatic) can be compared, although in the bulk experiments and pressure is maintained for long periods, unlike the transient compressions in the contact experiment.

The bulk flow or rupture stress of many organic polymers follows an equation nominally identical to Eq. (2), and Table 1 makes a comparison of material parameters for a number of such polymers [50]. The pressure coefficient α is similar for the two modes of deformation. However, the intrinsic shear stresses τ_0 for bulk deformation are about 10 times greater than those found in interfacial deformation. Part of the discrepancy may be attributed to the fact that during interfacial shear considerable changes in molecular orientation are produced. Again, in bulk experiments it is found that if the molecular or morphological units are oriented in the shear plane the value of τ_0 decreases significantly [51]. Although the discrepancy in the absolute values of τ_0 is too large to be explained away completely in these terms, the overall conclusion is that, roughly speaking,

TABLE 1 Values of τ_0 (Bulk)/τ_0 (Thin Film) and α (Bulk)/α (Thin Film) for Various Polymers[a]

Polymer	τ_0 (bulk) / τ_0 (thin film)	α (bulk) / α (thin film)
PMMA	4.6–53.0 (28.8)	0.5–2.9 (1.7)
PS	10.0–31.0 (20.05)	0.4–1.7 (1.05)
PET	3.1–11.6 (7.35)	0.3–1.0 (0.65)
PC	3.5–8.4 (5.95)	0.3–1.3 (0.8)
PP	3.2–4.0 (3.6)	0.7–1.2 (0.95)
HDPE	7.0–14.0 (10.5)	0.3–0.7 (0.5)
PVAC	2.9 (2.9)	1.5

[a]Parenthetic quantities are arithmetic means. The overall average values of α (bulk)/α (thin film) and τ_0 (bulk)/τ_0 (thin film) are, respectively, 0.9 and 11.3

the work done in continuous interfacial shear in the contact experiment is similar to that observed in the bulk experiment, where shear or rupture provides a single yield value associated with failure of the specimen. We consider this now in somewhat greater detail.

The similarity between thin-film, continuous shear, and bulk shear experiments raises questions concerning the detailed mode and location of the shear process. In bulk shear, quite apart from overall deformation, we may observe shear bands (due to shear thinning), adiabatic shear bands, or localized cracks. It is indeed possible that the true failure mechanisms may not be too different from those obtaining in the thin-film experiments. However, the significant difference is that the thin-film experiment provides the prospect of

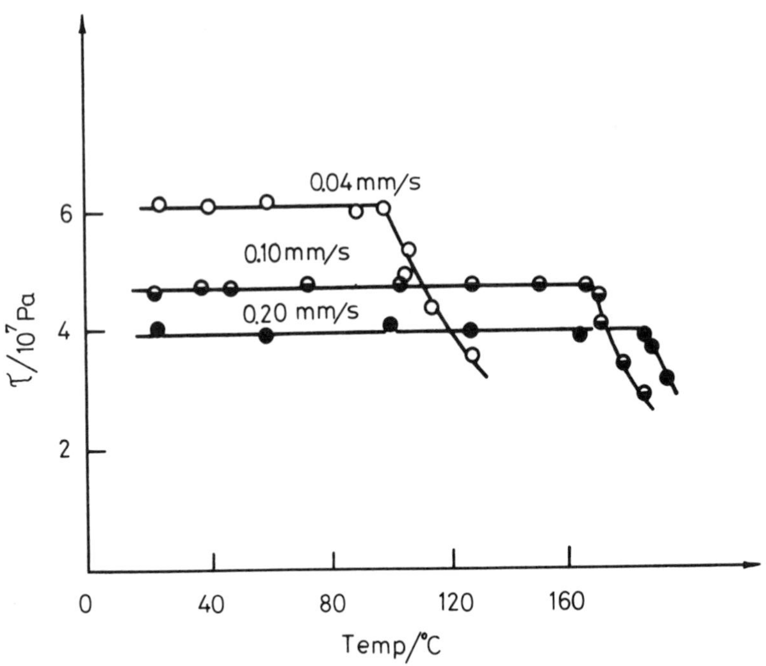

FIG. 7 Influence of temperature on the interfacial shear stress of poly(methyl methacrylate) thin films of thickness ∿100 nm. Contact pressure is 4.3×10^7 Pa; the variation with sliding velocity is shown [49]. The discontinuity is the glass transition temperature T_g, which increases with rate of strain. The fact that τ decreases with increasing velocity below T_g is evidence for viscoelastic retardation in compression.

continuous interfacial failure. Rather than producing a homogeneous shear zone, failure is localized at a substrate-film interface. This appears to occur with certain glassy polymer films such as poly-(methyl methacrylate). A number of analytical models have appeared in the literature which attempt to model energy dissipation in molecular planes; they are essentially molecular roughness models. Dejaguin [52a] produced a model of this type many years ago, and more recently Tabor [44] and Smith and Cameron [53] have elaborated this approach.

Figure 7 shows data for τ against temperature for a film of poly(methyl methacrylate) sheared in a glass contact. The film is deposited on only one surface and has a thickness of about 100 nm. Below the T_g of about 120°C there is strong evidence of interface slip and τ is not a function of temperature. Above the T_g the shear seems to be accommodated in the whole of the film and now τ decreases with increasing temperature. The value of Q [Eq. (3)] changes from near zero to about 16 kJ mol^{-1} at the T_g. This is evidence for a brittle-ductile transition, and a brittle failure may be a wall slip process of the sort described in the discussion of elasto=hydrodynamic lubrication. Figure 7 also indicates the presence of some effects due to retardation in compression [54]. Below the T_g, τ decreases with increasing velocity, as the contact time for the application of the normal stress decreases as the velocity increases. In the rubber state, above T_g, τ increases with increasing velocity.

V. SOME IMPLICATIONS OF THIN-FILM SHEAR

The shear properties of thin films provide a very convenient approach to our understanding of the friction between *unlubricated* bulk specimens. Suppose we have studied the shear of a thin film 5 to 500 nm thick of a given material confined between hard substrates. Consider now the behavior of this material as a bulk specimen sliding over a smooth rigid surface such as clean glass. The adhesion model of friction assumes that strong adhesion occurs at the regions of real contact and that the frictional work is produced by the sequential formation and shearing of these junctions during sliding (originally the rupture was envisaged as being ductile but this limitation is not necessary). The shear work is done in a very thin interfacial layer and the magnitude of this work is defined by the material and the contact conditions such as contact pressure, i.e., by an appropriate value of τ which is comparable with the bulk flow or rupture stress. Recalling Eq. (4)

$$\tau = \tau_0 + \alpha p \qquad (4)$$

and also noting that shear can occur only at areas of real contact A, the frictional force F is

$$F = \tau A = A(\tau_0 + \alpha p) \tag{6}$$

The coefficient of friction μ is F/W and hence (since A = W/p)

$$\mu = \frac{\tau_0}{p} + \alpha \tag{7}$$

This is a useful result [55]. For the experiment where the organic solid is slid over the glass plate the contact pressure p will be less than or equal to a normal flow stress (approximately the hardness), p_0. So

$$\mu = \frac{\tau_0}{p_0} + \alpha \tag{8}$$

In practice $\alpha > \tau_0/p_0$, so the friction coefficient is numerically close to the pressure coefficient α. If we return to the boundary lubrication experiment, where the dissipation zone is confined between much harder substrates, then μ is found to be almost exactly equal to α as the contact pressures are greater.

This result has potentially important practical consequences. The response described so far for organic solids is clearly applicable to the highly compressed liquid layers in elastohydrodynamic lubrication, where Eq. (4) applies at high values of strain rate. A very important application of these types of lubricated contact is in traction drives. The motion between contacting surfaces is transmitted via the film, and so the maximum force which can be transferred is given by a coefficient of traction (friction) which is nearly numerically equal to α. The pressure coefficient α is a material property, and its prediction from molecular models is clearly valuable in the design of traction fluids. Although the Eyring theory provides hints as to how the chemical structure of fluids might be modified to give improved traction, these ideas have not provided particularly fruitful in practice [19]. Often the molecular architecture required to produce large values of α is incompatible with effective protection of the metal surfaces.

A. The Three Regimes of Lubrication

We have shown that the three main regimes of lubrication can all be explained in terms of the rheological properties of the lubricant film.

In hydrodynamic lubrication the overall pressures are relatively low and the lubricant behaves as a Newtonian fluid. In elastohydrodynamic lubrication the extremely high contact pressures produce an enormous increase in the effective viscosity of the lubricant and convert it into a non-Newtonian fluid; in the extreme case it may be regarded as a waxy solid. Finally, in boundary lubrication the lubricant is initially in a solid form and responds to contact pressures in a manner resembling an elastohydrodynamic film. One remarkable feature is that the behavior in the three regimes is described surprisingly well by an Eyring-type equation involving the flow of molecules or molecular segments over one another. The intermolecular forces in all regimes are predominantly van der Waals forces. It is indeed possible to provide a simple molecular model for elastohydrodynamic and boundary lubrication in terms of molecular forces and to obtain shear strengths of the right order of magnitude. The approach is, however, descriptive rather than quantitative.

We now turn from problems of lubrication to the inverse—the problem of lubricant failure.

VI. LUBRICATION FAILURE

Lubricating films fail when the extent of direct solid-solid contact is unacceptable; the criteria are arbitrary ones. There are important modes of failure which do not involve solid-solid contact, where the stress transmitted through the fluid is sufficient to cause fatigue failure in the solids [56]. We will not discuss these. Although the criteria for failure are arbitrary, they are of two sorts [57,58]. There is a short-time failure caused by gross welding, sometimes called "scuffing," which when maintained for long periods produces extensive and irreversible damage. The other type of failure produces "steady-state" wear of a much more gentle type in which the surfaces are slowly eroded away by multiasperity contacts. There are oil additives which are chosen specifically to attenuate both processes, although it seems that some additives serve both purposes. Antiscuffing additives are often organometallics, often containing zinc and phosphorus, or organohalides and chalogenides, which are considered to form durable surface platings. The "antiwear" additives are mainly functionalized alkyl species containing oxygen, chlorine, and sulfur which are effective at lower surface and bulk temperatures.

A useful comprehensive review of surface chemistry in commercial boundary lubrication has been produced by Godfrey [59]. This review also describes the types of surface analytical tools which may be applied to surfaces worn in various lubricating environments. While high-vacuum techniques provide some indications of possible surface reactions, naturally they cannot provide direct information on the chemistry of the contact region *during* sliding.

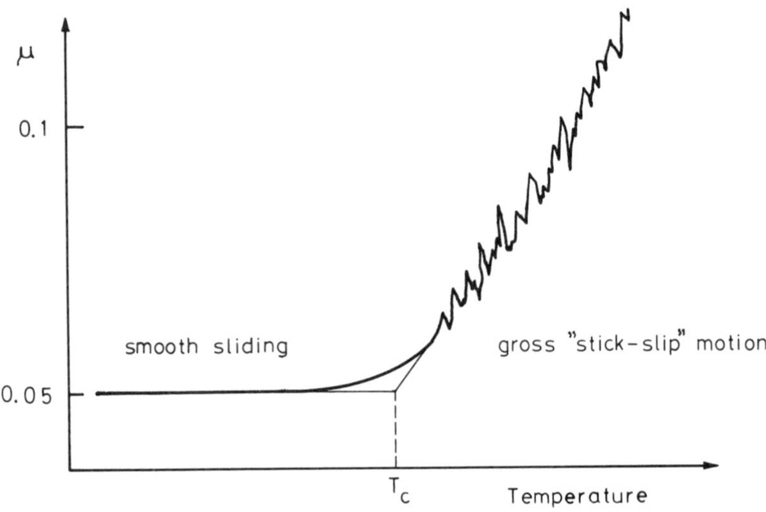

(a)

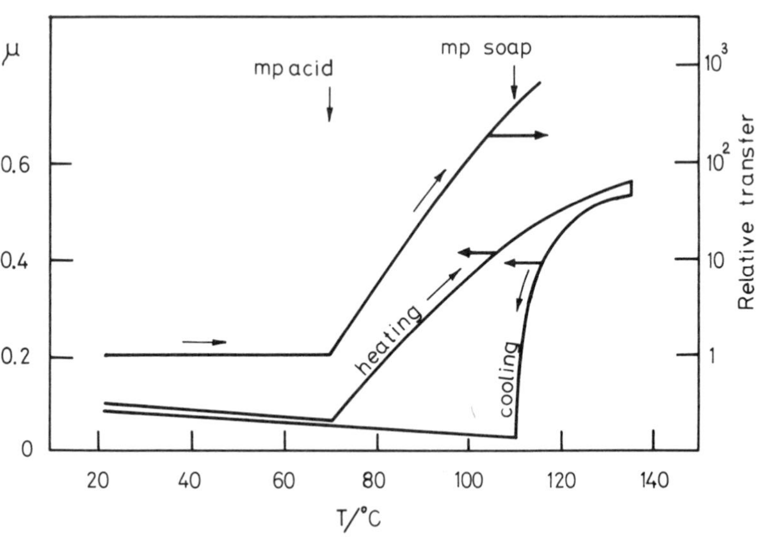

(b)

FIG. 8 (a) Schematic diagram of a coefficient of friction-temperature trace for a lubricating fluid containing a boundary lubricant. T_c is the scuffing temperature. (b) Précis of the data of Rabinowicz and Tabor for cadmium sliding on cadmium in the presence of palmitic acid, showing (upper curve) the fricitonal transfer as a function of temperature. Initially the sliding is smooth. As the temperature

Hydrodynamic lubrication fails if the film thickness is less than the height of surface asperities (low Dawson number). These will then tear or gouge out material from the other surface. However, in principle the tips of the asperities should be able to generate elastohydrodynamic films or, in the limit, to retain a boundary film. Thus the failure of a hydrodynamic film really involves the failure of an elastohydrodynamic or boundary film. There are two main ways in which this may occur. First, the film may be displaced, worn away, extruded, or even fractured (if the film material is brittle); these involve mechanical processes. The second mechanism is one in which surface temperatures produce desorption, dissolution, or evaporation of the film. We deal first with the effect of temperature [60] and the early experiments of Bowden and Tabor [60] and Frewing [61,62] on scuffing.

In one important series of experiments Frewing slid a metal hemisphere over a metal plate in neutral hydrocarbon fluids which contained varying amounts of a boundary lubricant such as stearic acid. The sliding speed was very low so that frictional heating could be ignored. The frictional force was measured as the temperature of the system was increased, and a well-defined point was reached when smooth sliding was replaced by intermittent or stick-slip motion [60,62] (Fig. 8). This was accompanied by a significant increase in the severity of surface damage. The temperature at which the motion changes is called the scuffing temperature, T_c, and it is often remarkably reproducible and reversible. Frewing [62] observed that T_c increased as the concentration of the additive in the oil, c, increased and that a plot of the logarithm of the concentration of the additive against the reciprocal of T_c was linear (Fig. 9). Frewing and indeed many others since [63-66] have chosen to interpret this transition in terms of a reversible physical adsorption process in which a critical surface coverage of adsorbate is required to maintain surface protection. Kingsbury [65] called this value the critical film defect. There are many very unsatisfactory aspects in the published analyses of this problem, apart from the important question of whether the sorption of, say, stearic acid onto steels, where we know chemical reactions occur, is a reversible process.

---

is increased the coefficient of friction (lower curve) falls until at about 70°C (the melting point of the acid) the sliding changes to stick-slip and finally at about 120°C it becomes very erratic. The transfer (measured by autoradiography) is little affected up to 70°C, after which it shows a marked increase with further increase in temperature. On cooling, the friction (bottom curve) falls again until at 110°C (the melting point of the soap) it reaches a low level, after which it slowly increases until room temperature is restored.

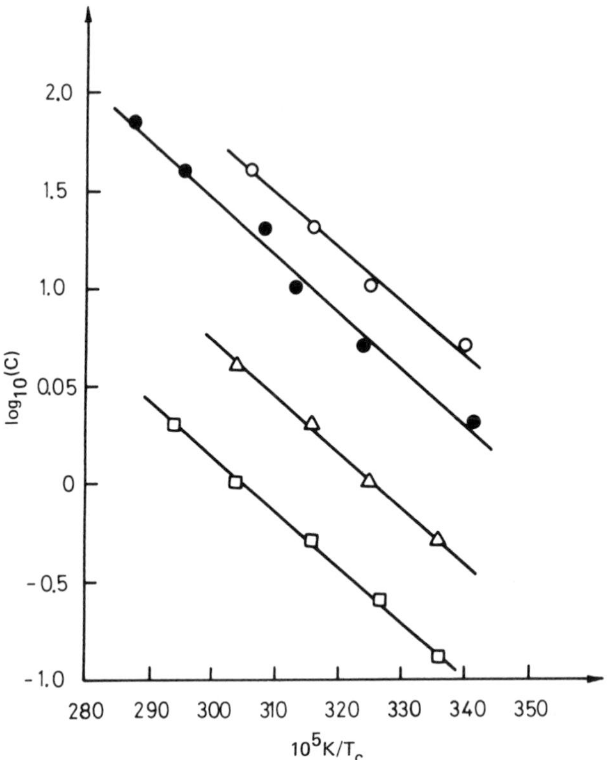

FIG. 9 Data obtained by Frewing [61,62] for the lubricating action of (○) capric acid, (△) myristic acid, (□) stearic acid, and (●) oleic acid in solution in a hydrocarbon fluid. The concentration (weight percent) of the solution is related to the scuffing temperature T_c by Eq. (9).

Published data show that up to 75% of the adsorption is reversible in some cases, for example, amines adsorbed from hexadecane onto iron powder. Most treatments simply derive the Langmuir adsorption isotherm and specify failure at a fixed surface concentration; a surface coverage of one-half is useful here as it usually enables a term in the final equation to be set to zero [64]. It is simpler to equate chemical potentials in a surface phase with that of the solution. For the solution

$$\mu_{i,b} = \mu_{i,b}^{\circ}(T) + RT \ln(\gamma_{i,b} x_{i,b})$$

where $\gamma_{i,b}$ is the activity coefficient of i where $\gamma_i \to 1$, $x_{i,b} \to 0$; $x_{i,b}$ is the mole fraction of i in solution; and $\mu_{i,b}°(T)$ is a hypothetical standard state. The surface species may be written, assuming a monolayer model, as

$$\mu_{i,s} = \mu_{i,s}°(T) + RT \ln(\gamma_{i,s} x_{i,s})$$

where $x_{i,s}$ is the surface mole fraction; $\gamma_{i,s}$ is a surface activity coefficient with $\gamma_{i,s} \to 1$, $x_{i,s} \to 0$; and $\mu_{i,s}°(T)$ is a hypothetical surface standard state. Equating the chemical potential, we obtain the usual expression

$$\Delta \mu_i°(T)_{s,b} = -RT \ln \frac{\gamma_{i,s} x_{i,s}}{\gamma_{i,b} x_{i,b}}$$

where $\Delta \mu_i°(T)_{s,b}$ is a standard free energy change for pure i in hypothetical states which can be written as

$$(\Delta h_i°)_{s,b} - T(\Delta S_i°)_{s,b}$$

where $(\Delta h_i°)_{s,b}$ and $(\Delta S_i°)_{s,b}$ are the standard enthalpy and entropy of the process. So we obtain

$$\frac{(\Delta h_i°)_{s,b}}{T} = R \ln(\gamma_{i,b} x_{i,b}) + R \ln(\gamma_{i,s} x_{i,s}) + (\Delta S_i°)_{s,b}$$

and if we neglect the temperature dependence of $\Delta S_i°$ and $\Delta h_i°$ and fix a critical value for surface coverage $x_{i,s}$, then we have a relationship between T_c and the critical bulk concentration $(x_{i,b})_c$

$$\frac{(\Delta h_i°)_{s,b}}{T} = R \ln(x_{i,b})_c + \text{constant} \tag{9}$$

The gradients in Fig. 9 are standard enthalpies of adsorption divided by R; these enthalpies are in terms of hypothetical standard bulk and standard surface states. Cameron and Spikes [64] carried out similar studies to evaluate $\Delta h_i°$. In addition, they prepared metal substrates in a finely comminuted form and carried out separate adsorption isothermal experiments on them. They found that the heats of adsorption calculated from adsorption isotherms were similar to those obtained from scuffing data; this is surprising in view of the fact that the surface chemistry of the comminuted specimen must have been substantially different from that of the plane surfaces. These authors also sought to unravel the constant in Eq. 9.

The thermodynamic equations produced are applicable if (1) the process involves reversible physical adsorption on a homogeneous surface, (2) the relevant critical condition is one of adsorbate coverage, and (3) the system is in chemical equilibrium. All three are questionable, but first we must review other experimental aspects of this subject.

Two pieces of work have a bearing on this problem. First, Tabor and co-workers [60,67] and also Frewing [62] have shown that with single-constituent lubricants scuffing corresponds to a physical transformation such as melting. For stearic acid on inert substrates smooth sliding occurs below the melting temperature and stick-slip motion above the melting temperature. On reactive metals the transition occurs at a temperature close to the melting or softening point of the appropriate soap [51] (Fig. 8). In the presence of a superincubent layer of paraffin oil the breakdown occurs at a lower temperature, presumably due to dissolution of the film. Failure appears to be due to a change in surface physical properties rather than surface composition, although the two may be related.

A more important series of observations was carried out later by Hirst and Hollander [68]. They investigated the influence of surface topography, load, and the stiffness of the measuring system. The latter is particularly interesting and was first reported by Bristow [69a] and from a different aspect by Tolstoi [69b]; the value of T_c may be changed by up to 100°C between 30 and 130°C by changes in the elasticity of the transducer used to monitor the frictional force. The value of scuffing temperature is therefore a function of the machine as well as the interface. This fact may not affect Δh_i°, but it certainly casts serious doubt on the significance of the constant term in Eq. (9).

Hirst and Hollander were able to make a good summary of the influence of roughness and load on their scuffing data by using a failure envelope (Fig. 10). To interpret these data we recognize that we need at least two parameters to describe the topography of rough surfaces: one is needed to express the fluctuations in height of the asperities, σ, the other to characterize the horizontal features, correlation distance β^*.

Hirst and Hollander assumed random roughness, which simplifies matters. The vertical distribution of asperity heights is Gaussian and a single parameter, σ, is sufficient to describe the variance. The horizontal description is more subtle; a correlation distance β^* is often used, assuming an exponential decay of correlation. For our purposes we may take σ/β^* as being proportional to the mean asperity slope. Returning to Fig. 10, we see that only certain combinations of σ and β^* produce a "safe" or nonscuffing condition. The upper failure lines depend upon the ratio σ/β^* and are also a function of the normal load. The strong indication is that this line is largely defined by the deformation characteristics of the substrates;

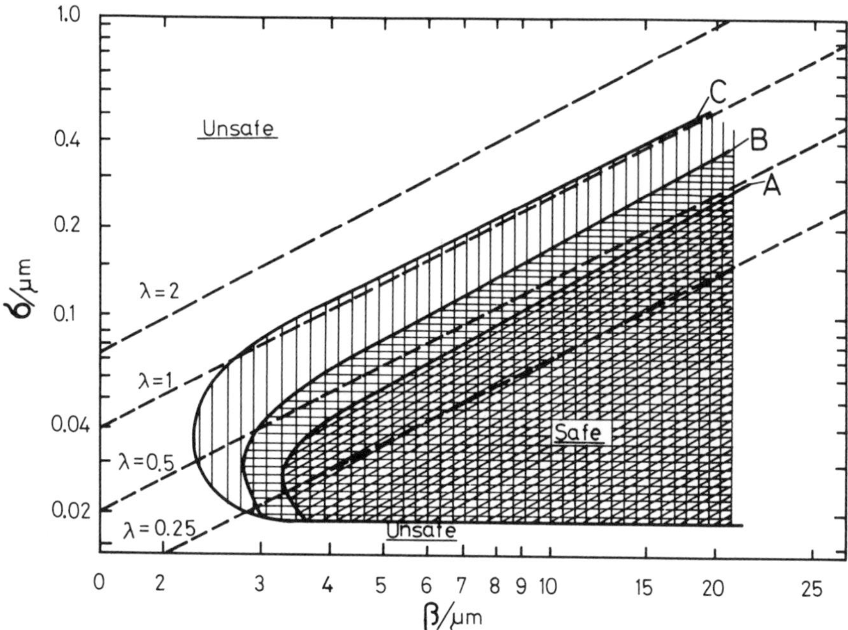

FIG. 10 Data of Hirst and Hollander [68] plotted to show the influence of surface topography on lubrication performance. "Unsafe" denotes lubrication failure; "safe" describes satisfactory operation. σ and β are, respectively, measures of the vertical and horizontal roughness scales. It is notable that a description of both scales is required in order to define the role of surface topography. The parameter λ is a constant which prescribes the likelihood of plastic deformation of the asperities. $\lambda = (E'/H)(\sigma/\beta)^2$, where E' and H are, respectively, the reduced Young's modulus and the hardness of the substrates. When $\lambda > 1$ extensive plastic deformation is probable. The upper bound of the envelope is clearly defined by the probability of plastic flow in the surface. A, B, and C refer to loads of 50, 10, and 2.5 N; the contact is of stainless steel with a smooth ball (12.7 mm in diameter) sliding on a rough plane. The lower bound indicates that a very smooth surface (superfinish) is difficult to lubricate. Lubricant is 1% by weight stearic acid in a hydrocarbon fluid.

gross plastic deformation of the asperities is more likely to produce scuffing than elastic deformation. In this connection it is significant that the likelihood of plastic deformation is proportional to σ/β^*. For very smooth surfaces (small σ) scuffing failure is found at all loads; Hirst and Hollander suggest that this is due to the fact that there are now "no longer any interruptions in the surface to prevent the growth of small adventitious welded functions." Rough surfaces can store pockets of trapped lubricant while smooth ones cannot. Hirst and Hollander did not explore the Frewing concentration variation experiment (they used a fixed concentration of stearic acid in a neutral oil), so we cannot compare their studies directly.

It now seems clear that in engineering practice the simple adsorption model has several weaknesses. We may also add a further problem, namely the interfacial temperatures at high sliding speeds. Flash temperatures of at least 800°C may be produced on sheared welds, and this may cause an appreciable rise in the local film temperature [46,60]. Certain calculations allow estimates of both temperatures and their contribution to thermally induced failure [70,71], but it is probably more useful to use a criterion based upon a critical energy dissipation [72,73].

We now consider some of the mechanical or physical causes of lubricant failure. We first ask whether pure liquids can form structured layers near solid surfaces and so constitute a useful barrier to asperity penetration. Early Russian workers thought they could and considered that these layers were very thick [74]. However, studies by the Australian school, using molecularly smooth mica as the solid surfaces, have shown them to be of shorter range [75,76]. They find, for example, that apolar liquids may develop a certain amount of structure, but it only extends a few nanometers away from the mica surface. Furthermore, even the rather low contact pressures generated between crossed mica cylinders can displace the liquid molecules from the mica surfaces. Thus pure apolar liquids are not effective as penetration barriers unless they are trapped at high pressures and become highly viscous materials.

By contrast, solid films such as Langmuir-Blodgett layers of stearic acid formed on mica or other surfaces can withstand much higher pressures both statically and *during* shear [43,65,77]. The enhanced durability of such solids may be due to strong lateral intermolecular forces, to their lower molecular mobility, to their higher energies of desorption, or to a combination of all three factors. The situation is even more involved if we consider the behavior of solid soap films formed in situ in metallic contacts; they are very complex in their chemistry and physical characteristics. The monolayer chemical composition may be important and can be changed by reversible adsorption as proposed by Frewing. Alternatively, the film may have a peculiar stability characteristic in relation to the ambient fluids; a Kraft point would produce some features of the observed behavior [78]. The role of fluid-additive and additive-additive interactions have not been stressed. *Additive interference* is a practical problem which occurs when a "package" of additives produces a result inferior to that anticipated from the behavior of the single components; a commercial apolar lubricating oil may contain a dozen or more additives. The converse is clearly desirable, and certain combinations of additives are incorporated in fluids in order to serve particular operating ranges. A combination of an aliphatic carboxylic acid and an organosulfur compound may produce a borader range of surface protection. At low temperatures a solid soap is formed and the sulfur species is relatively inactive; the lubrication is effective. At elevated temperatures the soap melts or desorbs, but now the sulfur compound is sufficiently unstable to react with the hot, clean, and potentially catalytic substrates to form a protective coating. The fact that warm metal surfaces may catalyze unusual decomposition has been recognized for many years. Attempts [79, 80] to simulate the surface reactions produced in sliding contacts in "test tubes" provide useful insight, but their relevance to lubrication in the contact is uncertain. The sliding interface region is a special chemical reactor. The temperautres and pressures are high and exist as transients. The reactants are possibly close to highly strained clean surfaces, which are thus highly catalytic in nature. Clearly, such conditions cannot be generated by conventional means. There are also more subtle effects arising from netural fluid-additive interactions.

Several authors, in particular Cameron, have claimed improved lubrication performance with aliphatic lubricants when the chain length of the alkane is the same as the alkyl chain length of the additive [81]. Some years ago workers at Battelle strongly favored the formation of smectic interface phases as a means of forming semi-solid protective films. These systems would have quite well-defined order-disorder transitions which could produce critical conditions of film breakdown and film rupture. The point about these films is that they are now no longer the monomolecular layers produced in physisorption.

We can now appreciate that no clear picture of lubrication failure is available. Solid layers produced by the adsorption or surface reactions of the boundary lubricants or the transient solid phases formed in elastohydrodynamic lubrication are more effective than weak liquid layers. There are many speculations on the reasons for the changes in interfacial phase behavior but none provides a complete description of the process of film breakdown.

VII. CONCLUSIONS

There are two aspects to interfacial phenomena in lubricated solid contacts: first, the origin of the friction when the solids are completely separated, and second, the mechanisms whereby the film is ruptured and extensive solid-solid contact occurs. The description of interfacial rheology and the prediction or rationalization of friction data is directly addressable. It demands a knowledge of the rheology of organic materials under rather special conditions. Many lubricants are van der Waals liquids or solids and hence, to a first order, show very similar rheological behavior. Their properties in interfaces are, with a few exceptions, what might be anticipated from an extrapolation of bulk response to conditions appropriate to those which exist in solid contacts. There are complications when the lubricant interacts chemically with the substrate, but again the surface organic product has a rheology not dissimilar from that of the parent lubricant. Liquid-solid and possibly solid-liquid transitions occur. Major changes in orientation are produced in the intense shear fields. There are transient time effects and adiabatic heating with the prospect of adiabatic shear zones developing. Wall or interface slip may occur as opposed to homogeneous shear. In spite of all these features we observe a commonality of response, and a simple and indeed crude model such as the Eyring stress-aided, thermally activated microscopic shear model provides a unifying approach. In contrast, the description of lubrication failure in either the boundary or the elastohydrodynamic regime remains poorly understood. Our better comprehension of interfacial rheology is a small, but nevertheless welcome, consolation.

REFERENCES

1. Count Benjamin Rumford, *J. Chem. 1*: 9 (1798).
2. N. Gane, P. Pheltzer, and D. Tabor, *Proc. R. Soc. London Ser. A 340*: 495 (1974).
3. J. N. Gregory, *Nature 157*: 443 (1946).
4. D. Dowson, *History of Tribology*, Longmans, New York, 1979.
5. O. Reynolds, *Philos. Trans. R. Soc. 177*: 157 (1886).
6. A. Cameron, *Basic Lubrication Theory*, Longmans, New York, 1971.
7. T. E. Stanton, *Proc. R. Soc. London Ser. A cii*: 241 (1923).
8. B. Towers, *Proc. Inst. Mech. Eng.* 532 (November 1883).
9. L. Rozeanu and N. Tipei, *Wear 28*: 298 (1974).

10a. H. S. Cheng, in *Fundamentals of Tribology*, N. P. Suh and N. Saka, eds., MIT Press, Cambridge, Mass., 1978.

10b. B. J. Hamrock, in *Tribology for the 80's*, NASA Conf. Rep. 2000, 503 (1983).

11. A. D. Roberts and D. Tabor, *Proc. R. Soc. London Ser. A 325*: 323 (1971).
12. H. S. Nagaraj, D. M. Sanbourn, and W. O. Winer, *Wear 49*: 43 (1978).

13. C. Barus, *Am. J. Sci. 45*: 87 (1893).
14. D. R. Adams and W. Hirst, *Proc. R. Soc. London Ser. A 332*: 523 (1973).
15. K. L. Johnson and A. D. Roberts, *Proc. R. Soc. London Ser. A 337*: 217 (1974).
16. W. Hirst and A. J. Moore, *Proc. R. Soc. London Ser. A 360*: 403 (1978).
17. R. S. Fein, *Trans. ASLE 8*: 59 (1965).
18. G. Harrison and E. G. Trachman, *J. Lubr. Technol. Trans. ASME F94*: 306 (1972).
19. K. L. Johnson, in *Elasto Hydrodynamic Lubrication Symposium, Leeds, 1979*, Institute of Mechanical Engineers, London, 1980.
20. G. Harrison, *The Dynamic Properties of Supercooled Liquids*, Academic Press, New York, 1976.
21. M. Alsadd, S. Bair, D. M. Dandborn, and W. O. Winer, *J. Lubr. Technol. Trans. ASME* F100 (1978) 206.
22. K. J. Johnson and R. Cameron, *Proc. Inst. Mech. Eng. 182*: 307 (1967-68).
23. R. S. Fein and K. L. Krentz, in *Interdisciplinary Approach to Friction and Wear*, P. M. Ku, ed., NASA SP-181, Washington, D.C., 1968, p. 358.
24. T. E. Tallian, *Wear 21*: 49 (1972).
25. See, for example, Tribology in the 80's, NASA Publ. 2000, Cleveland, 1983.
26. W. B. Hardy, *Collected Works*, Cambridge Univ. Press, Cambridge, 1936.
27. C. H. Giles and S. D. Forrester, *Chem. Ind. (London)*, 1616 (1969); 80 (1970); 43 (1971).
28. K. B. Blodgett and I. Langmuir, *Phys. Rev. 51*: 964 (1937).
29. E. B. Greenhill, *Trans. Faraday Soc. 45*: 625 (1949).
30. A. J. Smith and A. Cameron, *Spec. Discuss. Faraday Soc.*, 221 (1970).
31. K. G. Allum and J. R. Ford, *J. Inst. Pet. London 51*: 145 (1965).
32. K. G. Allum and E. S. Forbes, *J. Inst. Pet. London 53*: 173 (1967).
33. R. B. Jones and R. C. Coy, *Trans. Am. Soc. Lubr. Eng. 24*: 91 (1980).
34. E. S. Forbes, *Wear 15*: 87 (1970).
35. F. D. Tingle, *Nature 160*: 710 (1947).
36. M. Tomaru, S. Hironaka, and T. Sakurai, *Wear 41*: 117 (1977).
37. F. F. Ling, E. E. Klauss, and R. S. Fein, *Boundary Lubrication: An Appraisal of World Literature*, ASME, New York, 1969.
38. M. J. Furey, *Wear 26*: 369 (1973).
39. B. J. Briscoe and E. Tabor, *J. Adhes. 9*: 145 (1978).
40. R. F. Deacon and J. F. Goodman, *Proc. R. Soc. London Ser. A 243*: 464 (1958).

41. B. J. Briscoe and A. C. Smith, *Rev. Deformation Behav. Mater. 3*(3): 151 (1980).
42. B. J. Briscoe, B. Scruton, and R. F. Willis, *Proc. R. Soc. London Ser. A 333*: 99 (1973).
43. B. J. Briscoe and D. C. B. Evans, *Proc. R. Soc. London Ser. A 380*: 389 (1982).
44. D. Tabor, in *Microscopic Aspects of Adhesion and Lubrication*, J. Georges, ed., Elsevier, Amsterdam, 1982.
45. K. L. Johnson, in *Friction and Traction*, D. Dowson et al., eds., Westbury House, 1980.
46. F. P. Bowden and D. Tabor, *The Friction and Lubrication of Solids*, Oxford Press, London, 1950.
47. B. J. Briscoe and D. Tabor, *ACS Prepr. 21*(1): (1976).
48. B. J. Briscoe and A. C. Smith, *J. Appl. Polym. Sci. 28*(12): 3827 (1983).
49. J. Amuzu, B. J. Briscoe, and D. Tabor, *Trans. Am. Soc. Lubr. Eng. 20*(2): 152 (1977).
50. B. J. Briscoe and A. C. Smith, *Polymer 22*: 158 (1981).
51. L. A. Simpson and T. Hinton, *J. Mater. Sci. 6*: 558 (1971).
52a. B. V. Dejaguin, *Zh. Fiz. Khim.* 5: 1165 (1936).
52b. A. S. Akhmatov, *Molecular Physics of Boundary Lubrication*, Israel Program for Scientific Translations, Jerusalem, 1966.
53. A. J. Smith and A. Cameron, *Spec. Discuss. Faraday Soc.*, 221 (1970).
54. B. J. Briscoe and A. C. Smith, *J. Phys. D Appl. Phys. 15*: 579 (1982).
55. J. Amuz, B. J. Briscoe, and D. Tabor, *Trans. Am. Soc. Lubr. Eng. 20*(40): 354 (1977).
56. F. T. Barwell, *Fundamentals of Tribology*, N. P. Suh and N. Saka, eds., MIT Press, Cambridge, Mass., 1978, p. 401.
57. W. Hirst, *Proc. Conf. on Lubrication and Wear*, Institutions of Mechanical Engineers, London, 1957, p. 674.
58. J. K. Lancaster, *J. Phys. D Appl. Phys. 13*: 468 (1962).
59. D. Godfrey, in *Fundamentals of Tribology*, N. P. Suh and N. Saka, eds., MIT Press, Cambridge, Mass., 1980.
60. F. Bowden and D. Tabor, *The Friction and Lubrication of Solids*, Vol. II, Oxford Press, London, 1964.
61. J. J. Frewing, *Proc. R. Soc. London Ser. A 181*: 23 (1942).
62. J. J. Frewing, *Proc. R. Soc. London Ser. A 182*: 270 (1943).
63. A. Cameron and W. J. S. Grew, *Proc. R. Soc. London Ser. A 186*: 179 (1972).
64. A. Cameron and H. Spikes, *Proc. R. Soc. London Ser. A 336*: 407 (1974).
65. E. P. Kingsbury, *Trans. Am. Soc. Lubr. Eng. 3*: 30 (1960).
66. C. N. Rowe, *Trans. Am. Soc. Lubr. Eng. 9*: (1966).
67. J. W. Menter and D. Tabor, *Proc. R. Soc. London Ser. A 204*: 514 (1950).

68. W. Hirst and A. E. Hollander, *Proc. R. Soc. London Ser. A 337*: 379 (1974).
69a. J. R. Bristow, *Proc. R. Soc. London Ser. A 189*: 88 (1947).
69b. D. M. Tolstoi, *Wear 10*: 199 (1967).
70. H. Czichos and K. Kirschke, *Wear 22*: 321 (1972).
71. R. M. Matveevsky, P. B. Vipper, A. A. Markov, I. A. Buyanovsky, and V. Lashkhy, *Wear 45*: 143 (1977).
72. J. C. Bell and A. Dyson, Elastohydrodynamic Lubrication Symposium, 1972, *Proc. Inst. Mech. Eng.*, 61 (1972).
73. A. Dyson, *Tribiol. Int. 8*: 77, 117 (1975).
74. G. I. Fuks, *Res. Surf. Forces 2* (1964), translation, Consultants Bureau, New York, 1966, p. 79.
75. R. G. Horn and J. N. Israelachvili, *J. Chem. Phys. 75*: 1400 (1981).
76. H. K. Christenson, *J. Chem. Phys. 78*: 6906 (1983).
77. J. N. Israelachvili and D. Tabor, *Wear 24*: 386 (1973).
78. R. S. Fein, K. L. Kreuz, and S. J. Rand, *Wear 23*: 393 (1973).
79. D. Moorecroft, *Wear 18*: 333 (1971).
80. C. H. Bovington and B. Dacre, *Trans. Am. Soc. Lubr. Eng. 25*: 44 (1981).
81. T. C. Askwith, A. Cameron, and R. F. Crough, *Proc. R. Sòc. London Ser. A 291*: 500 (1966).
82. C. M. Allen and E. Draughlis, *Wear 14*: 363 (1969).

9

Pigment Dispersion in Apolar Media

R. B. McKAY Ciba-Geigy Pigments, Paisley, Scotland

I. INTRODUCTION

Dispersion of pigments in apolar media is a topic of importance to gravure printing ink and paint technologies. Pigments are finely divided solids supplied usually in powder or granular form. They are essentially insoluble in the media in which they are used and have to be dispersed therein by mechanical means. Most gravure inks are based on toluene or aliphatic hydrocarbon or mixtures of both. Most nonaqueous decorative paints are based on white spirit. Some industrial paints cured at around 80°C (low bake) are based on white spirit or white spirit/aromatic hydrocarbon mixtures. There is relevance also to the technology of plastics, particularly polyolefins, where pigment is dispersed in the molten polymer, but such viscous systems are not considered here.

Pigments can be classified in various ways, for example, in terms of chemical constitution (inorganic or organic), in terms of crystal lattice type (ionic, graphitic, or other molecular), in terms of crystal size, or in terms of surface polarity. General background on these various aspects has been given elsewhere in relation to surface characterization and dispersion of inorganic [1–4] and organic [5,6] pigments. The present chapter makes particular reference to copper phthalocyanine pigments, which are the most important organic pigments in the blue to green color region; they consist of finely divided molecular crystals (minimum dimension <0.05 μm; Fig. 1a–c) with a high degree of anisotropy, and have moderate to low surface polarity. In contrast are the inorganic titanium dioxide pigments, the most important white pigments, consisting of less finely divided ionic crystals (0.15 to 0.25 μm; Fig. 1d), with relatively high surface polarity unless specially treated. With the exception of carbon blacks, no other types of pigment have been subjected to such extensive investigative work as copper phthalocyanine and titanium dioxide pigments. Together they can be used to illustrate the various types of interfacial phenomena important in dispersion of pigments in apolar media.

Pigments are used to impart color or opacity. Being insoluble, their effectiveness depends on how finely divided they are, particle size and shape being the important parameters. Ideally, the particles in the final ink or paint are individual pigment crystals, as is largely the case with titanium dioxide pigments. Organic pigments, however, such as copper phthalocyanines, tend to differ in this respect. They consist of much smaller crystals (see Fig. 1) that tend to aggregate strongly during pigment manufacture unless special measures are taken. The particles in the final ink or paint are therefore usually small aggregates of crystals. Thus size, shape, and state of aggregation of crystals are factors of critical importance to dispersion properties.

Copper phthalocyanine pigments have a complex refractive index dominated by their strong absorption of visible light. Optimum absorption (corresponding to optimum color strength) occurs at a particle size well below 0.1 μm, and this is a primary objective in the manufacture of these pigments. Where particles are not spherically symmetrical it is the minimum dimension that is the most relevant [7]. In contrast, titanium dioxide pigments have a much higher refractive

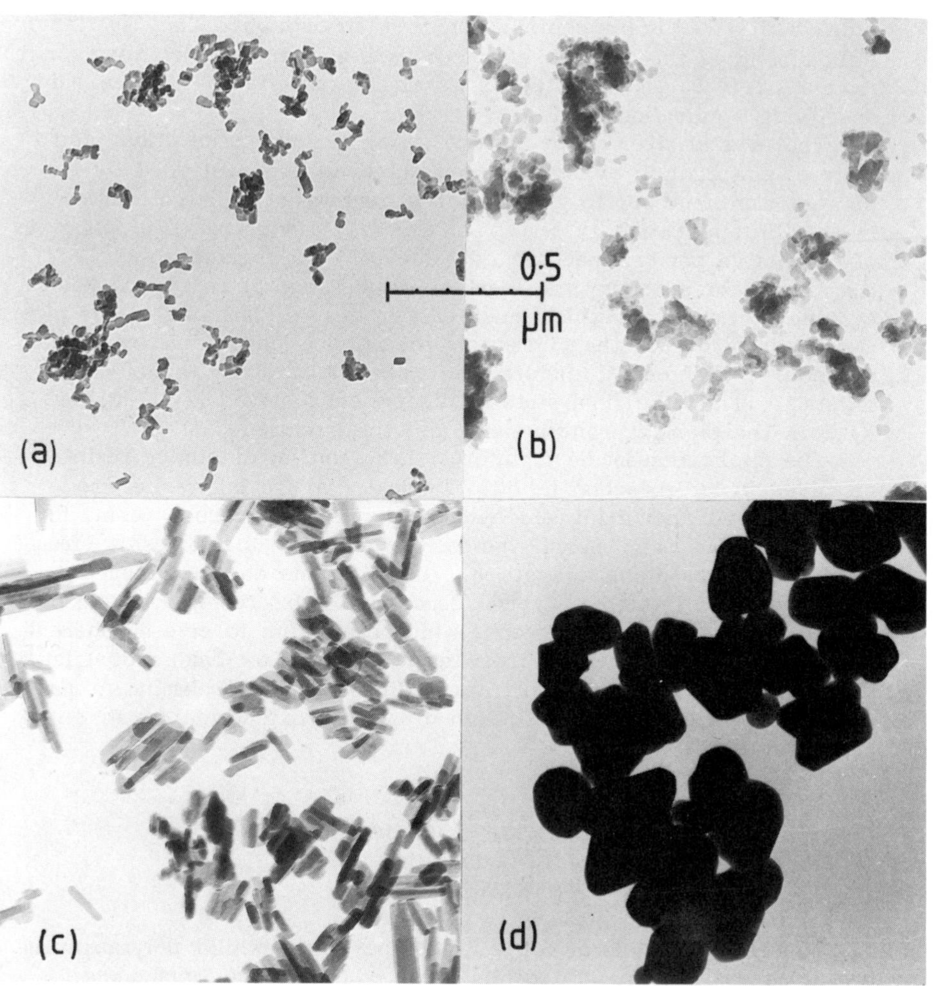

FIG. 1 Transmission electron micrographs of pigments: (a) α-CuPc (method 2, Table 1); (b) β-CuPc (method 3, Table 1); (c) β-CuPc (method 4, Table 1); (d) coated titanium dioxide pigment.

index with a negligible contribution from absorption. They scatter light strongly, the optimum effect occurring in the particle size range from 0.2 to 0.25 μm, and this is a primary objective of their role in opacifying application media.

The crystal size, shape, lattice type, surface composition, and state of aggregation of pigments are achieved and controlled by the pigment manufacturer to give required performance characteristics in specific application systems. In this way a wide range of pigments of distinctive performance can be achieved from a given chemical compound, for example, the Irgalite® Blue range of copper phthalocyanine pigments, on which much of this chapter is based. The pigment user disperses the pigment in the application medium by milling procedures designed to disintegrate aggregates rather than fracture crystals. Thus the final state of dispersion achieved is determined by both the pigment manufacturer and the customer.

The application media of interest here consist of binder resin dissolved in hydrocarbon liquid. Typical gravure ink resins are rosin-modified phenol-formaldehyde resins or hydrocarbon resins for toluene systems, and mixed zinc/calcium salts of abietyl resins (from wood rosin) for aliphatic systems. These resins do not undergo oxidative cross-linking. Typical decorative paint resins are long-oil alkyds incorporating drying oils, which cross-link to give a coherent film when exposed to air at room temperature. Low-bake industrial paints use, for example, short- to medium-oil alkyd/melamine-formaldehyde type resin mixtures, which cross-link oxidatively in air on baking at 80°C.

II. PHYSICAL CHARACTER OF COPPER PHTHALOCYANINE AND TITANIUM DIOXIDE PIGMENT CRYSTALS

The molecular crystals of copper phthalocyanine exhibit polymorphism, two of several known polymorphs being of commercial importance, namely the greenish blue β-form and the reddish blue α-form. The principal lattice features are shown in Fig. 2 (based on [8]). Note the stacking of the large planar molecules that leads to anisotropic character. The crystallographic properties and their consequences have recently been reviewed [9]. The more compact β-form is the thermodynamically stable form, to which other forms tend to revert at high temperatures or in contact with organic liquids. The α-form can be stabilized by partial chlorination in the 4-position (see Fig. 2), the equivalent of just over one-half chlorine atom per molecule being sufficient to prevent α→β-form conversion in hydrocarbon liquids. The stabilized α- and β-forms are both widely used in paints, but the β-form is predominant in inks.

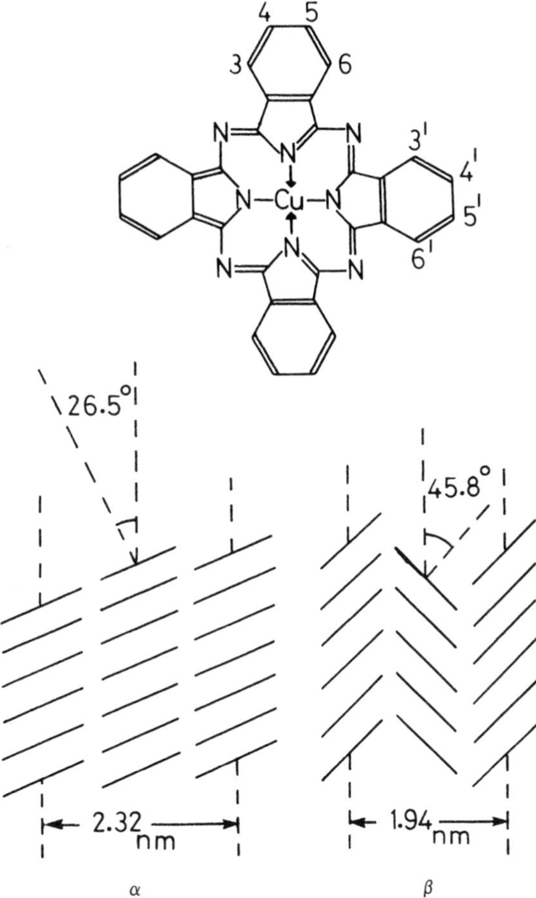

FIG. 2 Chemical constitution and crystal lattice features of copper phthalocyanine.

As normally synthesized, copper phthalocyanine consists of non-pigmentary β-form crystals in the shape of rods up to 100 μm long. Reduction to pigmentary size is achieved as detailed in Table 1 and produces the various types of crystal described.

Figure 1 shows some examples. β-form conversion in liquids (including hydrocarbons such as benzene) occurs by a complex series of processes involving Ostwald ripening [9,10]. On the other hand, the stabilized α-form remains in the α-form irrespective of the method of size reduction and with crystal size and shape comparable to those of the α-form.

The crystals of titanium dioxide also exhibit polymorphism, two of the three known polymorphs being of commercial importance, namely anatase and rutile. Both forms absorb ultraviolet light strongly and exhibit photochemical activity which is deleterious in application. Anatase is particularly prone to these problems, so that rutile tends to be preferred in paints, especially those for exterior use.

TABLE 1 Preparation Methods for Copper Phthalocyanine Pigments

Method of size reduction	Crystal form	Average crystal dimensions
1. Precipitation from solution in concentrated sulfuric acid	α	Bricks or platelets <0.03 μm in minimum dimension
2. Dry attrition in presence or absence of salts	>95α, or α,β mixtures depending on conditions	Bricks down to 0.02 μm in minimum dimension
3. Dry attrition in presence of salts, but with a small amount of crystallizing agent	β	Bricks, typically 0.02 to 0.04 μm in minimum dimension, axial ratio <2:1
4. Dry attrition in presence or absence of salts, followed by treatment with a crystallizing liquid	β	Rods, minimum dimension 0.02 to 0.04 μm, length up 0.3 μm
5. Shearing in a crystallizing liquid	β	Bricks, typically 0.03 μm in minimum dimension, axial ratio around 2:1

The crystals of commercial titanium dioxide pigments have surface coatings containing titanium oxide, silica, and alumina [1,2]. Silica counteracts photochemical activity; alumina improves dispersion properties. The oxides are applied as aftertreatments ideally to give continuous, coherent coatings. Although enlarging the particles by about 5 to 7% w/w in general-purpose pigments, the coatings actually increase the specific surface area determined by nitrogen adsorption (by about 50%). This is mainly due to porosity of the surface coating, the pores being penetrable by nitrogen and water molecules but inaccessible to larger molecules. Some grades are further treated with organic agents to improve compatibility with application media. Generally, however, the surface consists of a layer of water molecules adsorbed to hydrated oxide.

III. AGGREGATION OF COPPER PHTHALOCYANINE PIGMENT CRYSTALS

The state of aggregation of powdered solids can be characterized by the well-established gas adsorption method, using nitrogen as adsorbate. Details are given elsewhere [11]. Typical adsorption/desorption isotherms obtained with copper phthalocyanine pigments are given in Fig. 3.

Crystal size reduction by dry attrition or precipitation processes (methods 1, 2, and 3 of Table 1) produces copper phthalocyanine crystals in an aggregated state. Specific surface areas determined from nitrogen adsorption data by the BET method [11] (S_{BET}) are much lower than the corresponding geometric specific surface areas determined from electron micrographs (S_{TEM}). Nitrogen clearly does not gain access to all the crystal surfaces, only the outer regions of the aggregate being accessible. Aggregation also gives rise to the narrow hysteresis loops in the adsorption/desorption isotherms, the hysteresis extending to very low nitrogen pressures ($P/P_0 < 0.1$) [9,12,13]. Figure 3c and d show specific examples with β-form and α-form pigments, respectively. This type of hysteresis is well established with other solids and is associated with swelling of the pore structure of solid surfaces [11]. In the present case it indicates that nitrogen molecules penetrate between the crystals in the outer regions of the aggregate structure and prise the crystals apart to some extent. Once thus wedged in, the nitrogen molecules are difficult to remove [9]. The aggregate structure is generally not of a rigid type, where hysteresis would be expected only at $P/P_0 > 0.4$.

Much less aggregated are the rod-shaped β-form crystals produced by dry attrition of nonpigmentary material to give pigmentary, α,β-form mixtures, which are then treated with a crystallizing liquid (method 4 of Table 1). Indeed, optimization of processing conditions

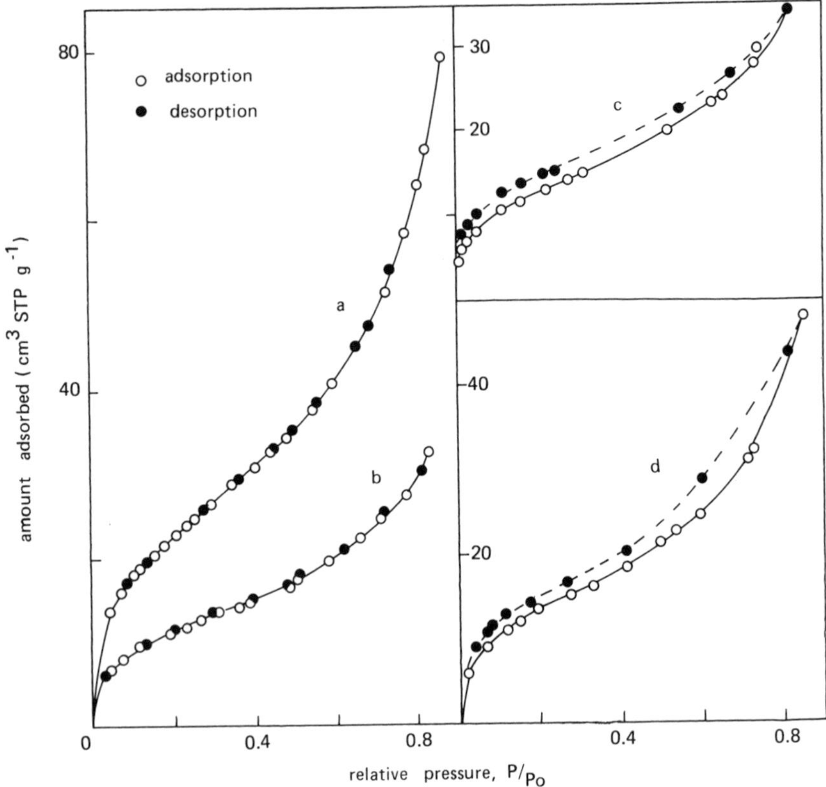

FIG. 3 Nitrogen adsorption isotherms: (a) β-CuPc-III; (b) β-CuPc-I; (c) β-CuPc-IV; (d) α-CuPc-II. (Data for a, c, and d are from Ref. [13].)

can produce pigments that are substantially unaggregated ($S_{BET} \rightarrow S_{TEM}$). In such cases there is little or no hysteresis in the nitrogen adsorption/desorption cycle [12,13]. Figure 3b and a show specific examples (β-CuPc-I, S_{BET} = 43 m^2 g^{-1}, S_{TEM} = 44 m^2 g^{-1}; β-CuPc-III, S_{BET} = 89 m^2 g^{-1}, S_{TEM} = 94 m^2 g^{-1}). Pigments can be identified from Table 2, columns 1 and 2.

The α- and stabilized α-forms are isomorphous and are more prone to aggregation than the β-form [9]. The α-lattice is the more loosely coherent, less efficiently packed, and more susceptible to plastic deformation—thus enhancing the area of contact between crystals forced together. The effectiveness of aftertreatment with liquids in reducing aggregation in milled β-form mixtures is therefore partly

TABLE 2 Morphology and Heats of Immersion of Copper Phthalocyanine Pigments[a]

Pigment[b]	Preparative method[c]	Crystal shape	S_{BET} ($m^2\ g^{-1}$)	Heats of immersion, h_i ($mJ\ m^{-2}$)		
				Toluene	*n*-Heptane	Dodecane
β-CuPc-I	4	Rod	43	77	—	—
β-CuPc-II	4	Rod	55	93	84	89
β-CuPc-III*	4	Rod	89	109	91	89
β-CuPc-VI*[d]	4	Rod	70	104	96	97
β-CuPc-IV*	3	Brick	48	218	—	—
β-CuPc-V	3	Brick	73	230	179	175
α-CuPc-II[e]	2	Brick	66	(372)[f]	200	208
α-CuPc-III	1	Brick	47	(283)[f]	—	—
α-CuPcCl-II	2	Brick	79	354	285	—
α-CuPcCl-III	1	"Rod"[g]	52	106	85	—

[a]Data of Ref. [9].
[b]Pigments prepared from exhaustively Soxhlet-extracted crude material reduced to pigmentary size by idealized methods, except those marked with an asterisk, which were plant samples.
[c]See Table 1.
[d]New data (not given in Ref. [9]).
[e]90% α-form, 10% β-form.
[f]Possibly slightly high due to some α→β-form conversion during measurement.
[g]Irregular.

due to controlled crystal growth and partly to $\alpha \rightarrow \beta$-form conversion. The stabilized α-form, which does not undergo phase conversion and which is more difficult to grow, cannot be so effectively deaggregated in this way. Neither can β-form pigments with brick-shaped crystals prepared directly by dry attrition (method 3 of Table 1).

Control of aggregation is of critical importance to the ease and extent of dispersion achievable in inks and paints [5].

IV. WETTING AND AGING

Bringing a copper phthalocyanine pigment into contact with an ink or paint medium initiates a number of processes. Not only is there wetting of immediately exposed surfaces, but there is also adsorption of binder resin, penetration of the aggregate structure by liquid molecules, and the possibility of various ageing effects, ranging from crystal phase change to localized crystal fusion and Ostwald ripening. Additives that are present in some pigments may become solvated and their disposition may be altered. Application of shearing or impaction tends to separate the constituent particles in all but the most strongly coherent aggregates, thereby releasing entrapped air or additives, and exposing the wetted crystal surfaces to adsorption of binder resin or additives in the dispersion medium. Wetting is the primary stage, but thereafter the various processes may all be occurring at any given instant. Investigative work must therefore be carefully designed.

Gravimetric vapor sorption measurements at room temperature have proved to be a useful prelude to wetting investigations of specially prepared pigments free from additives. Thus adsorption/desorption of toluene vapor on copper phthalocyanine pigments is more dependent on state of aggregation of crystals than on crystal lattice type [9,13,14]. Adsorption isotherms are virtually reversible on a β-CuPc pigment (β-CuPc-I) with well-developed, rod-shaped crystals that are essentially unaggregated (Fig. 4a). Furthermore, there is little difference between successive adsorption/desorption cycles and no significant effect on S_{BET} measured independently by nitrogen adsorption.

The behavior pattern, however, is quite different with α- and β-form pigments consisting of aggregates of small brick-shaped crystals (α-CuPc-I and β-CuPc-IV, $S_{BET} \ll S_{TEM}$) (Fig. 4b and c). There are marked hysteresis effects in the adsorption/desorption cycle and a marked decrease in uptake between successive cycles. The toluene vapor penetrates the aggregate structure. At high vapor pressures condensed toluene in the capillaries between crystals promotes cementation of the crystals, closure of the aggregate structure, and causes marked reduction in S_{BET} determined independently by nitrogen adsorption. This localized ageing effect occurs without

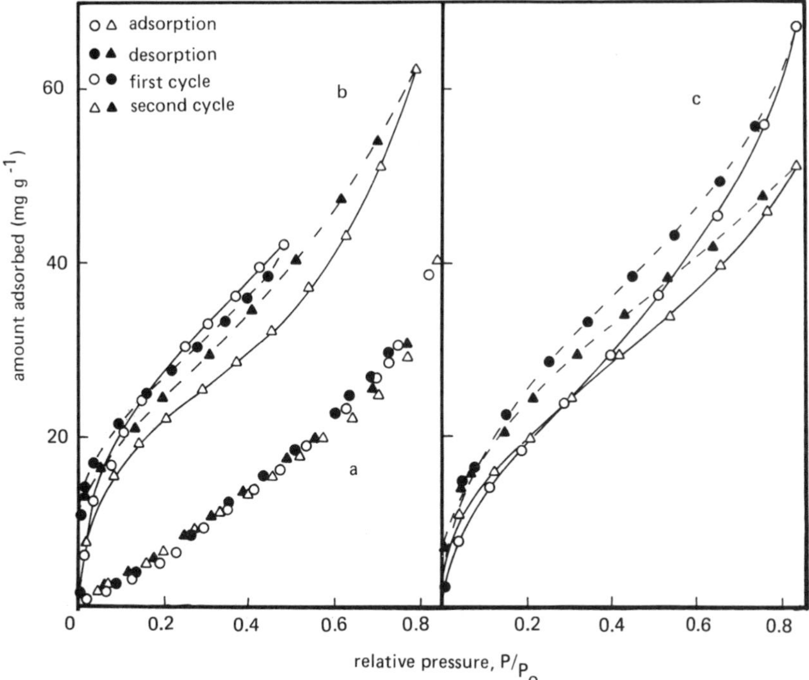

FIG. 4 Toluene vapor sorption isotherms: (a) β-CuPc-I; (b) α-CuPc-II; (c) β-CuPc-IV. (Data for b from Ref. [14], data for c from Ref. [13].)

detectable α→β-form conversion in the α-form pigments. Broadly comparable effects have been observed in independent studies of adsorption of benzene vapor on aggregated α- and β-form copper phthalocyanine pigments [15].

n-Heptane vapor on β-CuPc-IV shows significant hysteresis in the adsorption/desorption cycle, yet little difference in uptake between successive adsorption cycles (Fig. 5b). Thus, although *n*-heptane penetrates the aggregate structure to some extent, it does not cause significant ageing in the time span of the experiment. It is therefore milder in action than toluene. Like toluene, it gives virtually reversible adsorption on β-CuPc-III (Fig. 5a).

Wetting of copper phthalocyanine pigments in hydrocarbon *liquids* is also more influenced by aggregation of the crystals than by crystal lattice type [9]. Table 2 shows heats of immersion obtained with a range of β-, α-, and stabilized α-form pigments, all free from additives. The data are expressed as heat evolved per unit of S_{BET}

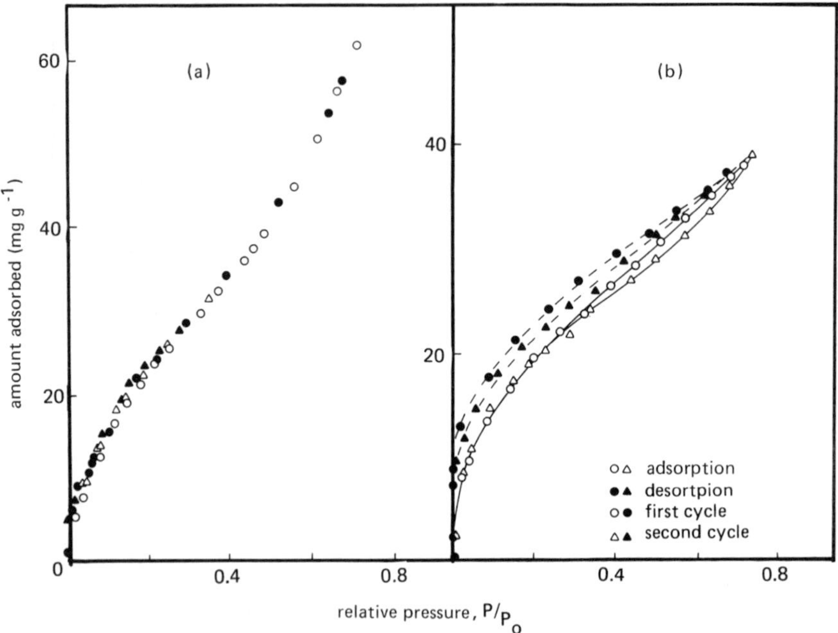

FIG. 5 *n*-Heptane vapor sorption isotherms: (a) β-CuPc-III; (b) β-CuPc-IV.

determined by nitrogen adsorption, and were obtained with a Calvet ambient microcalorimeter.

Heats in the range 77 to 109 mJ m^{-2} are given by the β-form pigments with rod-shaped crystals, all of which are essentially unaggregated ($S_{BET} \rightarrow S_{TEM}$). These heats represent wetting of a substantially open copper phthalocyanine surface. Larger heats in the range 175 to 372 mJ m^{-2} are given by the pigments with small brick-shaped crystals, whether α- or β-form. These pigments are extensively aggregated ($S_{BET} \ll S_{TEM}$). The liquids penetrate between the crystals in the aggregate structure to a much greater extent than nitrogen can and thus gain access to crystal surfaces that are inaccessible to nitrogen. Consequently, the surface area exposed to the hydrocarbon liquid is underestimated by nitrogen adsorption and the heats per unit of S_{BET} are correspondingly high. Thus a combination of heat of immersion and S_{BET} can be used to provide information on the state of aggregation of the pigment crystals.

An attempt has been made to take into account the anisotropic character of copper phthalocyanine crystals and resolve heats of immersion into the contributions from the ends (basal planes; ends of molecular stacks depicted in Fig. 2) and sides (prismatic planes) [6]. This requires comparison of unaggregated β-form crystals of different axial ratios. It has been found that specially prepared brick-shaped (isometric) and rod-shaped (acicular) crystals have given comparable heats of immersion in benzene (104 and 103 mJ m^{-2}, respectively) and in *n*-hexane (83 and 87 mJ m^{-2}), the comparability and actual magnitudes of the heats affirming that the pigments are essentially unaggregated. On the other hand, results in polar liquids, for example, *n*-butanol, *n*-butylamine, and nitrobenzene, are considerably higher with the brick-shaped crystals than with the rod-shaped crystals (e.g., 199 and 145, respectively, in *n*-butylamine). Since brick-shaped crystals have a higher proportion of end face area, it appears that the ends of the crystals have higher energy than the sides, where only the edges of the copper phthalocyanine molecules are exposed. Although this aspect of anisotropy had no significant effect on wetting by the hydrocarbon liquids themselves, it may affect adsorption of polar solutes dissolved in these liquids.

Some pigments are designed to undergo controlled ageing during dispersion in application media. So-called semifinished copper phthalocyanine pigments are aggregated α, β-form mixtures prepared by dry attrition in the absence of salts. During subsequent dispersion in toluene-based gravure ink media, $\alpha \rightarrow \beta$-phase conversion and crystal growth occur to give well-dispersed β-form pigment in the final ink. Small proportions of certain substituted copper phthalocyanine derivatives which adsorb on the pigment surfaces are incorporated by the pigment manufacturer to control the extent of crystal growth and interaction between crystals, thus achieving inks with high color strength and good rheological behavior [16].

V. DESORPTION/DISSOLUTION OF ADDITIVES FROM ORGANIC PIGMENTS

Pigments do not disperse spontaneously when they are immersed in organic liquids and effectively wetted. Input of mechanical energy is required to break aggregates and to release entrapped air that has been displaced from surfaces on wetting. The separation process is gradual and is commonly monitored in the case of organic pigments by observing color strength development as a function of time or amount of energy input. It is, of course, advantageous to develop full color strength in as short a time as possible with minimum energy input. Substantial advances in this direction have been achieved with organic pigments by coflocculating pigment with additives such as wood rosin or related abietyl compounds prior to isolation in the pigment manufacturing process [5]. Typical amounts of additive are in the range 5 to 40% by weight in the final pigment.

Copper phthalocyanine pigments seldom require more than 5%. The additives reduce crystal/crystal contact in the pigment by forming a mechanical barrier, but should dissolve away into the liquid phase during the dispersion process, thus releasing the pigment particles [17]. This principle can be demonstrated by simple sedimentation volume experiments in systems of low pigment concentration as follows. Table 3 shows how the sedimentation volume fraction V_S of copper phthalocyanine pigment dispersed in pure hydrocarbon liquids increases progressively with time of ball milling [18]. Although the pigment volume fraction is only 0.007, the V_S values are more than an order of magnitude higher due to flocculation.

The ratio of initial to final values of V_S, namely V_R, gives an indication of the ease of dispersion. Ideally, V_R should be unity, the pigment dispersing fully simply on shaking by hand so that subsequent ball milling should have no further effect. In the particular cases given in Table 3 the pigments are additive-free, give $V_R \ll 1$, and evidently have poor dispersibility. Table 4, however, shows V_R values for a range of organic pigments treated with large amounts of additive. It also shows the weight of additive that has dissolved into the liquid phase after dispersion—as determined by gravimetric analysis of the liquid phase separated from pigment by centrifugation.

TABLE 3 Sedimentation Data (Pigment Volume Fraction, c = 0.007)[a]

Surface-treated CuPcCl of type III in toluene			β-CuPc in n-heptane		
V_R	Milling time (hr)	V_S	V_R	Milling time (hr)	V_S
0.09	0	0.05	0.14	0	0.08
	0.25	0.11		1	0.19
	1	0.14		5	0.29
	3	0.17		10	0.40
	8	0.21		20	0.46
	16	0.32		40	0.56
	48	0.47		64	0.59
	67	0.53			

[a]Data of Ref. [18].

TABLE 4 "Desorption" of Resins from Organic Pigments

Pigment type	Wt. resin in 1 g pigment (g)	Toluene			n-Heptane		
			"Desorption"			"Desorption"	
		V_R	Pigment/liquid ratio	Wt. resin dissolved per gram pigment	V_R	Pigment/liquid ratio	Wt. resin dissolved per gram pigment
β-CuPc	0.20	0.60	0.008	0.19	0.28	0.004	0.09
			0.016	0.19		0.008	0.11
			0.029	0.20		0.026	0.12
Anilide yellow-I	0.21	>0.9	0.013	0.20	0.12	0.008	0.04
			0.029	0.20		0.017	0.05
			0.043	0.20		0.026	0.05
Anilide yellow-II	0.27	>0.9	0.016	0.28	0.12	0.009	0.07
			0.030	0.27		0.017	0.07
			0.045	0.26		0.026	0.07
Ca 4B toner	0.24	0.85	0.015	0.20	0.23	0.008	0.06
			0.030	0.20		0.017	0.05
			0.043	0.20		0.026	0.06

Clearly, ease of dispersion in toluene ($V_R \to 1$) is associated with complete or substantial dissolution of the additive. Difficulty of dispersion in *n*-heptane ($V_R \ll 1$) is associated with limited dissolution of additive. The limited dissolution is not simply due to saturation of the *n*-heptane, since the weight of additive that dissolves is virtually independent of the pigment/liquid ratios used. Instead, the cause of poor dispersibility is apparently the limited ability of *n*-heptane to penetrate the aggregate structure and release the resin.

VI. ADSORPTION OF BINDER RESINS

The binder resins in gravure ink or paint media adsorb to the pigment surface. Isotherms for apparent adsorption of resin from concentrated solution have the form shown in Fig. 6. Note especially the maximum and the linear region of negative slope. Determination is by the conventional procedure of equilibration, involving agitation

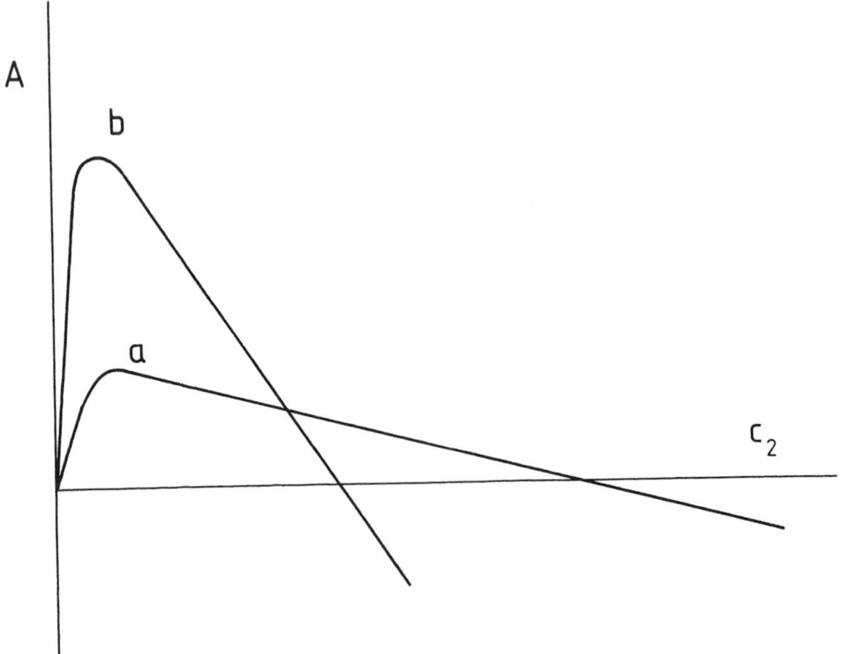

FIG. 6 General character of composite isotherms for adsorption of binder resin solutions to pigments: (a) titanium dioxide pigments; (b) copper phthalocyanine pigments.

of pigment in resin solution at constant temperature, followed by removal of pigment by centrifugation. In all cases control experiments free from pigment should be performed side by side and should be subjected to exactly the same procedures (including centrifugation). The concentrations of the control and test solution (weight fractions) after equilibration, c_1 and c_2 respectively, are determined by analysis. The apparent weight of resin adsorbed per unit weight of pigment is then deduced from $c_1 - c_2$. In this way the isotherm is constructed point by point.

The characteristic shape of the isotherms is the result of their composite nature (see Chap. 0). Both resin and solvent are adsorbed. At the high resin concentration used, uptake of solvent is a significant fraction of the total solvent present, increasingly so as the resin concentration is increased—hence the region of negative slope instead of a horizontal "plateau." Rehacek [19] has provided a method of analysis. Assuming that the bulk solution and the adsorbed layer can be treated as separate liquid phases and that the composition of the adsorbed layer remains constant throughout the linear region, then mass balances can be constructed as follows.

$$A = m_1(c_1 - c_2) \tag{1}$$

where A is the apparent weight of resin adsorbed per unit weight of pigment and m_1 is the weight of resin solution in contact with a unit weight of pigment.

$$m_1 = m_a + m_2 \tag{2}$$

where m_a is the weight of resin solution adsorbed per unit weight of pigment and m_2 is the weight of unadsorbed solution at equilibrium.

$$m_1c_1 = m_ac_a + m_2c_2 \tag{3}$$

where c_a is the concentration (weight fraction) of resin in the adsorbed layer.

Combining Eqs. (1), (2), and (3) gives

$$A = m_ac_a - m_ac_2 \tag{4}$$

which describes the isotherm. Thus m_a is the numerical value of the slope of the linear region; m_ac_a, the intercept on the A axis, is the true amount of resin adsorbed per unit weight of pigment; c_a is the intercept on the c_2 axis; and $1 - c_a$ is the proportion of solvent in the adsorbed layer.

Relative viscosity and sedimentation volume fractions of dispersions are constant over the region of resin concentration corresponding to the linear part of the isotherms. This supports the assumption that the composition of the adsorbed layer remains constant throughout the linear region [19].

The method of analysis was developed for adsorption of alkyd resins on titanium dioxide and other inorganic pigments in white spirit or xylene [19]. Copper phthalocyanine pigments in solutions of resins in hydrocarbon liquids give isotherms with the same general characteristics. Some examples are given in Fig. 7A (from [20]). The pigments used were free from resin or other additives designed

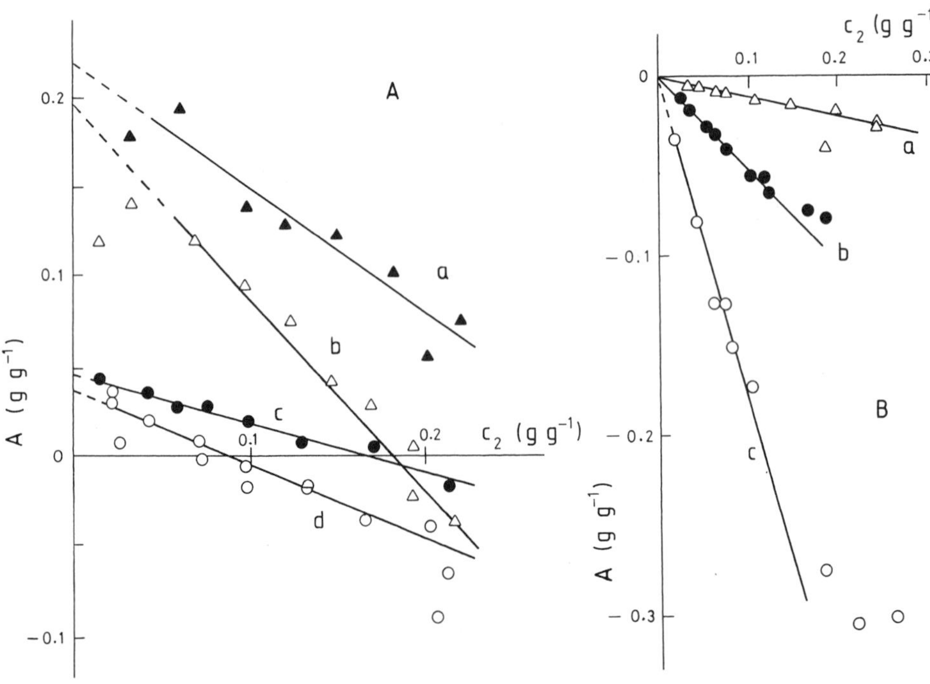

FIG. 7 Composite isotherms for adsorption of binder resin solutions on (A): a, stabilized α-CuPc of type 3 (long-oil alkyd in white spirit); b, c, and d, stabilized α-CuPc of type 1, β-CuPc-VI, stabilized α-CuPc of type 4, respectively (modified phenol-formaldehyde resin in toluene); and (B): a, b, and c, all Sephadex LH20 (respectively modified phenol-formaldehyde resin in toluene, short/medium-oil alkyd in xylene + butanol, Supronic E800 in water). (Data are from Refs. [20,29].)

to ease dispersion. Equilibration was conveniently achieved in cylindrical vessels rotated on rollers and containing a charge of steatite balls to ensure adequate dispersion. The formulation was designed such that the surface area of the balls in the vessel was negligible compared to that of the pigment; the use of control experiments as described earlier further ensured that no significant errors originated from this source.

Since copper phthalocyanine pigments are more finely divided than titanium dioxide pigments, their isotherms show greater uptake and steeper slopes, as indicated in Fig. 6. Moreover, they tend to cross the c_2 axis at lower c_2 values, thus indicating lower c_a and greater uptake of solvent. This is especially true for systems based on toluene. The stabilized α-form pigment of Fig. 7A,d is a good example. The particles of such pigments are aggregates of crystals that are penetrable by hydrocarbon liquids, especially toluene, as discussed earlier. Indeed, high proportions of liquid phase are associated with the particles in dispersion in hydrocarbon-based media. This has been deduced from determinations of limiting (or plastic [21]) viscosity made under high shear conditions, where weak flocculated structure is broken down [20].

The particles are thus part liquid phase (volume fraction v_i) and part pigment (volume fraction $1 - v_i$). Some v_i values are given in Table 5. The isotherms for such systems can be described by the modified equation

$$A = m_a c_a - (m_a + m_L) c_2 \tag{5}$$

where m_L is the weight of solvent-rich liquid which preferentially penetrates the interior of the aggregates per unit weight of pigment, large resin molecules being excluded [20].

When $m_a \to 0$, i.e., when only selective penetration of solvent occurs, there being no adsorption of resin solution, Eq. (2) reduces to

$$A = -m_L c_2 \tag{6}$$

Results with suspensions of the granular molecular sieve Sephadex LH20 (Pharmacia Fine Chemicals) instead of pigment conformed to Eq. (6) (Fig. 7B) [20]. This material is designed for use in gel permeation chromatography, where specific interaction with resin/ polymer solutes must be minimized. Values of m_L increased in the same order as did sedimentation volume fractions of the suspensions, the latter increase obviously resulting from increase in swelling of the gel granules by penetrating liquid. Whereas no pigment has shown this extreme behavior, the stabilized α-CuPc in Fig. 7A, d comes close.

TABLE 5 Volume Fractions of Liquid Associated with Organic Pigment Particles[a]

Pigment	System	v_i	v_a
Stabilized α-CuPc of type 1[b]	Phenolic/toluene	0.81	0.67
Stabilized α-CuPc of type 3[c]	Alkyd/white spirit	0.74	0.57
β-CuPc-VI[d]	Phenolic/toluene	0.93	0.33

[a]Data of Ref. [20].
[b]Treated with additive of type A (Table 7) to render flocculation-resistant.
[c]Similar to α-CuPcCl-II (Table 2).
[d]Prepared by method 4 of Table 1—essentially unaggregated.

Titanium dioxide pigments are less finely divided, and with them m_L is too small to be of much significance. Consequently, the isotherm data can be used to provide information on the degree of solvation of the adsorbed resin layer [22]. However, m_L cannot be ignored in the case of copper phthalocyanine pigments, especially α- and stabilized α-types, and β-types prepared by method 3 of Table 1, the particles of which are crystal aggregates. Neither can it be evaluated quantitatively, since in Eq. (5) the slope $-(m_a + m_L)$ cannot be resolved into its components. Consequently, with these pigments the isotherm data cannot be used to deduce the composition of the the adsorbed layer. If, however, it is assumed that $\rho_a = \rho_L$, which are respectively the densities of the adsorbed layer and the liquid phase, then the slope of the isotherm can be used to estimate the mean volume fraction of liquid phase v_a in the particles [20]. These values are high (see Table 5) and for the two stabilized α-form pigments are comparable to, but lower than, the corresponding v_i values deduced from rheological data. Whereas v_a detects uptake of liquid phase only when there is a change in resin concentration involved, v_i takes into account, additionally, liquid phase of the same concentration as the bulk liquid phase and simply trapped in pores and surface irregularities.

The high v_i value for the β-form pigment β-CuPc-VI in Table 5 indicates residual (strong) flocculation in the region of high shearing rate used in the rheological measurements (915 to 1818 sec^{-1}). β-CuPc-VI prepared by method 4 of Table 1 is free from additives and consists essentially of individually dispersed crystals ($S_{BET} = 70\ m^2\ g^{-1}$; $S_{TEM} = 76\ m^2\ g^{-1}$; $h_i = 104\ mJ\ m^{-2}$ in toluene, see Table 2).

Some studies with titanium dioxide pigments in decorative paint media have given values of adsorbed layer thickness estimated from:

$$\delta = \frac{m_a}{\rho_a S'} \quad [19,23] \tag{7}$$

or

$$\delta = \frac{c' - c}{S_{BET}} \quad [24] \tag{8}$$

where $S' = S_{BET}$ or an average specific surface area incorporating S_{BET}, c' is the effective pigment volume fraction in dispersion deduced from rheological measurements, and c is the actual pigment volume fraction. These estimates assume that S_{BET} represents the area available to binder resin. The assumption is not valid with pigments containing an alumina-rich surface coating. Such coatings are porous and part of the surface within the pore structure is inaccessible to resin molecules and even relatively small molecules such as stearic acid [2,25]. The values of δ are therefore underestimated. This is unlikely to affect the validity of comparisons of adsorption of different solutes on a given pigment (e.g., [26]), but it may invalidate comparison of adsorption of a given solute on differently modified pigments unless measures are taken to allow for porosity. Such measures are described elsewhere [25].

The binder resins of ink and paint systems are heterogeneous in terms of both chemical constitution and molecular size. Little has been reported on this aspect in the present context, but the feasibility of using gel permeation chromatography to separate resin size fractions has been demonstrated in relation to adsorption of solutions of alkyd resin in white spirit to titanium dioxide [27]. Use of radioactively labeled resin fractions in the same type of system is also feasible and has suggested that there is a preference for low molecular weight fractions [28].

VII. WEAK FLOCCULATION

Formation of an adsorbed layer of resin reduces the attraction between pigment particles. This stabilization effect is essential for achieving satisfactory levels of dispersion. It does not, however, eliminate interaction between particles altogether. Residual attraction gives rise to weak flocculation, which is manifest as gel-like structure formation, especially with copper phthalocyanine and other pigments that are extremely finely divided. The structure breaks and forms again on application and removal of shearing, and can be shown by rheological measurements to be truly reversible in a mechanical sense [29]. The flocculation is apparently of the weak or secondary minimum type.

Flocculation is generally undesirable in gravure inks and paints. The resulting structure formation limits the concentration of pigment usable in ink and paint concentrates. It adversely affects color strength of copper phthalocyanine (and organic pigments generally) in paints, which normally contain an excess of titanium dioxide pigment. Horizontal separation of blue and white pigments occurs to give a white-rich layer at the paint film surface (white flooding), which then appears coloristically weaker [1]. So the color strength becomes dependent on the degree of shearing during application of the paint to the substrate.

Flocculates are inherently ill-defined species and are difficult to characterize. There are, however, some methods that are useful for comparing dispersions and for investigating the effect that changes in the pigment/medium interface have on flocculation. One such method is settling.

A. Hindered Settling of Organic Pigment Dispersions

Flocculation causes rapid settling of the pigments, especially in dispersions with low pigment concentration. Dispersions of copper phthalocyanine pigments in hydrocarbon liquids, such as toluene, white spirit, and *n*-heptane, tend to exhibit hindered settling even in some cases at very low pigment concentrations, around 1% by weight [18]. Hindered settling is essentially characteristic of a crowded system. It differs from Stokes (normal) settling in that all the settling units, whether discrete particles or flocculates, settle at the same rate irrespective of their size. Hydrodynamic interactions between neighboring units cause the whole array of units to settle as a cloud at a rate dependent on their mean size. Furthermore, the rate of settling is restricted by the magnitude of the free space in the liquid phase between the units. This space influences the rate of upward flow of liquid being displaced by the settling units. The magnitude of the free space and hence the rate of settling are dependent on the concentration of pigment.

A boundary forms, ideally with clear liquid above and the cloud of settling units below. It moves downward with time according to the pattern shown in Fig. 8. Since flocculates are units of structure containing a large portion of liquid immobilized within, their mean density ρ_K is intermediate between that of the liquid phase ρ_L and that of the pigment ρ_S. It is given by

$$\rho_K = v_i \rho_L + (1 - v_i)\rho_S$$

so that

$$\rho_K - \rho_L = (\rho_S - \rho_L)(1 - v_i)$$

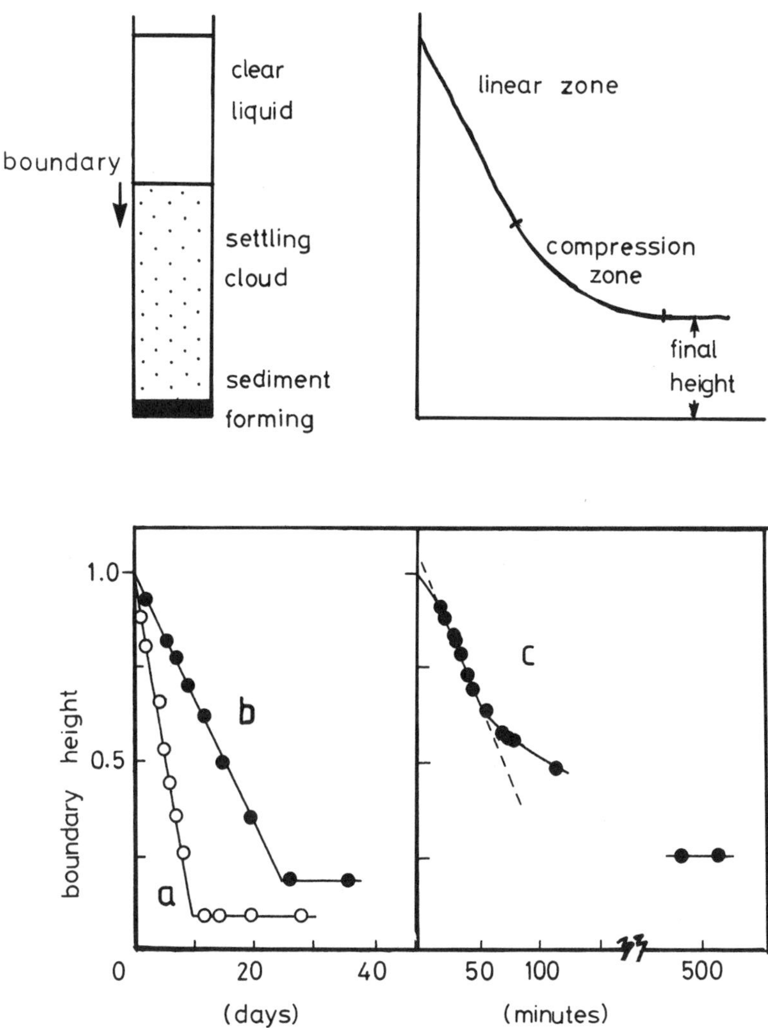

FIG. 8 Hindered settling characteristics. Top: Pictorial representation (left) with profile of boundary movement (right). Bottom: (a) stabilized α-CuPc of type 2; (b) stabilized α-CuPc of type 1; (c) stabilized α-CuPc of type 3. All in toluene containing modified phenol-formaldehyde resin. (Date of Ref. [18].)

where v_i is that fraction of the settling unit volume that is occupied by immobilized liquid phase [18].

Steinour's classic work on hindered settling took into account the effect of immobilized liquid on the volume fraction of settling units, but not the effect on their density [30]. This latter effect, however, can be allowed for by substituting $(\rho_K - \rho_L)$ for $(\rho_S - \rho_L)$ in Steinour's equation relating the rate of fall of the boundary U to the mean settling unit diameter 2r. Thus Steinour's equation

$$U = \frac{2g}{9\eta}(\rho_S - \rho_L)\varepsilon^2 r^2 10^{-1.82(1-\varepsilon)}$$

becomes

$$U = \frac{2g}{9\eta}(\rho_S - \rho_L)(1 - v_i)(\varepsilon')^2 r^2 10^{-1.82(1-\varepsilon')} \tag{9}$$

where g is the acceleration due to gravity, η is the viscosity of the liquid phase, and ε' is the volume fraction of free liquid in the dispersion, so that $(1-\varepsilon')$ is the volume fraction of the settling units [18].

Use of a simple model that considers flocculates as discrete units of structure that retain their identity in the final sediment leads to:

$$1 - \varepsilon' = pV_S \qquad \text{and} \qquad v_i = \frac{pV_S - c}{pV_S}$$

where V_S is the final sediment volume fraction when settling has ceased, p is a packing factor to allow for free space between close-packed units in the sediment, and c is the pigment volume fraction of the dispersion [18].

Equation (9) can then be rewritten as

$$U = \frac{2g}{9\eta}(\rho_S - \rho_L)\frac{c}{pV_S}(1 - pV_S)^2 r^2 10^{-1.82pV_S} \tag{10}$$

enabling 2r to be evaluated [18] from observations of U and V_S.

Well-defined linear settling regions are obtained with many copper phthalocyanine dispersions in hydrocarbon liquids (e.g., Fig. 8a–c). Apparently an equilibrium is established in this region where individual flocculate character is maintained. However, with strongly flocculating dispersions (large U, large V_S) there is nothing to prevent coalescence of flocculates into a coherent structure in the region of increasing concentration near the bottom of the container. The linear region is of short duration (e.g., Fig. 8c) and the concept of a packing factor is invalid (note the induction period while flocculates form and settling starts). In other cases that flocculate weakly (low U, low V_S) the linear region extends virtually all the way to the final sediment; there is no significant compression region (e.g., Fig. 8a and b). In these cases the settling units may retain their identity in the sediment, and indeed it has been possible to evaluate packing factors in some cases [18].

Equation (10) has been found satisfactory for sizing yeast cells in aqueous media—a type of system in which 2r and v_i can be determined independently by microscopy and analysis, respectively [18]. Furthermore, the principles have been found to apply well to aqueous kaolin dispersions [31]. Nevertheless, the main value of Eq. (10) is conceptual. Large V_S and U values indicate extensive flocculation; low V_S and U values indicate flocculation resistance; very low V_S and large U values indicate dense settling units indicative, for example, of poor separation of aggregates in the dispersion process; very large V_S with U indeterminate (due to merging of the linear and compression zones) indicates massive flocculation.

These principles can be used for comparisons with related series of dispersions with markedly different degrees of flocculation. Alternatively, Eq. (10) can be used by putting p = 1. Table 6 shows some typical results. Comparisons are made within series of dispersions of copper phthalocyanine pigments in toluene to determine the effect of phenolic resin on the state of flocculation. Each series has been prepared from a master dispersion in pure liquid so that all the constituent dispersions have the same pigment concentration (c = 0.005), but different resin concentrations ranging from 0 to 100 g l^{-1}.

It is evident from Table 6 that the presence of resin has a marked effect on the flocculation of some pigments, but not on others—even after correction for differences in density and viscosity of the liquid phase as embodied in calculations of 2r. The modified stabilized α-CuPc pigments of types 1 and 2 show enhanced flocculation resistance in the presence of resin. These pigments are modified with minor proportions of certain substituted copper phthalocyanine derivatives; the type 2 pigment has substantial flocculation resistance even in the absence of resin. Additive treatment is discussed at length below. Note that both β-form pigments show evidence of partly effective pigment/resin interaction. The interaction reduces flocculate size, but does not eliminate flocculation. This is typical of β-form pigments in general. At higher pigment concentrations they show clear evidence of flocculation in the phenolic/toluene ink system as determined from centrifugal hindered settling and rheological measurements discussed below.

Whereas hindered settling under gravity is useful with model systems of low pigment concentration (around 1% w/w), hindered settling using centrifugal acceleration instead of gravity can be applied directly

TABLE 6 Hindered Settling and Optical Microscopy Data for Copper Phthalocyanine Pigments Dispersed in Solution of a Phenolic Resin in Toluene

Pigment	Resin concn. ($g\ l^{-1}$)	U ($cm\ sec^{-1}$)	V_S	Qualitative deduction from U and V_S	2r, calculated from Eq. (10) with p = 1 (μm)	Brownian movement	Appearance
Modified stabilized α-CuPc of type 1[a]	0	4.7×10^{-5}	0.32	Effective pigment-resin interaction	19	None	Flocculated
		(4.6×10^{-5})[b]	(0.29)		(16)		
	10	2.9×10^{-6}	0.16		2	Vigorous	Discrete particles, 1 μm and less
		(2.5×10^{-6})	(0.15)		(2)		
	33	2.9×10^{-6}	0.13		2	Vigorous	
		(2.5×10^{-6})	(0.13)		(2)		
Modified stabilized α-CuPc of type 2[a]	0	1.2×10^{-5}	0.11	c	3	Just perceptible	Flocculated
	10	7.2×10^{-6}	0.06		1	Vigorous	Discrete particles, 2 μm and less
	33	5.7×10^{-6}	0.06		1	Vigorous	

Modified stabilized α-CuPc of type 3[a]	0	1.1×10^{-3}	0.22	No effective pigment-resin interaction	51	None	Extensive flocculation
		(1.0×10^{-3})	(0.19)		(42)		
	10	1.3×10^{-3}	0.22		52	None	
		(0.92×10^{-3})	(0.21)		(47)		
	33	1.1×10^{-3}	0.21		53	None	
		(0.88×10^{-3})	(0.24)		(57)		
A β-CuPc (77% β, 23% α) with brick-shaped crystals[a]	0	3.6×10^{-5}	0.30	Effective pigment-resin interaction	15	None	Flocculated
	10	3.3×10^{-6}	0.25		3	Partial[d]	Flocculated but only small flocculates
	33	2.7×10^{-6}	0.22		3	Partial	
	100	2.0×10^{-6}	0.22		3	Partial	
β-CuPc-VI with rod-shaped crystals	0	$>1 \times 10^{-5}$	0.22	Effective pigment-resin interaction	>5		
	10	5.7×10^{-6}	0.18		3		
	33	5.7×10^{-6}	0.15		3		
	100	4.4×10^{-6}	0.17		4		

[a] Data of Ref. [18].
[b] Dispersions prepared from a replicate master dispersion.
[c] Dispersion substantially resistant to flocculation even in absence of resin, but evidence of improved resistance in presence of resin.
[d] Some particles only.

to more realistic gravure ink and paint dispersions with pigment concentrations around 5% w/w [32]. A simple laboratory centrifuge with swing-out action is suitable for this purpose. Use of narrow tubes (e.g., 2 mm in diameter) minimizes radial effects. The same general principles apply as for gravity settling, but greater complexity arises from the fact that the acceleration increases (linearly) with distance from the axis of the centrifuge. Nevertheless, comparisons within related series of dispersions can still be made using the same concepts as for gravity settling. The comparative rate of settling U can be deduced from the time taken from the boundary to move a given distance, and V_S can be observed directly. Examples are shown in Fig. 9B. Where a large series of dispersions is being examined Eq. (10) with p = 1 can be used to simplify comparisons in terms of a flocculate size parameter 2r'. The average acceleration experienced is used in place of g. Details have been given elsewhere [32].

Hindered settling characteristics are dependent on the strength of interaction between the constituent pigment particles of settling units or flocculates. The stronger the interaction, the larger the flocculates, and vice versa.

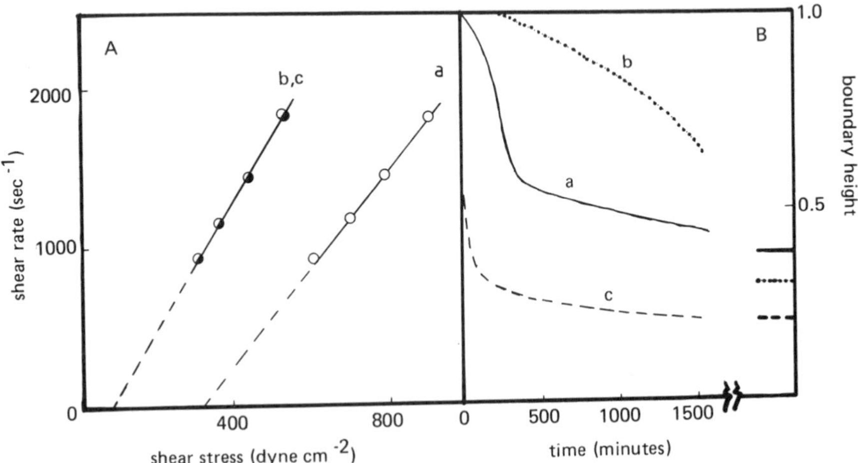

FIG. 9 (A) Effect of additives of type $CuPc(CH_2OOCR)_{2.5}$ on the rheological behavior of dispersions of β-CuPc-VI in the phenolic/toluene medium. (B) Effect on the corresponding centrifugal hindered settling behavior: a, no additive; b, R = $C_{11}H_{23}$; c, R = C_2H_5. Pigment volume fraction = 0.043. (Data are from Ref. [32].)

B. Strength of Flocculated Structure

The strength of flocculated structure can be estimated on a comparative basis from equilibrium shear stress measurements made with a suitable rotational viscometer, such as the Weissenberg rheogoniometer fitted with a Mooney cell [29,32]. The aim is to obtain plots of shear stress (measured) versus shear rate (applied) like those shown in Fig. 9A. The upper limit of shear rate usable is determined by the onset of marked frictional heating (about 2000 sec^{-1} in a gravure ink). The lower limit is determined by marked deviation from linearity. Care must be taken to eradicate effects of previous shear history by using an appropriate preshearing routine. The actual shear stress measurements should then be made rapidly in ascending order of shear rate to avoid long equilibration time for structure recovery. Rapid measurement is important with volatile systems such as gravure inks. The line of best fit and the intercept on the shear stress axis are then calculated by regression analysis. The intercept τ_0 is known as the extrapolated shear stress [21] and gives an indication of the strength of flocculated structure. The larger τ_0 is, the stronger the structure. The reciprocal of the slope is the limiting (or plastic [21]) viscosity prevailing when all weak flocculated structure has been broken down.

Intuitively

$$\tau_0 = \text{function}(n,f) \tag{11}$$

in which n is the number of particles per unit volume and f is the mean residual force of attraction of each particle for its neighbors. The number n increases with pigment concentration and degree of subdivision, which is dependent on crystal size and state of aggregation. Thus interpretation of differences in τ_0 between different dispersions requires additional information on at least n and f.

Combined use of such rheological measurement and centrifugal hindered settling data provides a means of comparing series of dispersions in terms of effects of pigment surface treatments [29,32]. Figure 9 provides an illustration. Dispersions a, b, and c are dispersions of a β-form copper phthalocyanine pigment in a phenolic/toluene gravure ink system. All three dispersions have the same concentration of pigment (c = 0.043) and resin and have been prepared in an identical manner. The only difference is that b and c each contain an additive of type B (Table 7) incorporated at 10% on pigment weight at the start of dispersion in the ink medium. The additives adsorb on the pigment surface [32]. They produce identical effects on the rheological properties, τ_0 being markedly reduced (Fig. 9A). Centrifugal hindered settling data, however, show that the reasons for the lowering of τ_0 are quite different (Fig. 9B); whereas the additive in c has reduced the level of dispersion (lower n),

TABLE 7 Some Types of Additive Used to Stabilize Copper Phthalocyanine Pigment Dispersions in Hydrocarbon Media

Type	Chemical constitution	n	R, etc.	Ref.
A[a]	$CuPc(CH_2NRR')_n$	2–4	$R = H$, $R' \leqslant C_6$, or typically $R + R' \leqslant C_8$	[36,37]
B	$CuPc(CH_2OOCR)_n$	2–3	$C_{12} \leqslant R \leqslant C_{18}$	[39]
C	$CuPc(CH_2OOC{-}C_6H_4{-}OC_{16}H_{33})_n$	2–3		[39]
D	$CuPc(CONHR)_n$	<2	$C_{12} \leqslant R \leqslant C_{18}$	
E	$CuPc(SO_2NH{-}C_6H_4{-}OC_{16}H_{33})_n$	2–3		[40]
F	$CuPc(S\bar{O}_3\overset{+}{N}HMeRR')_n$	2–3	$R = R' = C_{12}$ to C_{21}	[41]
G	$CuPc(CH_2NR_2 \leftarrow HO_3S{-}C_6H_4{-}C_{12}H_{25})_n$	2–3	$R = CH_3$ or C_2H_5	[16]
H	$CuPc(SO_2NH[CH_2]_mNR_2 \leftarrow HO_3S{-}C_6H_4{-}C_{12}H_{25})_n$	2	$R = CH_3$, $m \leqslant 4$	[16]

[a]Oligomeric types are also used; see Ref. [38].

the additive in b has enhanced the flocculation resistance (lower f). This simple approach is useful in designing additives for controlling strength of structure and is exemplified below.

It is useful to think in terms of flocculates with core and tentacles after the random flocculate model developed by Vold [33] using a computer simulation technique. This model shows that random flocculates are approximately spherically symmetrical with a central core and projecting tentacles. The proportion of constituent particles in the core increases with increase in flocculate size. Three-dimensional network structure is formed by interaction of the tentacles. Thus gentle shearing can be envisaged as breaking the intertentacle links and releasing the cores and tentacle fragments as settling units [32]. These units are believed to be the species that feature in centrifugal sedimentation. Thus weaker flocculates with small cores give small settling units and vice versa.

VIII. CONTROL OF FLOCCULATION

Control of flocculation of pigment dispersion in ink and paint media is achieved by modification of the surface character of the pigment particles. This, in turn, leads to modification of the properties of the pigment/medium interface and reduces the residual attraction between neighboring particles. In practice, it is not essential to eliminate flocculation altogether, substantial reduction in the strength of structure formed being sufficient.

There are two types of approach. One promotes interaction of the pigment surface with binder resin in the ink or paint medium in such a way as to form an effective adsorbed resin layer. This layer is necessarily different in composition and properties from the insufficiently effective layer formed at the surface of the unmodified pigment. The other approach features the use of additives that are preferentially adsorbed and themselves form an effective layer without necessarily involving the binder resin.

A. Flocculation Resistance Dependent upon Adsorption of Binder Resin

Commerical titanium dioxide pigments have a surface coating formed by treatment with silica and alumina as mentioned above. Increasing alumina content has been observed to progressively increase the fineness of dispersion and the opacification effect in decorative paints based on solutions of alkyd resin in white spirit [22].

Analysis of composite adsorption isotherms for such systems has shown that an alumina-rich coating [3% Al_2O_3, 1% T_iO_2) promotes adsorption of a layer with a high solvent content. This is indicative of well-solvated resin fractions (resin/solvent weight ratio in adsorbed layer = 0.22; solvent fraction, $1 - c_a = 0.82$) [22]. A mixed alumina/silica coating (1.5% Al_2O_3, 1.1% T_iO_2, 0.5% S_iO_2) gives an adsorbed layer with a lower proportion of solvent ($1 - c_a = 0.68$). A silica coating gives adsorption of poorly solvated resin fractions ($1 - c_a \to 0$), the isotherm being parallel to the c_2 axis in the wide range of c_2 used.

Alumina gives rise to positively charged sites in the pigment surface [22]. This has been shown by electrokinetic mobility measurements made directly on dispersions with realistic pigment concentrations by an electrodeposition method. Residual carboxylic acid groups in the alkyd resin interact with those sites, thus firmly anchoring the resin to the surface. Supporting evidence is that the alumina-rich surface shows a greater propensity to adsorb stearic acid from xylene than does the silica-rich surface [22]. Indeed, independent investigations have shown that although pure rutile adsorbs carboxylic acids [25,34], enhanced adsorption occurs in alumina-coated titanium dioxide, after correction for porosity [25]. Silica reduces adsorption [25].

Thus the presence of alumina in the surface of titanium dioxide pigments promotes adsorption of highly solvated resin fractions, firmly anchored to the surface, but with the solvated segments extending outward into the liquid phase. This conformation favors steric stabilization. Conversely, the resin has no specific attraction for a silica-rich surface coating, and the poorly solvated resin fractions adsorbed form a compact layer ineffective for steric stabilization [22].

The same broad principles apply to copper phthalocyanine pigments. Modification of the pigment surface is achieved by treatment with certain substituted derivatives of copper phthalocyanine at 5 to 10% on pigment weight. The classic example is type A of Table 7. Such additives adsorb on the pigment surface by virtue of the large planar copper phthalocyanine residue [35]. They may be introduced during pigment manufacture from aqueous or nonaqueous media or during dry attrition. In any case, adsorption equilibrium is established on subsequent dispersion in hydrocarbon liquids.

The adsorption of the additive introduces polar functional groups into the pigment surface. These functional groups interact with the binder resin in hydrocarbon-based gravure ink [32] or paint media [35]. Gravity hindered settling measurements given in Table 6 have already demonstrated the importance of interaction with the resin to flocculation resistance in toluene-based systems. Table 8 now gives some more specific examples. Note that stabilized α-form pigments have been used here, even though they are not normally used in practice in gravure inks. Unlike β-form pigments, however, they show no propensity to interact even in a partially effective manner with the phenolic resin, and thus flocculate severely, unless specially modified with particular additives. They are thus particularly suitable as substrates for examining the influence of additives by the gravity hindered settling technique.

TABLE 8 Gravity Hindered Settling Data for Stabilized α-Copper Phthalocyanine Pigment Dispersions—Effect of Additives on Pigment/Resin Interaction

Pigment	Additive	Resin concn. ($g\ l^{-1}$)	U ($cm\ sec^{-1}$)	V_S	2r calculated from Eq. (10) with p = 1 (μm)[c]	Effective pigment/resin interaction?
CuPcCl-IV[a]	None	0	Indeterminate	0.55		
		10		0.59	Indeterminate	No
		33		0.60		
CuPcCl-V[a]	Type A	0	5.3×10^{-4}	0.16	24	
	(Table 7)	10	3.8×10^{-6}	0.14	2	Yes
		33	2.5×10^{-6}	0.13	2	
CuPcCl-VI	b	0	1.0×10^{-3}	0.22	51	
		10	1.3×10^{-3}	0.22	52	No
		33	1.1×10^{-3}	0.22	55	
CuPcCl-VII	Tetrapyridino	0	2.5×10^{-4}	0.25	30	
	CuPc	10	7.3×10^{-6}	0.17	3	Yes
		33	5.0×10^{-6}	0.14	2	
CuPcCl-VIII	MnPc	0	9.1×10^{-4}	0.13	27	
		10	1.2×10^{-5}	0.08	2	Yes
		33	0.9×10^{-5}	0.09	2	

[a]Pigments prepared from exhaustively Soxhlet-extracted crude material by idealized size reduction methods.
[b]$CuPc(SO_3H)_2(SO_2N[CH_2CH_2OH]_2)_2$.
[c]From Eq. (10) with p = 1.

The unmodified pigment CuPcCl-IV shows massive flocculation whether or not resin is present. The additive of type A (Table 7) incorporated in CuPcCl-V induces flocculation resistance in the presence of resin. The acidic additive in CuPcCl-VI does not. Mono- to tetrapyridino copper phthalocyanine (N substituted for C in the 4, 4', etc. position; see Fig. 2) at 25% on pigment weight is also effective; CuPcCl-VII is an example. So too is manganese phthalocyanine at 10%, as in CuPcCl-VIII. The last two additives are not used in practice because of adverse cost and coloristic properties, respectively.

Very little has been reported on the mechanism of interaction with resin. There is evidence from cryoscopic measurements that the basic amino groups in additives of type A (Table 7) interact with carboxylic acids [35]. The particular interaction examined was between oleic acid and the additive with R = cyclohexyl, R' = cetyl, n = 3, in cyclohexane, the cetyl group being required to give sufficient solubility. There was little interaction with cetyl alcohol. Accompanying changes in the infrared absorption properties indicated hydrogen bonding between —COOH and —NRR'. It is conceivable that a similar mechanism could operate with, for example, the phenolic groups in the modified phenol-formaldehyde resin used in the above hindered settling experiments. The likelihood diminishes, however, with the weakly basic pyridino copper phthalocyanine derivatives, and how manganese phthalocyanine interacts has not been established, as it is of no commercial interest. All of these additives improve the rheological behavior of phenolic/toluene inks and are known to be effective also in aromatic alkyd/melamine-formaldehyde paint systems of low polarity [36,37,42,43].

B. Flocculation Resistance Independent of Adsorption of Binder Resin

Flocculation resistance can be achieved in the absence of large molecular species such as resins or polymers. It has been shown that dispersions of coated titanium dioxide in aliphatic hydrocarbon liquids can be stabilized by means of long-chain fatty acids [26]. The carboxylic acid group interacts with positively charged sites in the surface coating, thus firmly anchoring the aliphatic tail, which extends into the liquid phase. Investigations using oligomers of 12-hydroxystearic acid have shown that flocculation resistance increases with increase in chain length, at least up to the pentamer. This corresponds to an increase in adsorbed layer thickness up to around 70 to 90 Å, determined from rheological assessments of apparent pigment volume concentration by means of Eq. (8). The degree of flocculation or "flocculation factor" was determined by a rheological method—the lower the factor, the less the flocculation.

At constant adsorbed layer thickness corresponding to that of oleic acid (10–12 Å), flocculation resistance is enhanced by chain branching, especially near the free end of the hydrocarbon chain (Fig. 10). In this way branching increases solvated segment density where it matters most, namely in the outer region of the adsorbed layer. This is the region of overlap (interpenetration or deformation) between adsorbed layers of colliding particles, and it determines the effectiveness of steric stabilization. The valerate of 12-hydroxystearic acid (Fig. 10B) is the most effective agent and corresponds to the highest segment density in the outer regions of the adsorbed layer. A high degree of flocculation resistance is achieved. On the other hand, the valerate of 2-hydroxypalmitic acid (Fig. 10A) with branching in the inner part of the adsorbed layer is ineffective and shows no advantage over oleic acid. Figure 10 shows just how critically dependent the stabilization effect is on the location of branching.

Long-chain fatty acids are not effective with copper phthalocyanine pigments. The problem is in anchoring the aliphatic chains to the pigment surface, carboxylic acid groups being ineffective in this respect. Success, however, can be achieved by using certain types of copper phthalocyanine derivatives in which aliphatic chains are linked chemically to functional groups on the periphery of the phthalocyanine residue in the 4, 4', etc. positions. Examples are shown in Table 7 (types B, C, and E). The copper phthalocyanine residue is not appreciably soluble in hydrocarbon liquids, whereas the aliphatic chains are, thus conferring amphipathic character to the additives. These adsorb on the pigment, presumably with the phthalocyanine residue flat on the pigment surface and the fatty chains extending into the liquid phase. It is difficult to define the adsorption more precisely due to the limited solubility of the additives and their tendency to aggregate in solution in hydrocarbon solvents. For example, it requires prolonged high-speed centrifugation to achieve a steady concentration with solutions of additives of types B, C, and E (Table 7) [32]. The final solution is probably micellar. Although not in the present category in terms of stabilization mechanism, the additive of type A with R = cyclohexyl, R' = cetyl, n = 3, forms micelles four to five molecules in size in cyclohexane [35].

The mechanism of action of the additives has been investigated [32] by the combined use of rheological measurements and hindered settling described above. Figure 9 shows some results. Measurements were made directly on dispersions of β-CuPc-VI in pure toluene and in a gravure ink medium, consisting of a rosin-modified phenol-formaldehyde resin dissolved in toluene. The additive-free β-CuPc-VI has been fully discussed above (see Tables 2, 5, and 6) and is essentially unaggregated. Additives were introduced at the start of the dispersion process (overnight ball milling) at 10% on pigment weight.

Varying R in additives of type B (Table 7) from H, CH_3, C_2H_5, etc. to $C_{21}H_{43}$ produces the pattern of results shown in Fig. 11 in the phenolic/toluene ink system. All the additives reduce the strength

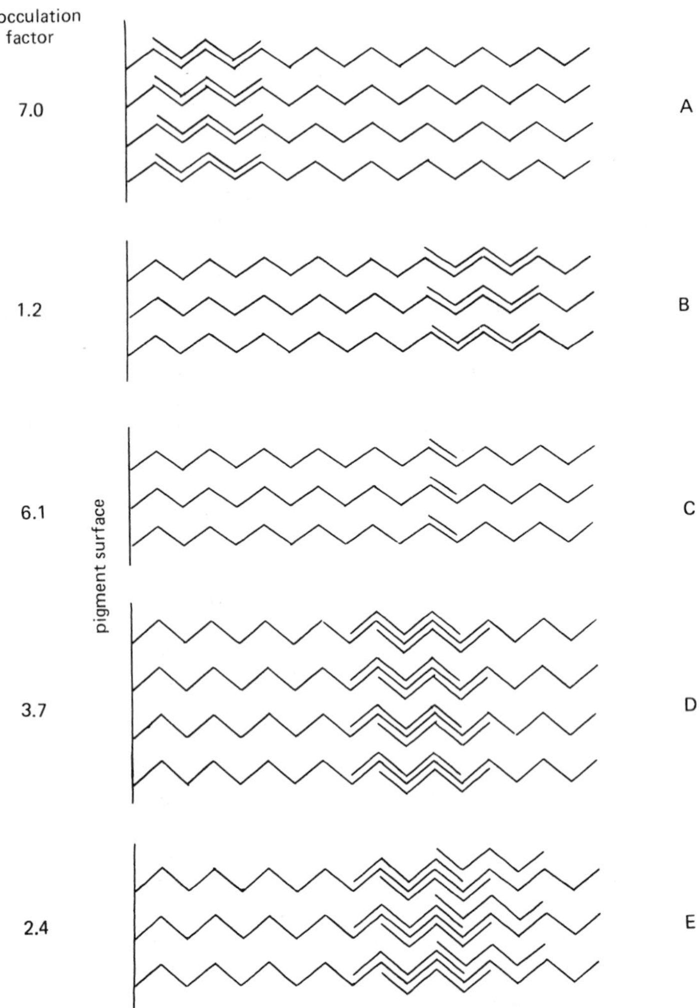

FIG. 10 Schematic representation of adsorbed fatty acids showing flocculation factor and regions of increased segment density (based on Ref. [26]). (A) Valerate of 2-hydroxypalmitic acid; (B) valerate of 12-hydroxystearic acid; (C) acetate of 12-hydroxystearic acid; (D) divalerate of 9,10-hydroxystearic acid; (E) trivalerate of 9,10,12-hydroxystearic acid.

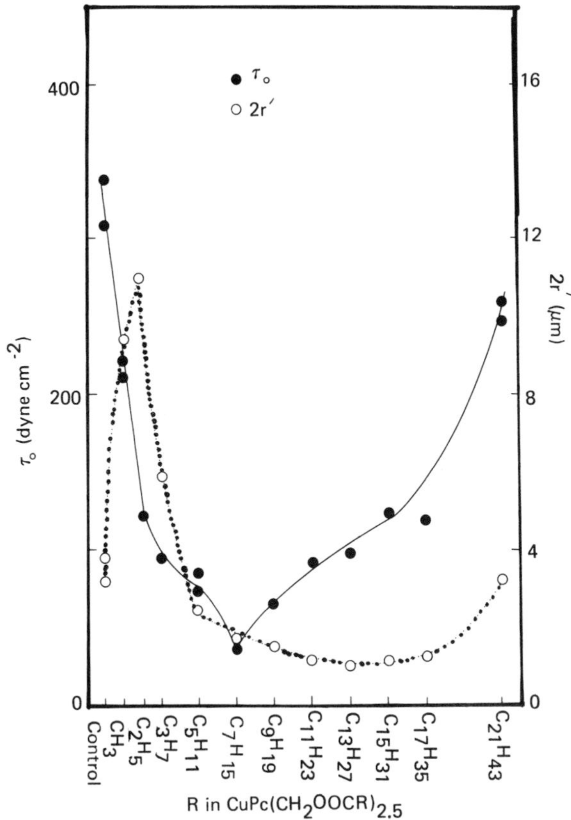

FIG. 11 Effect of varying R in the additive $CuPc(CH_2OOCR)_{2.5}$ on the properties of dispersions of β-CuPc-VI in the phenolic/toluene medium. (Data are from Ref. [32].)

of flocculated structure (represented by τ_0) below that of the additive-free control dispersion, but to extents that vary systematically with R. The most effective in this respect is $R = C_7H_{15}$. Enhanced flocculation resistance, however (as shown by lower 2r' values than that of the control dispersion), occurs at larger R values, in the range $C_7 < R < C_{17}$ (e.g., Fig. 9B,b), optimally at $C_{11} \leqslant R \leqslant C_{17}$. Additives with $R \leqslant C_3H_7$ give a poorer level of dispersion (e.g., Fig. 9B,c). Since a qualitatively similar effect occurred when the additives with $R \leqslant C_3H_7$ were introduced after the process of dispersion of the pigment, the poorer level of dispersion is likely to be the result of irreversible flocculation rather than incomplete separation during dispersion.

Additives with R = C_7H_{15} give the most attractive rheological behavior (lowest value of τ_0); in other words, the combined effect of n and f in Eq. (11) is at a minimum, even though neither n nor f is. The most attractive optical performance (highest color strength) is given when $C_{11} \leqslant R \leqslant C_{17}$, corresponding to optimum flocculation resistance.

The pattern of results in the pure toluene system is shown in Fig. 12. It is remarkably similar to that of the phenolic/toluene system, suggesting a common mechanism. Since there is no resin in the toluene system, the additives must be effective in their own right in that system, and therefore likewise by analogy in the phenolic/toluene system. Solubility data reinforce this view. Table 9 shows that there

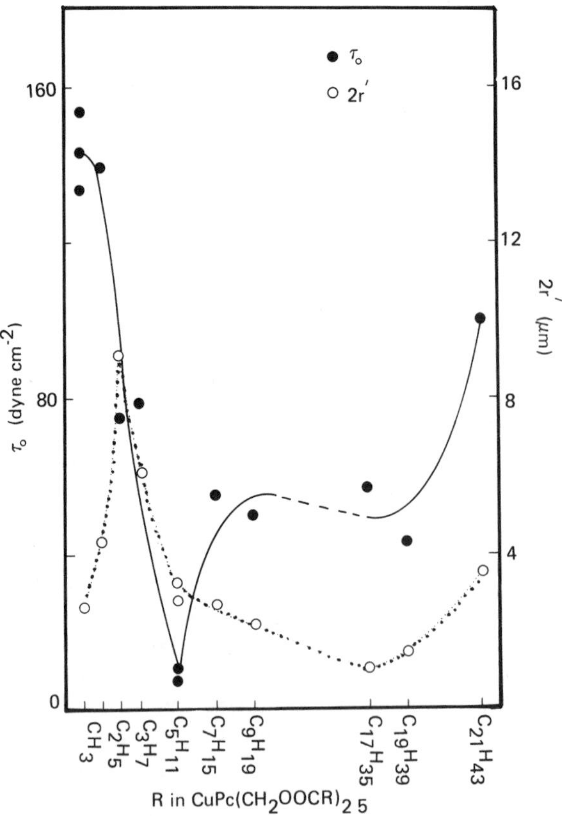

FIG. 12 Effect of varying R in the additive $CuPc(CH_2OOCR)_{2.5}$ on the properties of dispersions of β-CuPc-VI in toluene alone. (Data are from Ref. [32].)

TABLE 9 Solubility of $CuPc(CH_2OOCR)_{2.5}$ in Phenolic/Toluene Medium and Toluene Expressed as Optical Density (OD) of Saturated Solution (Actual Concentrations in Toluene Also Shown)[a]

R	Phenolic/toluene,	Toluene	
	OD	OD	Concentration (% w/w)
CH_3	0.06	0.08	0.019
C_2H_5	0.60	0.85	0.036
C_3H_7	1.17	1.34	0.057
C_5H_{11}	1.26	1.75	0.067
C_7H_{15}	2.12	2.25	0.168
C_9H_{19}	2.21	2.22	0.147
$C_{11}H_{23}$	1.99	2.23	0.146
$C_{13}H_{27}$	2.08	2.22	0.153
$C_{15}H_{31}$	2.06	2.21	0.168
$C_{17}H_{35}$	2.04	2.22	0.204
$C_{21}H_{43}$	0.64	0.94	0.065
(benzene ring)$OC_{16}H_{33}$	40.9	40.8	2.28

[a]Data of Ref. [32].

is no enhancement of solubility of the additives in the phenolic/toluene system—indeed, slightly the opposite. Thus there is no significant additive-resin interaction.

All the additives have finite solubility in both systems, so the ineffectiveness of some is not simply due to lack of access to the pigment surface. Therefore it is significant to the mechanism of stabilization that the most effective additives have the highest solubility. It follows that $C_{11} \leqslant R \leqslant C_{17}$ ensures a sufficiently thick and well enough solvated adsorbed layer to give effective steric stabilization.

Introduction of aromaticity into R (additive of type C, Table 7) results in a marked increase in solubility in both toluene and phenolic/toluene medium (Table 9), together with a marked increase in effectiveness in conferring flocculation resistance [32]. Table 10 compares

TABLE 10 Effect of Additives in Reducing Particle-Particle Interaction in β-Form Copper Phthalocyanine Pigment Dispersions in Phenolic/Toluene Ink and Toluene

Additive		Phenolic/Toluene			Toluene		
Type[a]	Specific formula	τ_0 (dyne cm^{-2})	2r' (μm)	Solubility (OD sat. solution)	τ_0 (dyne cm^{-2})	2r' (μm)	Solubility (OD sat. solution)[c]
B[b]	$CuPc(CH_2OOC_{15}H_{31})_{2.5}$	124	1.2	2.1	57	1.0	2.2
C[b]	$CuPc(CH_2OOC-C_6H_4-OC_{16}H_{33})_{2.5}$	4	0.2	40.9	6	0.4	40.8
E	$CuPc(SO_2NH-C_6H_4-OC_{16}H_{33})_2$	20	0.8	47.6	4	0.4	42.1
D	$CuPc(CONHC_{18}H_{37})_{0.9}$	13	0.4	1.1	48	2.4	0
F	$CuPc(\bar{SO_3}\overset{+}{N}HMe[C_{18}H_{37}]_2)_2$	18	0.6	4.4	54	2.7	0.2
A	$CuPc(CH_2NHcyclohexyl)_4$	26	0.6	High	4	4.5	High
	Control (no additive)	322	3.4		143	3.0	

[a]See Table 7.
[b]Data of Ref. [32].
[c]OD, optical density.

data for the additive of type C with data for one of the most effective additives of type B ($R = C_{15}H_{31}$). The enhanced flocculation resistance with type C is attributable to a thicker, better solvated adsorbed layer of thickness around 35 Å, as estimated from molecular models with extended chain conformation. It has been shown by adsorption isotherm determinations that both additive and resin are present at the pigment surface in the phenolic/toluene system [32], but there is no enhancement of resin uptake, indeed slightly the contrary. The presence of adsorbed resin does not impede the effectiveness of the additive.

C. Further Remarks

Table 10 widens the comparison to other types of additive. Whereas type E, the sulfonamido counterpart of type C, has properties comparable to type C, the remaining additives (types D and F) behave differently. They are effective in the phenolic/toluene system, but ineffective in toluene, in which they are much less soluble. They therefore depend on interaction with the phenolic resin for their effectiveness. In this they behave like type A, an example of which is included. Evidently there is no simple means of predicting the type of mechanism of action of an additive from its chemical constitution.

The type of mechanism of stabilization is important to performance in practice. Additives that depend on interaction with binder resin have two major disadvantages. First, they show marked system dependence. For example, type F (Table 7) is more effective in phenolic/toluene gravure ink systems than in rosinate/aliphatic systems. Second, as pigment concentration in the system is increased, effectiveness of the additive fails at relatively low pigment concentrations, when there is not enough of the relevant resin fractions present to form an effective adsorbed layer. On the other hand, additives effective in their own right are much less system-dependent and remain effective to much higher pigment concentrations. For example, in an alkyd/white spirit paint system containing the same β-form pigment (β-CuPc-VI), varying R in additives of type B (Table 7) gives a pattern of behavior broadly comparable to those shown in Figs. 11 and 12 for the phenolic/toluene ink and toluene systems. Minor differences are that the minimum in τ_0 occurs at $R = C_9H_{19}$ and optimum flocculation resistance at $C_{15} \leqslant R \leqslant C_{17}$.

ACKNOWLEDGMENT

The author thanks Ciba-Geigy for permission to publish this chapter.

REFERENCES

1. H. D. Jefferies, in *Dispersion of Powders in Liquids*, 3rd ed., G. D. Parfitt, ed., Applied Science, London, 1981, p. 395.
2. T. J. Wiseman, in *Characterisation of Powder Surfaces*, G. D. Parfitt and K. S. W. Sing., eds., Academic Press, London, 1976, p. 159.
3. E. Herrmann, in *Characterisation of Powder Surfaces*, G. D. Parfitt and K. S. W. Sing, eds., Academic Press, London, 1976, p. 209.
4. A. I. Medalia and D. Rivin, in *Characterisation of Powder Surfaces*, G. D. Parfitt and K. S. W. Sing, eds., Academic Press, London, 1976, p. 279.
5. R. B. McKay and F. M. Smith, in *Dispersion of Powders in Liquids*, 3rd ed., G. D. Parfitt, ed., Applied Science, London, 1981, p. 471.
6. R. Sappok and B. Honigmann, in *Characterisation of Powder Surfaces*, G. D. Parfitt and K. S. W. Sing, eds., Academic Press, London, 1976, p. 231.
7. B. Felder, *Helv. Chim. Acta 51*: 1224 (1968); *J. Color Appearance 1*: 9 (1971).
8. B. Honigmann and D. Horn, *FATIPEC Congr. XII*: 181 (1974).
9. J. R. Fryer, R. B. McKay, R. R. Mather, and K. S. W. Sing, *J. Chem. Technol. Biotechnol. 31*: 371 (1981).
10. B. Honigmann and D. Horn, in *Particle Growth in Suspension*, A. L. Smith, ed., Academic Press, London, 1973, p. 283.
11. S. J. Gregg and K. S. W. Sing, *Adsorption, Surface Area and Porosity*, 2nd ed., Academic Press, London, 1982.
12. R. R. Mather and K. S. W. Sing, *J. Colloid Interface Sci. 60*: 60 (1977).
13. R. R. Mather, *FATIPEC Congr. XIV*: 433 (1978).
14. C. R. S. Dean, R. R. Mather, and K. S. W. Sing, *Thermochim. Acta 24*: 399 (1978).
15. V. Y. Davydov, A. V. Kiselev, and T. V. Silina, *Kolloidn. Zh. 36*: 945 (1974).
16. BASF, British Patent 1,544,839 (1979).
17. J. Moilliet and D. A. Plant, *J. Oil Colour Chem. Assoc. 52*: 289 (1969).
18. R. B. McKay, *J. Appl. Chem. Biotechnol. 26*: 55 (1976).
19. K. Rehacek, *Farbe Lack 76*: 656 (1970).
20. R. B. McKay, in *Particle Size Analysis*, M. J. Groves, ed., Heyden, London, 1978, p. 421.
21. British Standards Institute, London, *Glossary of Rheological Terms*, BS5168, 1975.
22. M. J. B. Franklin, K. Goldsbrough, G. D. Parfitt, and J. Peacock, *J. Paint Technol. 42*: 740 (1970).

23. F. Biglieri and V. Di Paolo, *Double Liaison—Chim. Peint.* (No. 325): 381 (Oct. 1982).
24. A. Doroszkowski and R. Lambourne, *J. Colloid Interface Sci. 26*: 214 (1968).
25. D. Urwin, *J. Oil Colour Chem. Assoc. 52*: 697 (1969).
26. A. Doroszkowski and R. Lambourne, *Faraday Discuss. Chem. Soc. 65*: 252 (1978).
27. V. T. Crowl, *J. Oil Colour Chem. Assoc.* 55: 388 (1972).
28. K. Goldsbrough and J. Peacock, *J. Oil Colour Chem. Assoc. 54*: 506 (1971).
29. R. B. McKay, *Br. Ink Maker 19*: 59 (1977).
30. H. H. Steinour, *Ind. Eng. Chem. 36*: 618, 840, 901 (1944).
31. J. I. Bhatty, L. Davies, and D. Dollimore, *Surf. Technol. 11*: 269 (1980).
32. R. B. McKay, *6th International Conference Organic Coatings Science Technology, Athens, 1980*, G. D. Parfitt and A. V. Patsis, eds., Technomic, Westport, Conn., 1982, p. 173.
33. M. J. Vold, *J. Colloid Sci. 18*: 684 (1963).
34. R. H. Ottewill and J. M. Tiffany, *J. Oil Colour Chem. Assoc. 50*: 844 (1967).
35. W. Black, F. T. Hesselink, and A. Topham, *Kolloid Z. Z. Polym. 213*: 150 (1966).
36. BASF, British Patents 949,739 (1964) and 985,620 (1965).
37. ICI, British Patent 972,805 (1964).
38. ICI, British Patents 1,082,967 (1967) and 1,149,778 (1969).
39. Ciba-Geigy, British Patent 1,569,418 (1980).
40. Ciba-Geigy, British Patent 1,535,434 (1978).
41. Ciba-Geigy, British Patent 1,541,599 (1979).
42. KVK, British Patent 1,174,114 (1969).
43. Ciba, British Patent 971,093 (1964).

Index

D

T